Finals   Wed June

Finals   Wed June

BASIC CONCEPTS
OF PHYSICS
SECOND EDITION

ARTHUR BEISER

BASIC CONCEPTS
OF PHYSICS
SECOND EDITION

ADDISON-WESLEY PUBLISHING COMPANY
Reading, Massachusetts · Menlo Park, California
London · Amsterdam · Don Mills, Ontario · Sydney

This book is in the
ADDISON-WESLEY SERIES IN PHYSICS

ISBN 0-201-00491-7
GHIJKLMNOP-HA-89876543210

Permissions for quotations introducing sections are
as follows:

Page 2: From Bertrand Russell, *History of Western
Philosophy,* New York: Simon and Schuster, 1945.
Reprinted by permission of the publisher.

Page 53: From a lecture given by Max Born in 1928
and published in that year in Nachrichten von der
Gesellschaft der Wissenschaften zu Gottingen.
Reprinted in *Physics in My Generation* by Max Born,
c. 1969. Reprinted by permission of Springer-Verlag,
New York.

Page 264: From *The Feynman Lectures on Physics,*
Vol. II, pp. 1–11, by R. P. Feynman, R. B. Leighton
(California Institute of Technology), and M. L.
Sands (Stanford University), published by Addison-
Wesley, c. 1964.

Page 304: From Max Born's Nobel Prize lecture of
1954, published in *Science,* **122** (No. 3172), 675–679.
Copyright ©, The Nobel Foundation, 1955. Reprinted by permission of Elsevier Publishing
Company, Amsterdam, The Netherlands.

Page 339: From John C. Slater, *Quantum Theory of
Matter,* New York: McGraw-Hill, 1951. Reprinted
by permission of the publisher.

Page 349: From Victor F. Weisskopf, "The Three
Spectroscopies," *Scientific American,* May 1968,
p. 29. Reprinted by permission.

PREFACE

Basic Concepts of Physics is a concise introduction to its subject on an elementary level.
In preparing this edition, the original text was
both reorganized and rewritten. Modern
physics is discussed at greater length than
before, though classical physics is not skimped
within the limitations set by the nature of the
book. Special relativity is now treated early,
which permits the insights it offers into the
relationships between matter and energy and
between electricity and magnetism to be
considered when these topics are taken up.
Underlying these and other changes was the
desire to bring out more clearly the power and
beauty that reside in physics and whose
appreciation is one of the chief rewards offered
to its students. The practical aspects of physics
are no less significant, to be sure, but their
existence hardly needs emphasis today.

In an effort to communicate more effectively
with the reader, a large number of illustrations
have been integrated with the text so that
words and pictures jointly carry the flow of
ideas. Though frivolity has crept into some of
the pictures, the intent is always serious, and
there is no harm in being reminded that
physics can be fun. The book is divided into
forty-two relatively brief sections, each con-
centrating on a single theme, in place of the
fourteen chapters of the previous edition; the
same meat, but cut into what I hope are more
digestible chunks. Only the simplest mathematics is used, and a review of the elementary
algebra required is given in an appendix. The
exercises range from easy questions to
elaborate (but seldom difficult) problems that
serve to extend the text as well as the reader.

I wish to thank the students and teachers
who sent me comments on the first edition of
Basic Concepts of Physics; the criticisms and
suggestions provided useful guidance in carrying out the revision, and their friendly spirit
provided welcome encouragement.

Lovisa, Finland   A.B.
August 1971

CONTENTS

PART 1

SPACE AND TIME

1

INTRODUCTION

Physics is the science of matter, and seeks to understand its fundamental structure, properties, and behavior. As a science, physics is in the service of truth rather than utility, but modern technology as well as all the other sciences depends upon the insights it offers to the workings of the universe. To study physics is therefore to be introduced to the basic ideas involved in all of man's attempts to comprehend and control the natural world.

SCIENCE

Each of us is a member of the community of man, and each of us participates in some way in the continuing evolution of this community. Problems of politics and economics involve us all, and few people would wish to be ignorant of them.

Each of us is also part of the physical universe, whose evolution we can no more escape than we can escape the actions of our fellow men. We are made of atoms linked into molecules, liquids, and solids, and we live on a planet bound to a star which is a member of one of the galaxies that occupy the reaches of space. Atoms, galaxies, and everything in between have certain regularities of behavior in common, which are the "laws of nature." These same laws of nature underlie the technology that today dominates the lives of people and nations. An acquaintance with science is as vital to understanding man's place in modern civilization as it is to understanding his place in the universe.

"The scientist does not study nature because it is useful; he studies it because he delights in it." Henri Poincaré (1854–1912)

"Almost everything that distinguishes the modern world from earlier centuries is attributable to science." Bertrand Russell (1872–1970)

The scientist explores the natural world directly by experiment and observation and indirectly by abstract reasoning. Experiment and observation yield numbers, the raw material of science. Abstract reasoning yields theories, which are relationships that not only generalize the results of many measurements but also permit inferences to be drawn about phenomena either as yet undiscovered or impossible to examine directly. Nobody will ever visit the sun's interior or the interior of an atom, yet we know a great deal about what goes on in both places. The evidence is indirect but persuasive.

Every scientist dreams of illuminating some dark corner of the natural world—or, perhaps, of finding a dark corner where none had been suspected. The most accurate measurements, the most meticulous calculations, are not necessarily the most fruitful. Here is where creative imagination enters the picture, which is why the greatest advances in science have so often been the product of young, nimble, curious minds.

Science consists of more than just statements regarding certain features of the natural world—the ways in which these statements are obtained are equally significant. The two are intertwined. The results of an experiment are not independent of how the experiment was carried out, and the results of a theory cannot be divorced from the assumptions upon which the theory is based. The essence of science is that its cards are always face up on the table; nothing is hidden that bears upon the full meaning of what is being said.

What has made science such a powerful tool for investigating nature is the constant testing of its findings, both experimental and theoretical. Nothing is accepted on anybody's personal authority, on the basis of "common sense," or because it is part of a religious or political doctrine. Because every generalization in science is subject to modification in the light of fresh evidence, science is a living body of information and not a collection of

dogmas. Challengers of currently accepted ideas are not scorned—if their views are well-founded, anyway—nor is disgrace attached to a scientist whose contributions do not survive indefinitely. To rock the boat is part of the game; to overturn it is one way to win.

Poetry obtains its effects by suggestion rather than by bald statement. Poets delight in words rich in overtones of meaning, and a fine poem usually shimmers with ambiguity. Most poems evoke a variety of responses that depend upon who reads them and how the reader feels at the time.

In contrast science is objective, not subjective. A scientific statement that means different things to different people is unacceptable. But in recompense for the clarity of his language, the scientist is granted a view of the world far surpassing that of the poet in breadth, detail, variety, elegance, subtlety—and beauty. It may be the beauty of order in the structure of a crystal, or the beauty of disorder in the helter-skelter scurrying of gas molecules; the beauty of symmetry in the existence of an antiparticle for every elementary particle, or the beauty of asymmetry in the second law of thermodynamics; the beauty of certainty in the circling of the planets around the sun, or the beauty of uncertainty in the circling of electrons around an atomic nucleus. And many of the equations of science are beautiful too, though it may take a practiced eye to appreciate them.

PHYSICS

Broadly speaking, the science of physics is concerned with the elementary particles of which all matter is composed, the ways in which these particles interact with one another, and the behavior of the composite bodies (atoms, molecules, liquids, and solids) formed by elementary particles. Physics is thus the master science whose concepts and principles are drawn upon by such other sciences as chemistry, biology, astronomy, and geology in their exploration of narrower aspects of the world around us. Physics is the natural starting point in any serious study of science.

There is more than one way to approach physics. We can think of physics as a triumph of the human mind and examine its history and philosophy for their own sakes as part of our culture and also in the hope that its methods can be adapted to disciplines outside the natural sciences (a hope that has proved vain thus far). Or we can regard physics as a body of knowledge whose purpose is to be mined, as it were, for its applications to other sciences and to technology.

There is a third point of view as well. We inhabit a universe which has certain definite properties and regularities of behavior, and we ourselves are products of the unfolding of this universe. The insights physics offers into the fundamental character of the universe fit together into a pattern of the utmost intellectual and esthetic appeal, and it is this pattern which is emphasized in this book. It is not idle curiosity to ask why the sun shines, or what light is, or why the sky is blue, or what atoms are made of. These are profound questions, and to know their answers is profoundly enriching to one's personal life.

PARTICLES AND INTERACTIONS

As a sample of what lies in store, let us look at the picture physics provides of ultimate matter.

Consider a sample of matter of any sort—a drop of water, a grain of sand, a blade of grass —and imagine that it is somehow chopped up into finer and finer pieces. How far can this go on? Eventually each sample is reduced to its constituent atoms, of which there are 92 different varieties. If now the atoms themselves are broken up, we are left with just three kinds of *elementary particles* which cannot be subdivided further: protons, neutrons, and electrons. (There are other kinds of elementary particles, but these three are the ones primarily involved in ordinary matter.)

Elementary particles interact with one another in only four different ways, which are responsible for the structure and behavior of the entire universe. The four *fundamental interactions* are the strong nuclear, the weak nuclear, the electromagnetic, and the gravitational. Protons and neutrons join together by virtue of the strong nuclear interaction to form atomic nuclei, whose exact compositions are partly determined by the weak nuclear interaction as well. Electromagnetic forces hold electrons to nuclei to form atoms. Electromagnetic interactions between nearby atoms yield stable clumps of them called molecules and larger aggregates that constitute liquids and solids. On a larger scale, gravitation pulls matter together into the planets, stars, and galaxies that populate space.

Three particles and four interactions—these hardly seem enough to account for the span of things and events in everyday life, let alone for the evolution of the universe. Yet nothing else is needed—apparently—to explain phenomena ranging from the falling of a dropped stone to the development of plants and animals from inanimate matter early in the history of the planet earth.

To perceive such exquisite unity in the diversity around us took four hundred years of effort. The time scale is not arbitrary, for modern science began with Galileo (1564–1642). Before Galileo explanations of the natural world were sought in terms of self-evident principles, ideas supposed to be so clearly correct that experiments were unnecessary. Such a supposedly self-evident principle was Aristotle's assertion that a heavy object falls faster than a light one; another was the belief that all motion must be sustained by an applied force. The trouble with these principles, and many others like them, is that they are wrong—in the real world, moving bodies behave quite differently. Galileo's greatest contribution, overshadowing his specific discoveries in mechanics and astronomy, was the notion that statements about nature must be tested by observation.

The pages that follow present the various lines of evidence developed from Galileo's day to this which have led to the identification of the elementary particles and their interactions. The story is complicated, partly because it is impossible to analyze happenings on an atomic scale in terms of ordinary experience. To try to discuss atomic structure in everyday language is like trying to fix a watch while wearing boxing gloves. What are needed are precisely defined quantities (such as mass and energy) and precisely defined concepts (such as force field and wave), and a large part of the study of physics consists of mastering the vocabulary of these quantities and concepts.

To be sure, physics is concerned with a lot besides elementary particles. The table of contents of this book is a fair guide to the topics traditionally considered part of physics. What certain of these topics lack in relevance to the natural world, they make up for in practical utility. Ohm's law has nothing to say about atoms or planets, but it is an aspect of the behavior of matter that comes into play every time we switch on an electric lamp. Ohm's law is not one of the reasons to study physics in the first place, but on the other hand no apology is needed for its inclusion.

MODELS

At this point it is appropriate to mention a characteristic device physicists commonly use in their work.

It is not easy to get a firm intellectual grip on nature in the rough. Nothing is as simple. as it seems. We think of the earth as being round, but in fact it is more like a grapefruit with a bumpy skin than a perfect sphere. We say that the earth moves in an elliptical orbit about the sun, but in fact the orbit has wiggles no ellipse ever had. In order to extract the essence of a phenomenon, the physicists must idealize reality with the help of *models*. By choosing a sphere as a model for the actual earth and an ellipse as a model for its actual orbit, the physicist simplifies his task of analysis by isolating the most important features of his subject. If instead he had to deal from the start with a squashed, corrugated earth moving in an irregular path, he would find it hard to make any progress.

Another kind of model arises when the physicist tries to understand the microscopic world of atoms and molecules. For instance, a useful model of a gas pictures it as a collection of myriad tiny particles like billiard balls that fly about in all directions. This model is quite successful in accounting for many aspects of the behavior of gases. But it is still a model and not the whole story, since if we could somehow look directly at the structure of a gas, we would find that the particles of which it is composed are not at all like miniature billiard balls and in certain respects do not even act like particles in the usual sense.

There is no clear line between a model and a theory. Usually the term theory is reserved for a large-scale generalization that not only accounts for existing data but from which predictions about the outcome of new experiments can be made. A theory may be based upon a particular model, or it may not. Thus the kinetic theory of gases is a logical structure based upon the model of a gas as a collection of particles in rapid, random motion. On the other hand, there is no model directly associated with Newton's theory of gravitation, which deals with concepts that do not need to be pictured in order to be related to one another. Nevertheless, the key to formulating this theory was Newton's choice of spherical models for the sun, moon, and planets and of ellipses as models for the planetary orbits.

More than one model may be convenient or necessary in some cases. Light is a notable example. In many practical applications it is quite sufficient to imagine that light consists of "rays," thin pencils of something-or-other that are perfectly straight unless reflected or refracted, when they are bent through definite angles. A more sophisticated picture of light capable of explaining many more aspects of its behavior is the wave model. Without even specifying what kind of waves are involved, the wave model accounts for the interference, diffraction, and polarization of light, and interprets reflection and refraction in a straightforward way.

The next level of model-building brings even more rewards of understanding, but there is an unexpected (and at first glance disturbing) problem: *two* models are needed, of equal validity but different areas of application. One of them is a further development of the simple wave model, with the waves now identified as electromagnetic in character. The other model employs no waves at all but regards light as a stream of tiny particles. Certain phenomena can only be interpreted on the basis of the electromagnetic wave model, others only on the basis of the particle model.

Then what is light? The answer is that light is what it is, and no model that we can visualize in terms of our everyday experience can encompass its ultimate nature. But a purely abstract theory of light does exist whose conclusions are in superb agreement with experiment, and which makes comprehensible the need for two models. These models are not wrong, but each is limited in scope, which is true of all the other models in physics. Models are useful devices, but seldom are the last word.

Nearly always in this book we shall be considering models rather than actual things. It is a good habit to try to pick out those aspects of the phenomenon under study that are incorporated in its model and those that are left out, and to ask ourselves what is gained and what is lost in each case.

2

SCALARS AND VECTORS

"Philosophy is written in that great book which ever lies before our eyes—I mean the universe—but we cannot understand it if we do not learn the language and grasp the symbols in which it is written." Galileo Galilei (1564–1642)

Physics is rooted in observation, and few observations mean very much unless they are quantitative. Nor can the theories of physics have any significance unless they are quantitative statements that can be tested by measuring something and comparing the result with the predicted value. Before we actually take up the study of physics, then, we must review how physical quantities are expressed and treated mathematically.

UNITS

The process of measurement is essentially one of comparison. A certain standard quantity of some kind, called a *unit*, is first established, and other quantities of the same kind are compared with it. When we say a ladder is 2.6 meters long, we mean that its length is 2.6 times a certain distance called the meter whose magnitude is fixed by international agreement. Every measurement of a quantity must therefore have at least two parts, a number to answer the question "How many?" and a unit to answer the question "Of what?"

Nearly all quantities in the physical world can be expressed in terms of only four fundamental measurements, those of length, time, mass, and electric current. Thus every unit of area (the square foot, for instance) is the product of a length unit and a length unit; every unit of speed (the mile per hour, for instance) is a length unit divided by a time unit; every unit of force (the newton, for instance) is the product of a mass unit and a length unit divided by the square of a time unit; and every unit of electric charge (the coulomb, for instance) is the product of a unit of electric current and a time unit.

A *system of units* is a set of specified units of length, time, mass, and electric current from which all other units are to be derived. All units are arbitrary, and in fact thousands of different ones have been adopted at various times and places in the course of history. Even today a half dozen or so systems of units are in common use. Physics is complicated enough without having to worry about different systems of units, so we will confine ourselves to a single system in this book. Our choice is the MKSA system in which the units of length, time, mass, and electric current are respectively the meter, the second, the kilogram, and the ampere (Table 2.1). A brief discussion of the British system of units, also widely used at present, is given in Appendix C.

Because the various basic units in a system are not always convenient in size for a given measurement, other units have come into use within each system. Thus long distances are usually given in miles rather than in feet or in kilometers rather than in meters. The great advantage of the various metric systems is that they are wholly decimalized, which facilitates calculation, whereas the British system is quite irregular in this respect (1 mile = 5280 feet, for example, whereas 1 kilometer = 1000 meters).

It is so important that standard units be absolutely constant in magnitude that several of them have been redefined in recent years in terms of quantities in nature that are regarded as invariant under all circumstances and not subject to change in time. The standard meter, for instance, is now defined as exactly 1,650,763.73 wavelengths of one of the spectral lines of the krypton isotope Kr^{86} instead of as the distance between two scratches on a certain platinum–iridium bar kept at Sèvres, France. Measuring rods can be directly compared with the new standard meter in any well-equipped laboratory in the world.

Table 2.1   Some metric units

Length
1 meter (m) = the standard unit = 39.37 inches =
3.28 feet
1 centimeter (cm) = 0.01 meter
1 millimeter (mm) = 0.1 centimeter = 0.001 meter
1 kilometer (km) = 1000 meters = 0.621 mile

Time
1 second (s) = the standard unit
1 minute (min) = 60 seconds
1 hour (hr) = 60 minutes = 3600 seconds
1 day = 24 hours = 86,400 seconds

Mass
1 kilogram (kg) = the standard unit
1 gram (g) = 0.001 kilogram
1 milligram (mg) = 0.001 gram = 0.000,001 kg
(1 kilogram corresponds to 2.21 pounds in the sense
that the weight of 1 kilogram is 2.21 pounds.)

Electric Current
1 ampere (A) = the standard unit
1 milliampere (mA) = 0.001 ampere

SCALAR AND VECTOR QUANTITIES

Some physical quantities require only a number and a unit to be completely specified. It is quite sufficient to say that the mass of a man is 85 kg, that the area of a farm is 160 acres, that the frequency of a sound wave is 660 cycles/s, and that a light bulb consumes electrical energy at the rate of 100 watts. These are examples of *scalar quantities*.

It is not always possible to confine quantitative statements to numbers and units alone if ambiguity is to be avoided. A man's mass is the same whether he stands on his feet or on his head, but the motion of a car traveling north at 60 mi/hr is hardly the same as that of a car traveling in a circle at 60 mi/hr. A quantity whose direction is significant is called a *vector quantity.* Vector quantities occur frequently in physics, and their arithmetic is different from the arithmetic of scalar quantities.

Table 2.2

Subdivisions and multiples of metric units are widely used, and each is designated by a prefix according to the corresponding power of ten.* The most common prefixes are listed below.

Prefix	Power of ten	Abbreviation	Example
pico-	10^{-12}	p	1 pF = 1 picofarad = 10^{-12} farad
nano-	10^{-9}	n	1 ns = 1 nanosecond = 10^{-9} second
micro-	10^{-6}	μ	1 μA = 1 microampere = 10^{-6} ampere
milli-	10^{-3}	m	1 mg = 1 milligram = 10^{-3} gram
centi-	10^{-2}	c	1 cm = 1 centimeter = 10^{-2} meter
kilo-	10^{3}	k	1 kW = 1 kilowatt = 10^{3} watts
mega-	10^{6}	M	1 MJ = 1 megajoule = 10^{6} joules
giga-	10^{9}	G	1 GeV = 1 gigaelectron-volt = 10^{9} electron volts

* See Appendix B for a review of powers-of-ten notation.

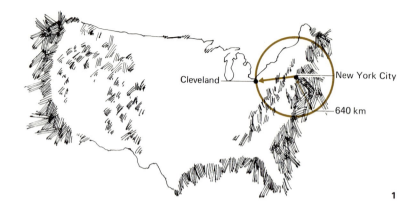

Cleveland

New York City

640 km

1

The simplest example of a vector quantity is *displacement*, which is a change in position. If we are told that a certain airplane leaves New York City and flies for 640 km (a scalar statement), we know nothing about its actual path or final destination; it could land anywhere within a circle of radius 640 km whose center is New York. If we are told instead that the airplane flies for 640 km to the west (a vector statement), we know where it is headed and can identify its destination as Cleveland. (1)*

Two other vector quantities of special significance are *velocity* and *force*. The velocity of a body incorporates both its speed (a scalar quantity) and its direction: the speed of a car might be 100 km/hr, whereas its velocity is 100 km/hr to the north. This may seem a minor distinction, but it is not. Given its speed and direction, we can readily calculate where a moving body will be after a certain time interval, but without knowing the direction we cannot determine the path of the body at all.

_____

* The boldface numbers in parentheses are a key to the figures.

A force is often spoken of as a "push" or a "pull." Though there is more to the concept of force than this description indicates, it is adequate for the time being and we shall postpone a more elaborate discussion until later. Forces are important because they are responsible for all changes in motion: a force is needed to start a stationary object in motion and to deviate or stop a moving one. Evidently the direction as well as the magnitude of a force must be known in order to assess its effects, and vector methods must be used to treat forces as well as to treat displacements and velocities.

VECTORS

The most straightforward way to represent a vector quantity is to draw a straight line with an arrowhead at one end to indicate the direction of the quantity. The length of the line is proportional to the magnitude of the quantity. A line of this kind is called a *vector*.

Here is how a 640-km westward displacement might be shown, with a compass rose to establish orientation and a distance scale (here 1 cm = 100 km) to establish the relationship

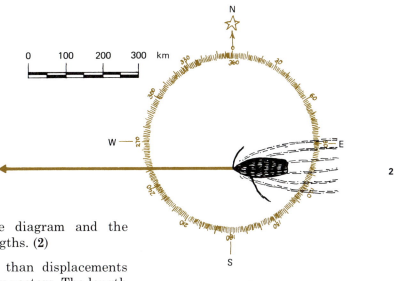

between lengths in the diagram and the corresponding actual lengths. (**2**)

Vector quantities other than displacements can also be represented by vectors. The length of the vector in each case is proportional to the magnitude of the quantity it represents, and its direction is that of the quantity. Thus a velocity of 20 m/s is represented on a scale of 1 cm = 5 m/s by an arrow 4 cm long, and a force of 350 lb is represented on a scale of 1 in. = 1000 lb by an arrow 0.35 in. long. (**3**)

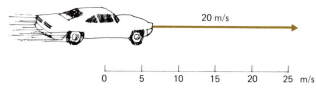

The symbols of vector quantities are customarily printed in boldface italic type (**F** for force), while lightface italics are used both for scalar quantities (*m* for mass) and for the magnitudes of vectors. Thus we might denote a 640-km westward displacement by the symbol **A** and its magnitude of 640 km by the symbol *A*. In handwriting, vector quantities are indicated by placing arrows over their symbols, for instance $\vec{A}$.

VECTOR ADDITION

Ordinary arithmetic is used to add two or more scalar quantities of the same kind together. Ten kg of potatoes plus 4 kg of potatoes

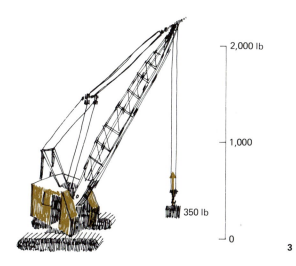

equals 14 kg of potatoes. The same is true for vector quantities of the same kind whose directions are the same. Thus a net upward force of 200 newtons (N) is required to lift a 200-N weight, regardless of whether the force is applied by one, two, or ten men. It is the net force on a body, not the individual forces, that affects the motion of the body. (4)

What do we do when the directions are different? A man who walks 7.0 km north and then 4.0 km west does *not* undergo a net displacement of 11 km from his starting point, even though he has walked a total of 11 km. A vector diagram provides a convenient method for determining his actual displacement. The procedure is to make a scale drawing of the successive displacements *A* and *B*, and to join the starting point and the terminal point with a single vector *R*. The required net displacement is *R*, whose length corresponds to 8.1 km and whose direction corresponds to 30° west of north. (5)

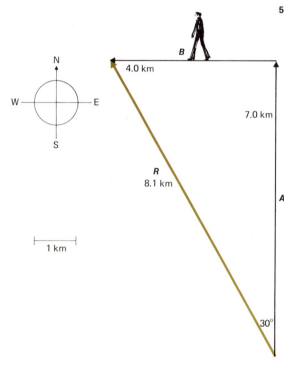

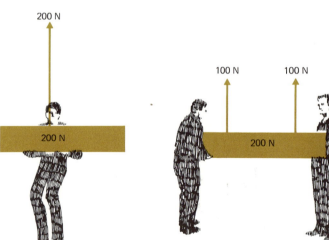

4

The general rule for adding vectors is as follows. To add **B** to **A**, shift **B** parallel to itself until its tail is at the head of **A**. In its new position, **B** must have its original length and direction. The vector sum $A + B$ is a vector **R** (often called the *resultant*) drawn from the tail of **A** to the head of **B**. To find the magnitude R of the resultant **R**, measure the length of **R** on the diagram and compare this length with the scale. The direction of **R** with respect to **A** or **B** may be determined with a protractor. (6)

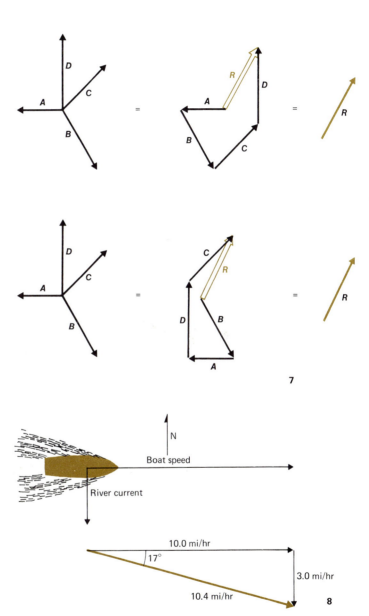

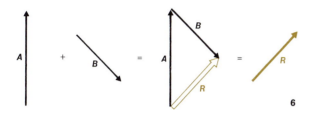

6

The same procedure may be used with any number of vectors: place the tail of each vector at the head of the previous one, keeping their lengths and original directions unchanged, and then draw a vector **R** from the tail of the first vector to the head of the last. **R** is the required resultant. The order in which the vectors are added does not affect the magnitude or direction of the resultant. (7)

7

Vector diagrams may be used to add vector quantities of any kind, not just forces and displacements. Here is an example of the vector addition of velocities. A boat is heading east at 10.0 mi/hr in a river flowing south at 3.0 mi/hr; what is the boat's velocity relative to the earth? Adding the vectors representing these velocities yields a single vector that represents a resultant velocity whose magnitude is 10.4 mi/hr and whose direction is about 17° south of east. (8)

8

VECTOR COMPONENTS

Just as we can add together two or more vectors to give a single resultant vector, so we can break up a vector into two or more others. The process of replacing one vector by two or more others is called *resolving* the vector, and the new vectors are called the *components* of the original one. Often the best way to analyze a physical problem is to resolve a vector into components.

When we add the vector A to the vector B, we get the vector C, which is the vector sum of A and B. If instead we start with the vector C, we can equally correctly resolve it into the two vectors A and B. A and B are the components of C. (9)

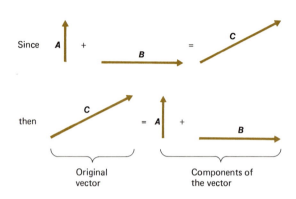

Since A + B = C

then C = A + B

Original vector — Components of the vector

9

When a boy pulls a wagon with rope that is at the angle θ above the ground, only part of the force he exerts is reflected in the motion of the wagon, since the wagon moves horizontally while the force F is not horizontal. We can resolve F into two component vectors, F_x and F_y, where

$$F_x = \text{horizontal component of } F,$$
$$F_y = \text{vertical component of } F. \text{ (10)}$$

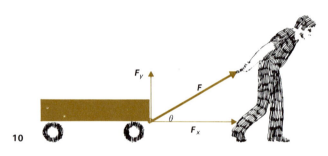

10

The horizontal component F_x is responsible for the wagon's motion, while the vertical component F_y merely pulls upward on it. F_x is the projection of F in the horizontal direction, and F_y is the projection of F in the vertical direction. Almost invariably the components of a vector are chosen to be perpendicular to one another, though not necessarily horizontal and vertical relative to the earth's surface.

If the force the boy exerts on the wagon has the magnitude $F = 15.0$ lb and $\theta = 24°$, from a vector diagram we find that

$$F_x = 13.7 \text{ lb,}$$
$$F_y = 6.1 \text{ lb. (11)}$$

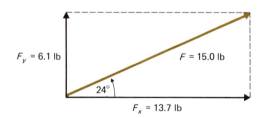

F_y = 6.1 lb F = 15.0 lb

24°

F_x = 13.7 lb

11

14   Scalars and vectors

We note that the algebraic sum of the magnitudes F_x and F_y is 13.7 lb + 6.1 lb = 19.8 lb, although the boy exerted a force of only 15 lb. Where is the mistake? The answer is that there is no mistake: F_x and F_y are vectors whose directions are different, so the *only* correct way to add them is vectorially. The sum of the magnitudes F_x and F_y has no meaning at all.

EXERCISES

1. Is it correct to say that scalar quantities are abstract, idealized quantities with no precise counterparts in the physical world, whereas vector quantities properly represent reality?

2. What kind of quantity is the magnitude of a vector quantity?

3. The resultant of three vectors is zero. Must they all lie in a plane?

4. What is the minimum number of unequal forces whose vector sum can equal zero?

5. A sled has a stick attached to it at an angle of 30° above the horizontal. Would you rather move the sled by pushing down along this stick or by pulling along it?

6. A force of 25 N and another of 35 N act on the same object. What is the maximum total force they can exert? the minimum total force?

7. Three 10-lb forces act on the same object. What is the maximum total force they can exert on it? the minimum total force?

8. A force of 3 N acts perpendicularly to a force of 4 N. What is the magnitude of their resultant?

9. Which of the following sets of displacements might be able to return a car to its starting point?
   a) 2, 8, 10, and 25 mi
   b) 5, 20, 35, and 65 mi
   c) 60, 120, 180, and 240 mi
   d) 100, 100, 100, and 400 mi

10. A force of 20 lb is applied to the lower end of a window pole that is 53° above the horizontal. How much vertical force is being applied to the window?

11. In going from one city to another, a car travels 120 mi west, 50 mi north, and 40 mi southeast. How far apart are the two cities?

12. The *Queen Mary* is heading due west at 20 mi/hr in the presence of a southeast wind of 15 mi/hr. In what horizontal direction will smoke from the liner's funnel go?

13. The ketch *Minots Light* is heading northeast at a speed of 7 knots through a tidal stream which is flowing southeast at 3 knots. Find the magnitude and direction of its velocity relative to the earth's surface.

14. Two tugboats are towing a ship. One of them exerts a horizontal force of 7.5 tons and the other a horizontal force of 9.6 tons; the angle between the towropes is 30°. Find the resultant force exerted on the ship.

15. Find the vertical and horizontal components of a 100-lb force that is directed 40° above the horizontal.

16. A body is moving in a plane at 12 m/s in a direction 40° clockwise from the $+x$-axis. What are the x- and y-components of its velocity?

17. Two particles are moving in a plane; one has the velocity components $v_x = 1$ m/s, $v_y = 5$ m/s, and the other has the velocity components $v_x = 4$ m/s, $v_y = 3$ m/s. If both started from the same point, what is the angle between their paths?

18. A length of pipe is mounted on a truck traveling at 30 km/hr. It is raining, and the raindrops fall vertically downward at 50 m/s. At what angle with the vertical should the pipe be tilted so the raindrops pass through it without touching its sides?

3

VELOCITY

"For every apparent change in place occurs on account of the movement either of the thing seen or of the spectator, or on account of the necessarily unequal movement of both. For no movement is perceptible relative to things moved equally in the same direction—I mean relative to the thing seen and the spectator." Nicolas Copernicus (1473–1543)

Everything in the physical world is in motion, from the electrons, protons, and neutrons within atoms to the largest galaxies of stars. Since the goal of physics is an understanding of the nature and behavior of the physical world, the physicist is directly concerned with things in motion in nearly all of his work.

FRAME OF REFERENCE

When something has changed its position with respect to its surroundings, we say it has *moved*.

There are two separate ideas here. One is that of *change*: when something has moved, the world is not exactly the same as it was before. The other idea is that of *frame of reference*: if we are to notice that something is moving, we must be able to check its position relative to something else. The choice of a frame of reference for reckoning motion depends upon the situation. For a car, the most convenient frame of reference is the earth's surface; for a sailor, it is the deck of his ship; for a planet, it is the sun; for an electron in an atom, it is the atom's nucleus.

The key to solving many problems in physics is the proper selection of a frame of reference. Newton was able to interpret the motions of the earth and planets in terms of the gravitational pull of the sun only because he recognized the sun as the center of the solar system. If instead he had considered the earth as the center, with the sun and the other planets moving around it, their paths would have appeared to be very irregular and it is unlikely that these seemingly complicated motions would have permitted him to find the law of gravitation.

CONSTANT VELOCITY

Let us imagine we have marked off 100-m intervals with chalk on a straight road, and that we are now standing beside the road with a watch, pencil, and paper in hand. A car comes along. (1)

We note the exact time at which the front of the car passes each of the marks we made on the road. The measurements are shown in Table 3.1.

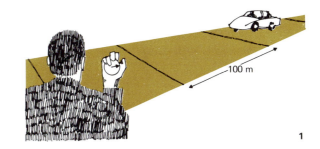

Table 3.1

Total distance, m	0	100	200	300	400	500
Elapsed time, s	0	4.55	9.09	13.6	18.2	22.7

To interpret these measurements a sheet of graph paper is helpful. Along the bottom of the sheet we establish a time scale running from 0 to 25 s, and on the left-hand side we establish a distance scale running from 0 to 500 m. Next we plot each of our measurements as a point on the graph, where the point is the intersection of the lines on the graph paper that pass through the time and distance figures on the two scales. The point representing 300 m and 13.6 s is shown here. (2)

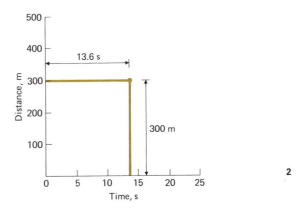

Each point on the complete graph represents the results of a measurement. The various points come very close to lying along a straight line. The line does not go through all the points, but we may reasonably attribute the deviations to experimental errors. We have no way of knowing from the graph itself whether additional measurements between those we did take would also fall on the same line, but let us suppose that if we had made such additional measurements, they too would conform to this line. (3)

When a graph of one quantity versus another results in a straight line, each quantity is *directly proportional* to the other. We can verify this proportionality from the graph,

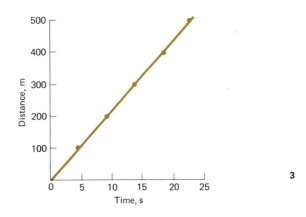

where we see that doubling t means that s also doubles, tripling t means that s also triples, and so on. Thus we can write

$$\text{Distance} = \text{velocity} \times \text{time},$$
$$s = vt,$$

where v, the *velocity* of the car, is the constant of proportionality. It is the rate at which the distance covered by the car changes with time. The greater the time t, the farther the car goes, in exact proportion to t; hence the car is said to move with *constant velocity*.

Velocity is a vector quantity and in general is denoted by the vector symbol $\mathbf{v}$. However, since we are concerned with straight-line motion for the time being, we can consider velocity to be a scalar quantity with the provision that a given direction (here the one in which the car is headed) is to be reckoned as positive and the opposite direction is to be reckoned as negative. Thus the car's velocity is positive when it is moving ahead and is negative when it is moving in reverse. In general, the magnitude of a velocity is called *speed*.

How can we find the value of v, the velocity of the car? What we do is rewrite the equation $s = vt$ in the form

$$\text{Velocity} = \frac{\text{distance}}{\text{time}},$$
$$v = \frac{s}{t},$$

and then calculate s/t for each of the measurements. The results, worked out to four figures, are given in Table 3.2. Only three of these figures are significant, however, since the original measurements had only three significant figures. When the values of v are rounded off, we find that

$$v = 22.0 \, \frac{\text{m}}{\text{s}}.$$

Table 3.2

Total distance, m	0	100	200	300	400	500
Elapsed time, s	0	4.55	9.09	13.6	18.2	22.7
Velocity, m/s		21.98	22.00	22.06	21.98	22.03

The advantage of knowing that the velocity of a particular car (or other body) is constant is that we can then predict exactly how far it will go in a given period of time; or, given the distance it is to travel, we can determine the time required.

Problem. A car moves at the constant velocity of 30 m/s. How far will it travel in exactly 1 hr?

Solution. First we note that

$$1 \text{ hr} = 60 \text{ min} \times 60 \, \frac{\text{s}}{\text{min}} = 3600 \text{ s}.$$

Hence we have

$$\text{Distance} = \text{velocity} \times \text{time},$$
$$s = vt$$
$$= 30 \, \frac{\text{m}}{\text{s}} \times 3600 \text{ s}$$
$$= 108{,}000 \text{ m}.$$

The proper way to express this result is

$$s = 1.1 \times 10^5 \text{ m},$$

since only two figures are significant here.

Problem. How long does it take the same car to travel 6.0 km?

Solution. For this problem we must rewrite the basic equation $s = vt$ in the form

$$\text{Time} = \frac{\text{distance}}{\text{velocity}},$$
$$t = \frac{s}{v}.$$

Since there are 1000 m in a kilometer, we have

$$t = \frac{s}{v}$$

$$= \frac{6.0 \times 1000 \text{ m}}{30 \text{ m/s}} = \frac{6000 \text{ m}}{30 \text{ m/s}}$$

$$= 200 \text{ s} = 2.0 \times 10^2 \text{ s}$$

since again the accuracy of the answer extends to two significant figures only.

INSTANTANEOUS VELOCITY

Not all cars have constant velocities. In Table 3.3 are the data that might be taken for a car that starts on the marked stretch of road from a stationary position at $t = 0$.

Table 3.3

Total distance, m	0	100	200	300	400	500
Elapsed time, s	0	28	40	49	57	63

When we plot these data on a graph, we find that the line joining the various points is not straight but shows a definite upward curve. In each successive interval of time (as marked off at the bottom of the graph) the car goes a greater distance; in other words, the car is going faster and faster. Because s is not directly proportional to t, the car's velocity is not constant. (4)

Even though v varies, at every instant it has a definite value. To find this *instantaneous velocity* at a particular time t, we draw a straight line tangent to the v–t curve at that value of t. (The length of the line does not matter.) Then we determine v from that straight line from the formula

$$v = \frac{\Delta s}{\Delta t}$$

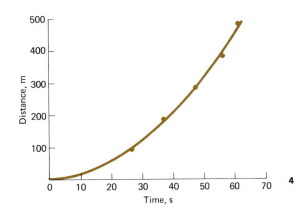

where Δs is the distance interval between the ends of the line and Δt is the time interval between them. The instantaneous velocity of the car at $t = 40$ s is, from the graph,

$$v = \frac{\Delta s}{\Delta t}$$

$$= \frac{100 \text{ m}}{10 \text{ s}} = 10 \frac{\text{m}}{\text{s}}. \quad \textbf{(5)}$$

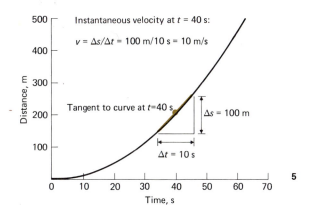

6

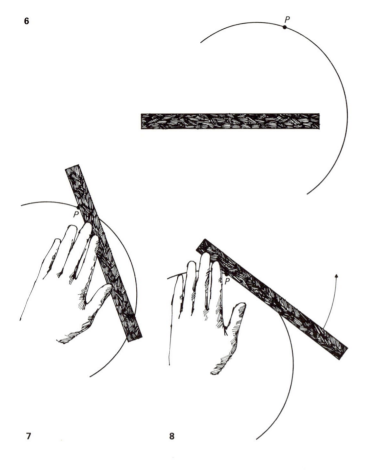

When we define instantaneous velocity as the tangent to the *s–t* curve, we are simply applying our previous procedure to a very short segment of the curve. The shorter the segment of the curve, the more nearly straight it is. In the limit of infinite shortness, the segment is a straight line whose slope is the same as that of the tangent to the curve there. Thus the tangent is equal to the velocity *v* in a time interval of vanishing duration, which is what is meant by instantaneous velocity.

In Table 3.4 are the instantaneous velocities of the car at 10-s intervals as determined from the preceding graph.

Table 3.4

Elapsed time, s	0	10	20	30	40	50	60
Instantaneous velocity, m/s	0	2.5	5.0	7.5	10	12.5	15

HOW TO DRAW A TANGENT

Here is how a tangent line is drawn to a curve at some point *P* with the help of a ruler. (**6**)

First the ruler is placed across the curve from *P* to some other nearby point. (**7**)

The ruler is then turned, using *P* as the pivot. (**8**)

7   **8**

When the ruler just barely touches the curve at *P*, a line drawn along the ruler's edge is the tangent. (**9**)

9

AVERAGE VELOCITY

In some situations we are interested in the instantaneous velocity of a moving body at a certain specific time, but very often our chief concern is with its *average velocity*. The average velocity, denoted $\bar{v}$, of a body during a period of time t is the total displacement s it has undergone in that period of time divided by t:

$$\text{Average velocity} = \frac{\text{total displacement}}{\text{total time}},$$

$$\bar{v} = \frac{s}{t}.$$

We note this is the same as the definition of v for a body moving at constant velocity, since in that case v and $\bar{v}$ are the same. For a body whose velocity varies, however, the instantaneous velocity is in general different from the average velocity.

Let us find the average velocity of the second car, whose instantaneous velocity went from $v = 0$ at the start to $v = 15$ m/s after 60 s had gone by. Since the car went a total distance of 450 m in the 60 s (which we find from the graph), its average velocity was

$$\bar{v} = \frac{s}{t}$$

$$= \frac{450 \text{ m}}{60 \text{ s}}$$

$$= 7.5 \frac{\text{m}}{\text{s}}.$$

Problem. An airplane takes off at 9:00 A.M. and flies in a straight path at 300 km/hr until 1:00 P.M. At 1:00 P.M. its velocity is increased to 400 km/hr and it maintains this velocity in the same direction as before until it lands at 3:30 P.M. What is its average velocity for the entire flight?

Solution. To calculate the average velocity of the airplane we must know the total distance s that is covered and the total time t involved. In the first part of the flight the airplane covers

$$s_1 = v_1 t_1$$

$$= 300 \frac{\text{km}}{\text{hr}} \times 4.00 \text{ hr}$$

$$= 1200 \text{ km},$$

and in the second part it covers

$$s_2 = v_2 t_2 = 400 \frac{\text{km}}{\text{hr}} \times 2.50 \text{ hr} = 1000 \text{ km}.$$

Hence we obtain for s and t the values

$$s = s_1 + s_2 = (1200 + 1000) \text{ km} = 2200 \text{ km},$$

$$t = t_1 + t_2 = (4.00 + 2.50) \text{ hr} = 6.50 \text{ hr},$$

and the average velocity is

$$\bar{v} = \frac{s}{t}$$

$$= \frac{2200 \text{ km}}{6.50 \text{ hr}}$$

$$= 338 \frac{\text{km}}{\text{hr}}.$$

The instantaneous velocity of some object may change in direction as well as in magnitude continuously, but the rule for finding average velocity is essentially the same as before, except that displacement rather than distance is involved. We compute the average speed from $\bar{v} = s/t$, where the s now represents the magnitude of the displacement s from the position of the object at the start of the time interval being considered to its position at the end of the time interval. The direction of v is the direction of the displacement vector

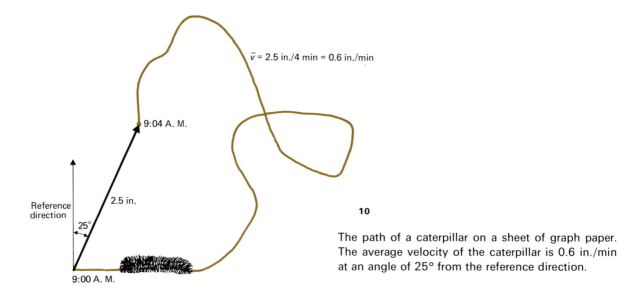

$\bar{v}$ = 2.5 in./4 min = 0.6 in./min

9:04 A. M.

Reference direction

2.5 in.

25°

9:00 A. M.

10

The path of a caterpillar on a sheet of graph paper. The average velocity of the caterpillar is 0.6 in./min at an angle of 25° from the reference direction.

from the starting point to the terminal point. As shown, this procedure yields the average velocity regardless of the vagaries of the object. (**10**)

The terms *speed* and *velocity* are often used interchangeably, since the direction in which something is moving may be understood by implication. However, it is good practice to reserve "speed" for the rate at which something covers distance and "velocity" for the complete picture of its motion in space. A boat traveling in a circle has a constant speed of 20 mi/hr, but its instantaneous and average velocities are both continually changing, with the latter equaling zero periodically. Each of these terms, then, has a different meaning and each has its own area of usefulness in describing motion.

EXERCISES

1. A ship travels 12 mi at constant speed in 40 min. What is its speed in mi/hr?

2. A car travels at 40 mi/hr for 2 hr, at 50 mi/hr for 1 hr, and at 20 mi/hr for $\frac{1}{2}$ hr. What is its average speed?

3. The speed of light is 3×10^8 m/s. How many minutes does it take light to reach us from the sun, which is 1.5×10^{11} m away?

4. Echoes return in 4 s to a man standing in front of a cliff. How far away is the cliff? (The speed of sound in air at sea level is about 330 m/s.)

5. A man hears the sound of thunder 6 s after seeing a lightning flash. How far away was the lightning? (The speed of light is so great, 3×10^8 m/s, that the man sees the flash a negligible time after it occurs.)

6. A woman observes lightning strike a building 1 km away from her. How long must she wait until she hears the thunder?

7. In climbing a wall 1.00 m high, a caterpillar slips down 1 cm after it has ascended 2 cm, repeating the process until it reaches the top. The caterpillar takes 10 min to get to the top of the wall. Find its average speed and average velocity in cm/s during the entire climb.

8. What is the air speed of an airplane that requires 3 hr and 30 min to go 1200 mi from one city to another when it has a 70 mi/hr tail wind?

9. (a) What is the speed with respect to the ground of an airplane whose air speed is 200 mi/hr when it is bucking a head wind of 56 mi/hr? (b) What distance can it cover relative to the ground in 5 hr?

10. Measurements of the light reaching us from distant galaxies of stars indicate that these galaxies are all moving outward with speeds that are proportional to their distance r from the earth, so that in general $v = Hr$ where H, called the *Hubble constant*, has a value of about $2.4 \times 10^{-18} \, \mathrm{s}^{-1}$. The farther a galaxy is from us, the faster it is receding; thus a galaxy $10^{25} \, \mathrm{m}$ away has a recession speed of about 2.4×10^7 m/s. These outward motions of the galaxies support the notion that the entire universe is expanding. Although the galaxies seem to be receding from the earth, this does not mean the earth is at the center of the universe. The situation is like that of a spotted balloon: as the balloon is inflated, the spots move farther apart from one another. Use the above value of H to estimate the age of the universe, that is, how long ago the expansion began. Express the answer in years (1 yr $\approx 3 \times 10^7$ s).

4

ACCELERATION

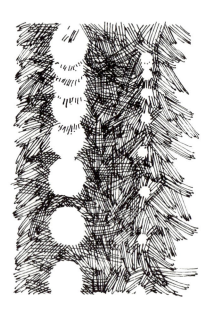

In the real world, few bodies move at constant velocity for very long. The velocity of a body changes only when it interacts with something else in some way, which means that such interactions occur almost all the time. For the moment our concern is with how changing velocities are described; much of the remainder of this book is devoted to the interactions responsible for them.

ACCELERATED MOTION

A body whose velocity is not constant is said to be *accelerated*. This term is applied regardless of whether v is increasing, decreasing, or changing in direction. Here are three examples of accelerated motion which show the successive positions of a car after equal periods of time:

Increasing velocity. The car goes faster and faster, so it travels a longer distance in each period of time. (**1**)

Decreasing velocity. The car goes slower and slower, so it travels a shorter distance in each period of time. (**2**)

Changing direction. The magnitude of the car's velocity is constant, but its direction changes. This car is also accelerated. (**3**)

Just as velocity is the rate of change of displacement with time, *acceleration* is the rate of change of velocity with time. The symbol for acceleration is a, since it is a vector quantity. For the time being we are discussing straight-line motion only, so we shall consider acceleration as a scalar quantity a.

If a body's velocity is v_0 to begin with and changes to v after a time interval of t, the body's acceleration is given by the formula:

$$\text{Acceleration} = \frac{\text{change in velocity}}{\text{time interval}},$$

$$a = \frac{v - v_0}{t}.$$

When the final velocity v is greater than the initial velocity v_0, the acceleration is positive, which signifies that the body is going faster and faster. When the final velocity is *less* than the initial velocity, the acceleration is negative, which signifies that the body is going slower and slower.

Just what is it that causes something to be accelerated? The answer is a force, which for the moment we can think of as a push or a pull. In the absence of a force, an object at rest remains at rest and an object in motion continues in motion at constant velocity. These matters are discussed in detail in Sections 7 to 9.

CONSTANT ACCELERATION

Let us return to the data on car No. 2 of the previous section and plot a graph of its instantaneous velocity v versus time t. All the

1

2

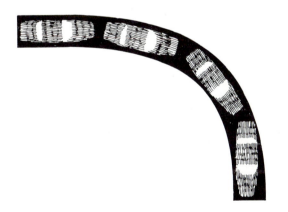

3

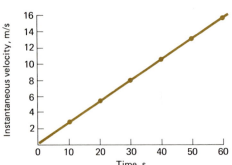

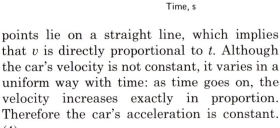

4

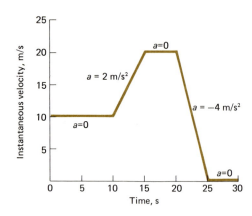

5

points lie on a straight line, which implies that v is directly proportional to t. Although the car's velocity is not constant, it varies in a uniform way with time: as time goes on, the velocity increases exactly in proportion. Therefore the car's acceleration is constant. **(4)**

From its definition, acceleration is expressed in terms of

$$\frac{\text{Velocity}}{\text{time}} = \frac{\text{distance/time}}{\text{time}} = \frac{\text{distance}}{\text{time}^2}.$$

A body whose acceleration increases by 10 m/s in each second would thus have its acceleration expressed as

$$10 \, \frac{\text{m/s}}{\text{s}} = 10 \, \frac{\text{m}}{\text{s}^2}.$$

Let us calculate the acceleration of car No. 2. The initial velocity of the car is $v_0 = 0$ and after, say, $t = 20$ s it is $v = 5.0$ m/s. Hence we have

$$a = \frac{v - v_0}{t}$$

$$= \frac{(5.0 - 0) \text{ m/s}}{20 \text{ s}}$$

$$= 0.25 \text{ m/s}^2.$$

If we make the same calculation at the later time $t = 40$ s, we have, since v is now 10 m/s,

$$a = \frac{(10 - 0) \text{ m/s}}{40 \text{ s}} = 0.25 \text{ m/s}^2.$$

The value of a is the same because the acceleration is constant. If the acceleration were not constant, different values of a would be obtained at different times.

Not all accelerations are constant, of course, but a great many real motions are best understood by idealizing them in terms of constant accelerations, and we shall accordingly restrict ourselves to the latter.

Problem. Discuss the motion of the car whose velocity–time graph is shown here. **(5)**

Solution. A horizontal line in a v–t graph means that v does not change, so the acceleration is 0 at first. A line sloping upward means a positive acceleration (v increasing), and a line sloping downward means a negative acceleration (v decreasing). Hence the car at first moves with the constant velocity of 10 m/s, then is accelerated at 2 m/s^2 to the velocity of 20 m/s, again travels at constant speed for 5 s, and finally undergoes a negative acceleration of -4 m/s^2 until it comes to rest.

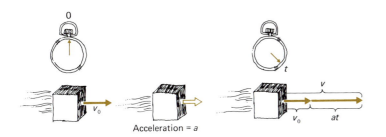

Acceleration = a

6

VELOCITY AND ACCELERATION

In the event a body starts to accelerate from some initial velocity v_0, its change in velocity at during the time interval t in which the acceleration a (assumed constant) occurs is added to v_0. Hence the final velocity v at a time t after the acceleration begins is the initial velocity v_0 plus the change in velocity at:

Final velocity = initial velocity +
                              velocity change,

$$v = v_0 + at. \ \textbf{(6)}$$

Problem. A car has an initial velocity of 30 m/s and an acceleration of -1.5 m/s^2. Find its velocity after 10 s and after 50 s have elapsed.

Solution. The car's velocity after 10 s is

$$v = v_0 + at$$

$$= 30\,\frac{m}{s} - 1.5\,\frac{m}{s^2} \times 10\ s$$

$$= (30 - 15)\,\frac{m}{s}$$

$$= 15\,\frac{m}{s}.$$

The velocity of the car is less than it was originally, since a negative acceleration means that v is decreasing. After 50 s the velocity of the car, if a stays the same, is

$$v = v_0 + at$$

$$= 30\,\frac{m}{s} - 1.5\,\frac{m}{s^2} \times 50\ s$$

$$= (30 - 75)\,\frac{m}{s}$$

$$= -\,45\,\frac{m}{s}.$$

The minus sign means that the car is now going in the opposite direction (we can think of it as going backward).

DISPLACEMENT AND ACCELERATION

A body has some initial velocity v_0 when it begins to be uniformly accelerated. As we have seen, its velocity after a time t is $v = v_0 + at$. The next question to ask is, How far does the body go during the time interval t?

We know that, in general, $s = \bar{v}t$, so that if we can determine the average velocity during the time interval t we can also find s, the distance through which the body moves.

Because the acceleration a is constant, v is changing at a uniform rate, and

$$\bar{v} = \frac{\text{initial velocity} + \text{final velocity}}{2}.$$

Here the initial velocity is v_0 and the final velocity is $v_0 + at$; hence

$$\bar{v} = \frac{v_0 + v_0 + at}{2} = v_0 + \tfrac{1}{2}at.$$

The distance traveled is accordingly

$$s = \bar{v}t = (v_0 + \tfrac{1}{2}at) \times t$$

which yields the very useful formula

$$s = v_0 t + \tfrac{1}{2}at^2. \; \textbf{(7)}$$

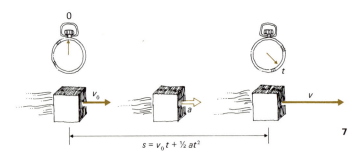

$s = v_0 t + \tfrac{1}{2} at^2$

7

Problem. A car has an initial velocity of 30 m/s and an acceleration of -1.5 m/s^2. Find how far it goes in the first 10 s and the first 50 s after the acceleration begins.

Solution. In the first 10 s the car travels

$$s = v_0 t + \frac{1}{2}at^2$$

$$= 30\,\frac{\text{m}}{\text{s}} \times 10\text{ s} + \frac{1}{2} \times \left[-1.5\,\frac{\text{m}}{\text{s}^2}\right] \times (10\text{ s})^2$$

$$= (300 - 75)\text{ m} = 225\text{ m},$$

and after 50 s,

$$s = v_0 t + \frac{1}{2}at^2$$

$$= 30\,\frac{\text{m}}{\text{s}} \times 50\text{ s} + \frac{1}{2} \times \left[-1.5\,\frac{\text{m}}{\text{s}^2}\right] \times (50\text{ s})^2$$

$$= (1500 - 1875)\text{ m}$$

$$= -375\text{ m}.$$

The latter result means that the car is 375 m *behind* its starting point after 50 s have elapsed.

It is not hard to find a relationship among s, v_0, v, and a that does not directly involve the time t. The first step is to rewrite the defining formula for acceleration, which is

$$a = \frac{v - v_0}{t},$$

in the equivalent form

$$t = \frac{v - v_0}{a}.$$

This gives the time during which the velocity of the body involved changed from v_0 to its final value of v. Now we substitute this expression for t into the formula for the distance traveled,

$$s = v_0 t + \frac{1}{2}at^2.$$

The result is

$$s = v_0 \frac{(v - v_0)}{a} + \frac{1}{2}a\frac{(v - v_0)^2}{a^2}$$

$$= \frac{v_0 v}{a} - \frac{v_0^2}{a} + \frac{v^2}{2a} - \frac{v_0 v}{a} + \frac{v_0^2}{2a}$$

$$= \frac{v^2}{2a} - \frac{v_0^2}{2a}.$$

Finally, by multiplying both sides of the last equation by $2a$ and rearranging the terms, we arrive at the formula

$$v^2 = v_0^2 + 2as.$$

Problem. How far will the car of the previous example have gone when it comes to a stop?

Solution. When the car is at rest, $v = 0$, and so, since

$$v_0 = 30 \text{ m/s}$$

and

$$a = -1.5 \text{ m/s}^2,$$

we have

$$v^2 = v_0^2 + 2as,$$

$$s = \frac{v^2 - v_0^2}{2a}$$

$$= \frac{0^2 - (30 \text{ m/s})^2}{2 \times (-1.5 \text{ m/s}^2)}$$

$$= \frac{900 \text{ m}^2/\text{s}^2}{3 \text{ m/s}^2} = 300 \text{ m}.$$

The car comes to a stop after it has gone 300 m and then, since its acceleration is negative, it begins to move in the negative (backward) direction.

These examples have been worked out not because they are, in themselves, especially significant, but because they illustrate the power of the mathematical approach to physical phenomena. By defining certain quantities and relating them to each other and to events that actually occur in the real world, a whole theoretical structure of equations may be built up. This structure is an instrument enabling us to solve problems which otherwise would each require a separate, perhaps difficult or impossible, experiment. We must remember that the validity of the theoretical structure depends upon its experimental basis; but once this is established, we may proceed to work out its consequences with pencil and paper.

THE ACCELERATION OF GRAVITY

Drop a stone, and it falls. Does the stone fall at a constant speed, or is it accelerated? Does the motion of the stone depend upon its weight, or its size, or its color? Over two thousand years ago questions such as these were answered by Greek philosophers, notably Aristotle, on the basis of "logical reasoning" only. To them it seemed reasonable that heavy things should fall faster than light things, for example. Almost nobody felt it necessary to perform experiments to seek information on the physical universe until Galileo revolutionized science by doing just that: performing experiments. Modern science owes its success in understanding and utilizing natural phenomena to its reliance upon experiment.

What Galileo found, as the result of measurements made on balls rolling down inclined planes, was that *all freely falling bodies have the same acceleration* near the earth's surface. This acceleration, which is called the acceleration of gravity (symbol g), has the value

$$g = 9.81 \text{ m/s}^2$$

to three significant figures. In most problems in elementary physics it is sufficient to let $g = 9.8 \text{ m/s}^2$. (In British units, $g = 32 \text{ ft/s}^2$.)

A body falling from rest thus has a velocity of 9.8 m/s after the first second, a velocity of 19.6 m/s after the next second, and so on. Hence the greater the distance through which a stone falls after being dropped, the greater

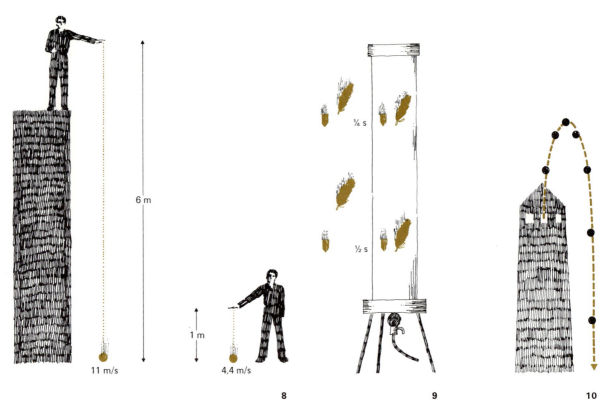

6 m

1 m

11 m/s

4.4 m/s

¼ s

½ s

8   9   10

its speed when it hits the ground. But the stone's *acceleration* is always the same. (**8**)

Another aspect of Galileo's work deserves comment. His conclusion that all things fall with the same constant acceleration is only an *idealization* of reality. The actual accelerations with which objects fall depend upon many factors: the location on the earth, the size and shape of the object, and the density and state of the atmosphere. For example, a bullet falls faster than a feather does in air because of the effects of buoyancy and air resistance. Galileo perceived that the essence of the phenomenon was a constant acceleration downward, with other factors acting merely to cause deviations from the constant

value. In a vacuum the bullet and feather fall with exactly the same acceleration. (**9**)

The acceleration of gravity is independent of the initial state of motion of a body. For example, when a ball is thrown upward from a point above the ground, the downward acceleration of gravity eventually causes its upward velocity to decrease to zero. At this moment the ball is at the maximum height it will reach, and it then starts to fall exactly as though it had been dropped from there. (**10**)

If instead the ball is thrown horizontally from a certain height, it will reach the ground at the same time as another ball that is simply dropped. In both cases the vertical component

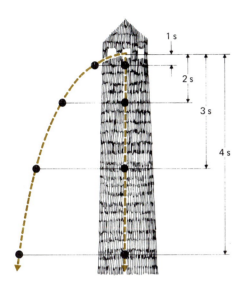

1 s
2 s
3 s
4 s

11

of the ball's velocity is zero at first, both undergo the same acceleration, and so both take the same period of time to fall. **(11)**

FREE FALL

We can apply the formulas we derived earlier for motion under constant acceleration to bodies in free fall. It must be kept in mind, of course, that the direction of g is always downward, no matter whether we are dealing with a dropped object or with one that is initially thrown upward.

Problem. A stone is dropped from the top of the Empire State Building in New York City, which is 450 m high. Neglecting air resistance,

how long does it take the stone to reach the ground? What is its speed when it strikes the ground?

Solution. The general formula for the distance traveled in the time t by an accelerated object is

$$s = v_0 t + \frac{1}{2}at^2.$$

Here $v_0 = 0$, since the stone is simply dropped with no initial velocity, and the acceleration is $a = g$. Hence we have

$$h = \frac{1}{2}gt^2,$$

where h represents vertical distance from the starting point. First we solve this formula for t, which yields

$$t = \sqrt{\frac{2h}{g}},$$

and then we substitute $h = 450$ m and $g = 9.8$ m/s² to obtain

$$t = \sqrt{\frac{2h}{g}}$$

$$= \sqrt{\frac{2 \times 450 \text{ m}}{9.8 \text{ m/s}^2}} = \sqrt{91.8 \text{ s}}$$

$$= 9.6 \text{ s}.$$

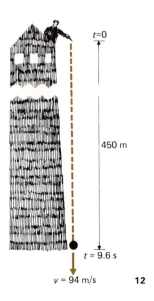

t=0

450 m

t = 9.6 s

v = 94 m/s   **12**

Here $v_0 = 16$ m/s and $a = -9.8$ m/s^2, and at the top of the path $s = h$ and $v = 0$. (We are reckoning up as $+$ and down as $-$ in this calculation.) Hence

$$h = \frac{0 - (16 \text{ m/s})^2}{2 \times (-9.8 \text{ m/s}^2)}$$

$$= \frac{256 \text{ (m/s)}^2}{19.6 \text{ m/s}^2}$$

$$= 13 \text{ m}.$$

When will the stone strike the ground? A body takes precisely as long to fall from a certain height h as it does to rise that high (provided that h is its maximum height, as it is here), just as a body's final velocity when dropped from a height h is the same as the initial velocity required for it to get that high. As we saw earlier,

$$h = \frac{1}{2}gt^2$$

is the distance a body dropped from rest will fall in the time t. Here we have

$$t = \sqrt{\frac{2h}{g}}$$

$$= \sqrt{\frac{2 \times 13 \text{ m}}{9.8 \text{ m/s}^2}}$$

$$= 1.6 \text{ s}.$$

Because the stone takes as long to rise as to fall, the total time it is in the air is twice 1.6 s, or 3.2 s.

To determine the position of the stone a given time after it has been thrown, we make use of the general formula

$$s = v_0 t + \frac{1}{2}at^2.$$

Knowing the duration of the fall makes it easy to compute the stone's final velocity:

$$v = v_0 + at$$

$$= 0 + gt$$

$$= 9.8 \frac{\text{m}}{\text{s}^2} \times 9.6 \text{ s}$$

$$= 94 \text{ m/s}. \text{ (12)}$$

Problem. A stone is thrown upward with an initial velocity of 16 m/s. What will its maximum height be? When will it strike the ground? Where will it be in 0.8 s? in 2.4 s?

Solution. To find the highest point the stone will reach, we rewrite the formula

$$v^2 = v_0^2 + 2as$$

in the form

$$s = \frac{v^2 - v_0^2}{2a}.$$

Here, again reckoning the upward direction as $+$, we have

$$v_0 = 16 \text{ m/s}$$

and

$$a = -9.8 \text{ m/s}^2.$$

For $t = 0.8$ s,

$$h = 16\,\frac{\text{m}}{\text{s}} \times 0.8 \text{ s} - \frac{1}{2} \times 9.8\,\frac{\text{m}}{\text{s}^2} \times (0.8 \text{ s})^2$$

$$= 10 \text{ m}.$$

When we substitute $t = 2.4$ s, we find that again

$$h = 16\,\frac{\text{m}}{\text{s}} \times 2.4 \text{ s} - \frac{1}{2} \times 9.8\,\frac{\text{m}}{\text{s}^2} \times (2.4 \text{ s})^2$$

$$= 10 \text{ m}.$$

All this apparently paradoxical result means is that at 0.8 s the stone is at a height of 10 m on its way *up*, then it goes on further to its maximum height of 13 m, and at 2.4 s it is again at a height of 10 m but this time on the way *down*. (13)

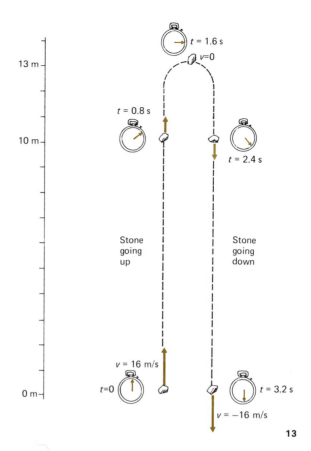

13

EXERCISES

1. A body whose speed is constant
   a) must be accelerated.
   b) might be accelerated.
   c) cannot be accelerated.
   d) has a constant velocity.

2. An example of a body whose motion is *not* accelerated is a car that
   a) turns a corner at the constant speed of 10 mi/hr.
   b) descends a hill at the constant speed of 30 mi/hr.
   c) descends a hill at a speed that increases from 20 mi/hr to 40 mi/hr uniformly.
   d) climbs a hill, goes over its crest, and descends on the other side, all at the constant speed of 30 mi/hr.

3. Can a rapidly moving body have the same acceleration as a slowly moving one?

4. The acceleration of a moving object is constant in magnitude and direction. Must the path of the object be a straight line? If not, give an example.

5. A hunter aims a rifle directly at a squirrel on a branch of a tree. The squirrel sees the flash of the rifle's firing. Should it stay where it is or drop from the branch in free fall at the instant the rifle is fired?

6. A man at the masthead of a sailboat moving at constant velocity drops a wrench. The man is 60 ft above the boat's deck at the time, and the stern of the boat is 36 ft aft of the mast. The

effect of air resistance on the motion of the wrench is negligible. Is there a minimum speed the sailboat can have such that the wrench does not land on the deck? If so, what is this speed?

7. Is it true that an object dropped from rest falls three times farther in the second second after being released than it does in the first second?

8. A movie is shown that appears to be of a ball falling through the air. Is there any way to find out from what appears on the screen if the movie is actually of a ball being thrown upward but is being run backwards in the projector?

9. A car's speed decreases from 18 m/s to 10 m/s in 4 s. Find its acceleration.

10. A car's speed increases from 10 m/s to 25 m/s in 10 s. Find its acceleration.

11. A passenger in an airplane flying from New York to San Francisco notes the times at which he passes over various cities and towns. With the help of a map he determines the distances between these landmarks, and compiles the table shown below. Plot the distance covered by the airplane versus time from these data, and describe the airplane's motion with the help of this graph.

Time (P.M.):	4:00	5:22	5:46	5:55	6:30	6:48	8:32	9:04
Distance (mi):	0	480	620	680	910	1030	1550	1710

12. A train passes successive mile posts at the times indicated below. Plot the data on a graph, and determine from this whether the train's speed is constant or not. If it is not constant, plot the train's speed in each time interval versus time, determine whether its acceleration is constant, and, if so, what the value of the acceleration is.

Distance (mi):	0	1	2	3	4	5	6	7	8
Time (min):	0	1.5	2.9	4.2	5.5	6.7	7.8	8.9	10.0

13. A car undergoes an acceleration of 8 m/s^2 starting from rest. How far does it go in the first 0.1 s? in the first 1 s?

14. The tires of a car begin to lose their grip on the pavement at accelerations greater than about 5 m/s^2. Express this acceleration in terms of the number of seconds required to reach a speed of 60 mi/hr (27 m/s) starting from rest.

15. An airplane takes 20 s and 400 m of runway to become airborne, starting from rest. What is its speed when it leaves the ground?

16. A stone is dropped from a cliff 490 m above the ground. How long does it take to fall?

17. Divers in Acapulco, Mexico, leap from a point 36 m above the water. If there were no air resistance, what would their final speed be?

18. A stone is thrown vertically upward at 9.8 m/s. When will it reach the ground?

19. A ball is thrown vertically downward with an initial speed of 10 m/s. What is its speed 1 s later? 2 s later?

20. A ball is thrown vertically upward with an initial speed of 10 m/s. What is its speed and direction 1 s later? 2 s later?

21. A man in an elevator drops an apple 2 m above the elevator's floor. With a stopwatch he determines how long the apple takes to fall to the floor. What does he find
   a) when the elevator is ascending with an acceleration of 1 m/s^2?
   b) when it is descending with an acceleration of 1 m/s^2?
   c) when it is ascending at the constant speed of 2 m/s?
   d) when it descending at the constant speed of 2 m/s?
   e) When the cable has broken and it is descending in free fall?

5

THE RELATIVITY OF TIME

"I am anxious to draw attention to the fact that this theory is not speculative in origin; it owes its invention entirely to the desire to make physical theory fit observed fact as well as possible." Albert Einstein (1879–1955)

Few physical theories represent so drastic an assault on traditional habits of thought as does the theory of relativity. Relativity links time and space, matter and energy, electricity and magnetism—and for all the seeming magic of its conclusions, most of them can be reached with the simplest of mathematics. The theory of relativity was proposed in 1905 by Albert Einstein, and little of physical science since then has remained unaffected by his ideas.

MEASUREMENTS AND RELATIVE MOTION

In the previous few sections no special point was raised about how measurements of length, time, mass, and electric current are carried out. It was simply assumed that these quantities could be determined in some way, and that since standard units exist for each of them, it doesn't matter who makes a particular determination—everybody ought to get the same figure. There is certainly no question of principle associated with, say, finding the length of an airplane on the ground: all we need do is place one end of a tape measure at the airplane's nose and note the number on the tape at the airplane's tail.

But what if the airplane is in flight and we are on the ground? Certainly it is not hard to find the length of a distant object with the help of a surveyor's transit to measure angles, a tape measure to establish a base line, and a knowledge of trigonometry. When the object is moving, however, things become more complicated, because now we must take into account the fact that light does not travel instantaneously from one place to another but does so at a definite, fixed velocity—and light is the means by which information is carried from a distant object to our measuring instruments. When a careful analysis is made of

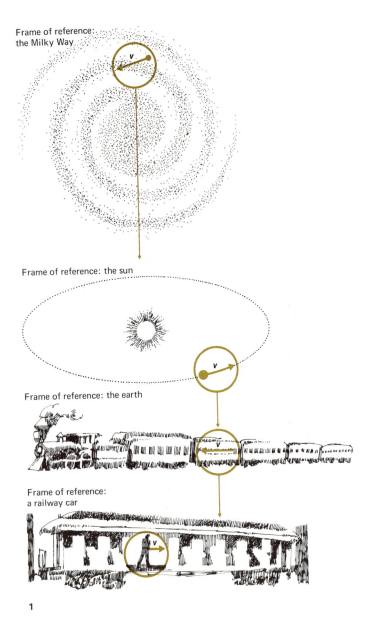

Frame of reference: the Milky Way

Frame of reference: the sun

Frame of reference: the earth

Frame of reference: a railway car

1

THE PRINCIPLE OF RELATIVITY

The theory of relativity rests upon the twin pillars of the principle of relativity and the principle of the constancy of the speed of light. These principles are the fruit of countless experiments and have survived both direct tests of their validity and tests of the validity of the conclusions that follow from them.

The principle of relativity states:

The laws of physics are the same in all frames of reference moving at constant velocity with respect to one another.

To appreciate the significance of the above statement, let us review what is meant by a frame of reference. When we observe something moving, what we actually detect is that its position relative to something else is changing. A passenger moves relative to a train; the train moves relative to the earth; the earth moves relative to the sun; the sun moves relative to the galaxy of stars (the Milky Way) of which it is a member; and so on. In each case a frame of reference is part of the description of the motion: it is meaningless to say that something is moving without specifying with respect to what the motion occurs. (**1**)

There is no universal frame of reference that can be used everywhere. If we see something changing its position with respect to us at constant velocity, we have no way of knowing whether *it* is moving or *we* are moving. If we were isolated from the rest of the universe, we would be unable to find out if we were moving at constant velocity or not—indeed, the question would make no sense. All motion is relative to the observer, and there is no such thing as "absolute motion."

At one time it was thought that the stars are stationary and so could serve as a universal

the problem of measuring physical quantities when there is relative motion between the measuring instruments and whatever is being observed, many surprising results emerge.

frame of reference. But modern observations show that there are no "fixed stars." All the stars in the universe move relative to one another, some of them at relative velocities of over 50,000 miles per second. The reason the motions of the stars are hard to detect is simply that they are extremely far away. Here is what the constellation Big Dipper (a) looked like 200,000 years ago, (b) looks like today, and (c) will look like 200,000 years from now. The arrows in (b) show the directions of motion of the stars that make up the Big Dipper. (**2**)

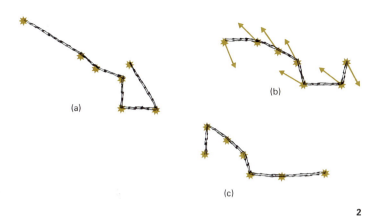

SPECIAL AND GENERAL RELATIVITY

The principle of relativity follows directly from the absence of a universal frame of reference. If the laws of physics were different for different observers in relative motion, they could infer from these differences which of them were "stationary" in space and which were "moving." But such a distinction does not exist in nature, and the principle of relativity is an expression of this fact.

Thus experiments of any kind whatsoever performed in, for instance, an elevator which is ascending at a constant velocity yield exactly the same results as the same experiments performed when the elevator is at rest or is descending at a constant velocity. On the other hand, an isolated observer *can* detect accelerations, as any elevator passenger can verify.

The theory of relativity is divided into two parts. The *special theory of relativity* is confined to situations that involve the motion of frames of reference at constant velocity with respect to one another; the *general theory of relativity* deals with situations that involve frames of reference accelerated with respect to one another. In this book we shall consider only the special theory.

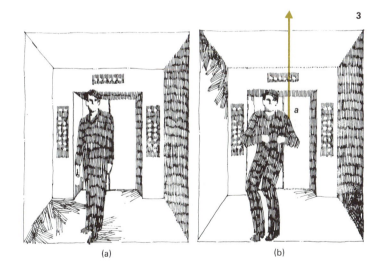

Facing the rear of an elevator, there is no way a passenger can tell whether he is ascending or descending at a constant rate or is at rest. (**3a**)

But he *can* detect an acceleration! (**3b**)

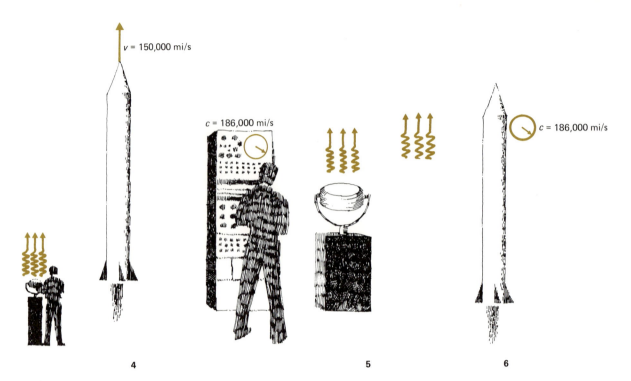

4   5   6

THE VELOCITY OF LIGHT

The principle of the constancy of the velocity of light states:

The velocity of light in free space has the same value for all observers, regardless of their state of motion.

This velocity, denoted by the symbol c, is equal to

$$c = 3.00 \times 10^8 \text{ m/s}$$

which is about 186,000 mi/s.

At first glance the constancy of the velocity of light may not seem so very extraordinary, which is a misleading impression indeed. Let us examine a hypothetical experiment in essence no different from actual experiments that have been performed in a number of ways.

Suppose I turn on a searchlight at the same moment you take off in a spacecraft at a speed of 150,000 mi/s. (**4**)

We both measure the speed of the light waves from the searchlight using identical instruments. From the ground I find their speed to be 186,000 mi/s, as usual. Common sense tells me you ought to find a speed of (186,000–150,000) mi/s or only 36,000 mi/s for the same light waves. (**5**)

But you also find their speed to be 186,000 mi/s, even though to me you seem to be moving parallel to the waves at 150,000 mi/s. As usual, common sense is wrong. (**6**)

There is only one way to account for the apparent discrepancy between the above results without violating the principle of relativity, and that is to conclude that measurements of space and time are not absolute but depend upon the relative motion of the observer and that which is observed. If your clock ticks more slowly than it did on the ground and your meter stick is shorter in the direction of motion of the spacecraft, then you will find 186,000 mi/s for a velocity that I think should be only 36,000 mi/s. To you, your clock and meter stick are the same as they were on the ground before you took off, but to me they are different because of the relative motion. Time intervals and lengths are relative quantities, not absolute ones.

THE RELATIVITY OF TIME

Measurements of time are affected by relative motion between an observer and what he observes. As a result, all moving clocks tick more slowly than clocks at rest do, and all natural processes that involve regular time intervals occur more slowly when they take place in a moving frame of reference.

We begin by considering the operation of the particularly simple clock shown here. In this clock a pulse of light is reflected back and forth between two mirrors. Whenever the light strikes the lower mirror, an electrical signal is produced that is registered as a mark on the recording tape. Each mark corresponds to the tick of an ordinary clock. (**7**)

Two clocks. One clock is at rest in the laboratory, and another identical clock is in a spaceship moving at the velocity v relative to the laboratory. An observer in the laboratory watches both clocks. Does he find that they tick at the same rate? (**8**)

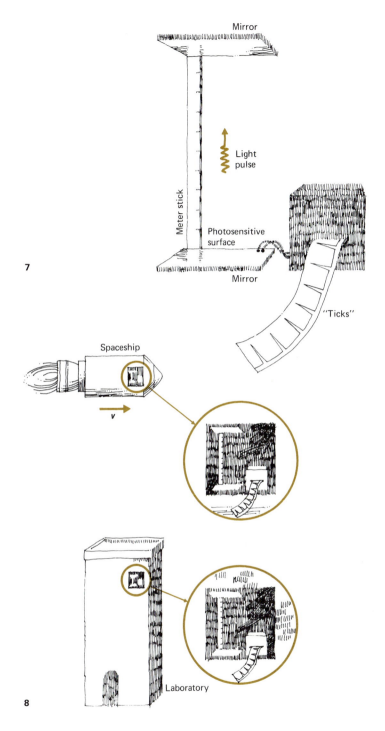

7

8

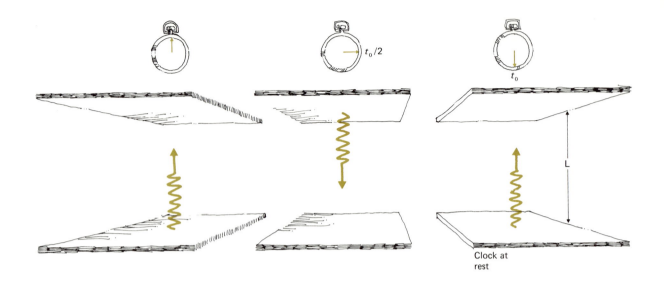

9

The laboratory clock. The mirrors are L apart, and the time interval between ticks is t_0. Hence the time needed for the light pulse to travel the distance L between the mirrors at the velocity c is $t_0/2$, and so

$$\text{Time} = \frac{\text{distance}}{\text{speed}},$$

$$\frac{t_0}{2} = \frac{L}{c}$$

$$t_0 = \frac{2L}{c}. \quad \textbf{(9)}$$

The moving clock as seen from the laboratory. The time interval between ticks is t. Because the clock is moving, the pulse of light follows a zigzag path in which it travels the distance $ct/2$ in going from one mirror to the other in the time $t/2$. From the Pythagorean theorem,

$$\left[\frac{ct}{2}\right]^2 = L^2 + \left[\frac{vt}{2}\right]^2. \quad \textbf{(10)}$$

Path taken by light pulse as seen from laboratory

Clock in motion v

10

How is t related to t_0? To find out, we first solve the preceding equation for t:

$$\left[\frac{ct}{2}\right]^2 = L^2 + \left[\frac{vt}{2}\right]^2,$$

$$\frac{t^2}{4}(c^2 - v^2) = L^2,$$

$$t^2 = \frac{4L^2}{(c^2 - v^2)}$$

$$= \frac{(2L)^2}{c^2(1 - v^2/c^2)},$$

$$t = \frac{2L/c}{\sqrt{1 - v^2/c^2}}.$$

But the quantity $2L/c$ is the time interval t_0 between ticks in the laboratory clock, as we saw. Hence

$$t = \frac{t_0}{\sqrt{1 - v^2/c^2}}. \qquad \text{Time dilation}$$

Here is a reminder of what the symbols in this important formula represent:

t_0 = time interval on clock at rest,

t = time interval on clock in relative motion,

v = velocity of relative motion,

c = velocity of light.

TIME DILATION

Because the quantity $\sqrt{1 - v^2/c^2}$ is always smaller than 1 for a moving object, t is always greater than t_0. *A clock moving with respect to an observer ticks more slowly than a clock that is stationary with respect to the same observer.* This effect is referred to as *time dilation* (dilate = to become larger).

Now let us turn the situation around and ask what an observer in a spacecraft finds when he compares his clock with one on the ground. The only change needed in the preceding derivation is the direction of motion: if the man on the ground sees the spacecraft moving to the east, the man in the spacecraft sees the laboratory on the ground moving to the west. To the man in the spacecraft the light pulse of the ground clock follows a zigzag path that requires a total time per round trip of

$$t = \frac{t_0}{\sqrt{1 - v^2/c^2}},$$

whereas the light pulse in his own clock takes t_0 for the round trip. Thus to the man in the spacecraft the clock on the ground ticks at a slower rate than his own clock does. A clock moving relative to an observer is *always* slower than a clock at rest relative to him, regardless of where the observer is located.

THE TWIN PARADOX

Since life processes occur with regular rhythms, a person constitutes a biological clock and must behave in the same way as any other clock when in motion relative to an observer. There is no difference in principle between heartbeats and the ticks of a clock. This means that the life processes of a man in a spacecraft occur at a slower rate than they do on the ground, so he ages more slowly than somebody back on the ground does.

The celebrated case of the twin brothers Alphonse and Gaston illustrates the consequences of time dilation in space travel. Alphonse is 20 yr old when he embarks on a

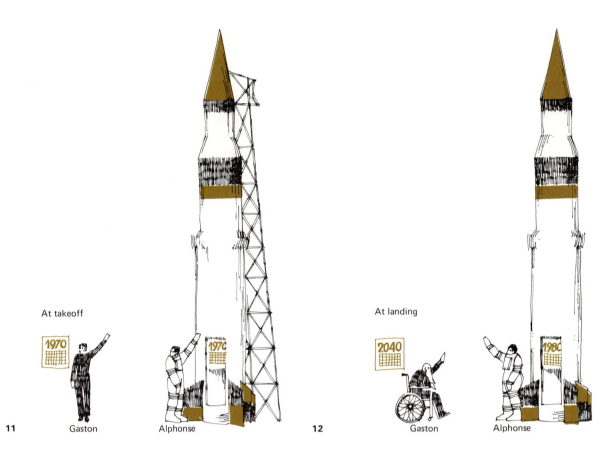

At takeoff

11   Gaston   Alphonse

At landing

12   Gaston   Alphonse

space voyage at a speed of 2.97×10^8 m/s, which is 99% of the velocity of light. **(11)**

To Gaston, who has stayed behind, the pace of Alphonse's life is slower than his own by a factor of

$$\sqrt{1 - \frac{v^2}{c^2}} = \sqrt{1 - \frac{(0.99c)^2}{c^2}} = 0.141 \approx \frac{1}{7}.$$

Alphonse's heart beats only once for every 7 beats of Gaston's heart; Alphonse takes only one breath for every 7 of Gaston's; Alphonse thinks only one thought for every 7 of Gaston's. Eventually Alphonse returns after 70 years

have elapsed by Gaston's calendar—but Alphonse is only 30 yr old, whereas Gaston, the stay-at-home twin, is 90 yr old. **(12)**

Gaston is baffled by the youth of his brother the astronaut. "After all," he argues, "according to the principle of relativity, *my* life processes should have appeared 7 times slower to Alphonse, so by the same reasoning I ought to be 30 and Alphonse ought to be 90."

But advanced age has dulled Gaston's powers of reasoning. The two situations are *not* interchangeable at all. Alphonse, in his spacecraft, had experienced enormous and prolonged accelerations in reaching the velocity of 99% of the velocity of light, in turning around, and finally in coming in to a landing. In each of these accelerations Alphonse changed from one frame of reference moving at constant velocity relative to the ground to another. Gaston, on the ground, has not been accelerated and hence has stayed in the same frame of reference all the time. What Gaston measured is in accord with the formulas of special relativity. Alphonse's accelerations affect his own measuring instruments, and indeed his own life processes, and the theory of general relativity must be applied to his journey; the conclusion of this theory is that Alphonse is indeed younger than Gaston on his return, and by the amount expected on the basis of Gaston's measurements.

In situations of this sort, where one frame of reference is accelerated and another is not, the measurements of an observer in the latter frame give the correct results when interpreted with the help of special relativity.

It is important to keep in mind that the slowing down of a moving clock is significant only at relative velocities not far from the velocity of light. Such velocities are readily attained by elementary particles, and most of the experiments that have confirmed time dilation and have verified the above interpretation of the twin paradox have employed such particles. Today's spacecraft are far too slow for their crews to return perceptibly younger than if they had stayed on the earth. The highest velocity reached by the Apollo 11 spacecraft on its way to the moon was only about 10,840 m/s, or 0.0036% of the velocity of light. At this velocity clocks on the spacecraft differ from those on the earth by less than 1 part in 10^9.

EXERCISES

1. Can an observer in a windowless laboratory determine whether the earth is traveling through space with a constant velocity or is stationary? whether the earth is accelerated? whether the earth is rotating?

2. If the speed of light were smaller than it is, would relativistic phenomena be more or less conspicuous than they are now?

3. Until the sixteenth century the earth was widely believed to be the center of the universe. According to the *Ptolemaic system*, which was a detailed picture of this idea, the sun and moon revolve around the earth in circular orbits while the planets each travel in a series of loops. The stars are supposed to be fixed to a giant crystal sphere that turns once a day. In 1543 Copernicus published the hypothesis that the sun is the center of the solar system, with only the moon revolving around the earth. The stars are far away in space, and the earth rotates daily on its axis. The *Copernican system* thus refers planetary motions to the sun rather than to the earth.
   a) Would you characterize the Ptolemaic system as incorrect and the Copernican system as correct?
   b) What considerations would you suppose led to the adoption of the Copernican system?
   c) Today the sun is known not to be the center of the universe. Why is the Copernican system still taught?

4. A certain process requires 10^{-6} s to take place in an atom at rest in the laboratory. How much time will this process require to an observer in the laboratory when the atom is moving at 5×10^7 m/s?

5. A spacecraft leaves the earth at a speed of 2×10^8 m/s. How much time (as measured on the earth) must elapse before a clock in the spacecraft and one on the earth differ by 1 min?

6. How fast would a spacecraft have to travel relative to the earth in order that each day on

the spacecraft correspond to two days on the earth?

7. An airplane is flying at 300 m/s (672 mi/hr). How much time must elapse before a clock in the airplane and one on the ground differ by 1 s?

8. A light-pulse clock is a more exotic timepiece than most of us are accustomed to. Suppose that a cuckoo clock is installed next to a light-pulse clock in a spacecraft. Will it also run slow relative to a clock on the ground?

[*Note:* The quantities

$$\sqrt{1 - \frac{v^2}{c^2}} \quad \text{and} \quad \sqrt{\frac{1}{1 - \frac{v^2}{c^2}}}$$

occur often in special relativity. When the relative velocity v is much smaller than the velocity of light c, a lot of tedious arithmetic can be avoided by using the approximate formulas

$$\sqrt{1 - \frac{v^2}{c^2}} \approx 1 - \frac{1}{2}\left(\frac{v^2}{c^2}\right) \qquad v \ll c,$$

$$\frac{1}{\sqrt{1 - \frac{v^2}{c^2}}} \approx 1 + \frac{1}{2}\left(\frac{v^2}{c^2}\right) \qquad v \ll c.$$

The symbol $\approx$ means "is approximately equal to." When v is $0.5c$ or less, these formulas are accurate to within at worst a percent or so. When v is greater than $0.5c$, however, they should not be used.]

6

THE RELATIVITY OF SPACE

"The influence of motion (relative to the coordinate system) on the form of bodies and on the motion of clocks, also the equivalence of energy and inert mass, follow from the interpretation of coordinates and time as products of measurement." Albert Einstein (1879–1955)

Measurements of space as well as of time are affected by relative motion. A moving object is always shorter in the direction of motion than it is when at rest. As in the case of time intervals, the relativistic length contraction is significant only when very high velocities are involved.

MUON DECAY

A good illustration of time dilation occurs in the decay of the unstable elementary particles called *muons*.

Muons have masses 207 times that of the electron and may have positive or negative electric charges. A muon decays into an electron an average of 2.0×10^{-6} s after it comes into being. Muons are created in the atmosphere thousands of meters above sea level as the ultimate result of collisions between air atoms and fast cosmic-ray particles (largely protons) that reach the earth from space; a muon passes through each cm^2 of the earth's surface a little more often than once a minute. The muon velocities are observed to be about 2.994×10^8 m/s, or $0.998c$ where c is the velocity of light. But in $t_0 = 2.0 \times 10^{-6}$ s, the average muon lifetime, they can travel a distance of only

$$vt_0 = 2.994 \times 10^8 \, \frac{m}{s} \times 2.0 \times 10^{-6} \, s$$

$$= 600 \, m,$$

whereas they actually come into being at elevations 10 or more times greater than this.

The key to resolving this paradox is to note that the average muon lifetime of 2×10^{-6} s is what an observer at rest with respect to a muon would find. If we could collect some muons at the instant of their creation and

45

time their decays when they are at rest, we would find an average of $t_0 = 2 \times 10^{-6}$ s. **(1)**

However, when we are on the ground and the muons are hurtling toward us at the considerable speed of 0.998c, we find instead that their lifetimes have been extended by time dilation to

$$t = \frac{t_0}{\sqrt{1 - \dfrac{v^2}{c^2}}}$$

$$= \frac{2 \times 10^{-6} \text{ s}}{\sqrt{1 - \dfrac{(0.998c)^2}{c^2}}}$$

$$= \frac{2 \times 10^{-6} \text{ s}}{\sqrt{0.004}} = \frac{2 \times 10^{-6} \text{ s}}{0.0632}$$

$$= 31.6 \times 10^{-6} \text{ s.} \textbf{ (2)}$$

The fast muons have lifetimes almost 16 times longer than those of their brethren at rest. In a time interval of $t = 31.6 \times 10^{-6}$ s, a muon whose velocity is 0.998c can cover the distance

$$vt = 2.994 \times 10^8 \, \frac{\text{m}}{\text{s}} \times 31.6 \times 10^{-6} \text{ s}$$

$$= 9500 \text{ m.}$$

Despite its brief lifespan of $t_0 = 2.0 \times 10^{-6}$ s in its own frame of reference, a muon is able to reach the ground from quite considerable altitudes because in the frame of reference in which these altitudes are measured, the muon lifetime is $t = 31.6 \times 10^{-6}$ s.

THE LORENTZ CONTRACTION

As we saw, the arrival of cosmic-ray muons at sea level from high altitudes is not in conflict with the brevity of their lives (in their frames of reference) since these lives are increased 16-fold (in our frame of reference) by their

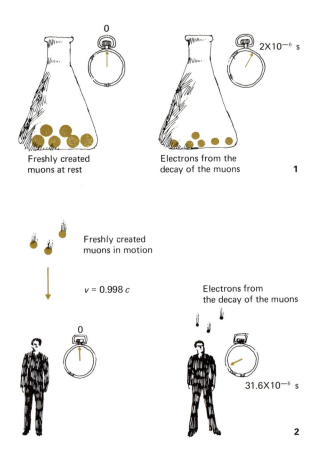

Freshly created muons at rest

Electrons from the decay of the muons

1

2×10^{-6} s

Freshly created muons in motion

$v = 0.998\,c$

Electrons from the decay of the muons

31.6×10^{-6} s

2

relative motion. But what if somebody could accompany the muons downward at the same velocity of 0.998 c, so that to him the muons are at rest? Both the muons and the observer are now in the same frame of reference, and the muon lifetime is only 2.0×10^{-6} s in this frame. The question is, does the moving observer find that the muons reach the ground, or does he find that they decay beforehand?

The principle of relativity states that the laws of physics are the same in all frames of reference moving at constant velocity with respect to one another. An observer on the ground and an observer moving with the muons are

As seen from the ground, the muon altitude is L_0.

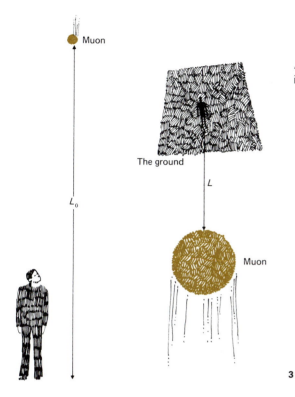

Muon

The ground

L

L_0

Muon

3

As seen by the muon, the ground is L below it, which is a shorter distance than L_0.

out to be!) enables us to infer at once the extent of the shortening—it must be by the same factor of $\sqrt{1 - v^2/c^2}$ that the muon lifetime is extended from the point of view of a stationary observer. Thus a distance we on the ground measure to be L_0 will appear to the muon as the abbreviated distance

$$L = L_0\sqrt{1 - v^2/c^2}.$$

In our frame of reference, the maximum distance a muon can go at the velocity $v = 0.998c$ before it decays is $L_0 = 9500$ m. The corresponding distance in the muon's frame of reference is

$$L = L_0\sqrt{1 - v^2/c^2}$$

$$= 9500 \text{ m} \times \sqrt{1 - \frac{(0.998c)^2}{c^2}}$$

$$= 9500 \text{ m} \times \sqrt{0.00400} = 9500 \text{ m} \times 0.0632$$

$$= 600 \text{ m}.$$

This is precisely how far a muon traveling at $0.998c$ can go in 2.0×10^{-6} s. Both points of view—from the frame of reference of someone on the ground, to whom the muon lifetime is dilated, and from the frame of reference of the muon itself, to which the distance to the ground is contracted—give the same result.

in relative motion at a constant velocity, namely $0.998c$, and if we on the ground find that the muons reach our apparatus before they decay, then the moving observer must also find the same thing. Though the appearance of an event may be different to different observers, the fact of the event's occurrence is not subject to dispute.

The only way an observer in a muon's frame of reference can reconcile its arrival at sea level with the lifetime of 2.0×10^{-6} s he finds is if the distance the muon travels is shortened by virtue of its motion. (3) The principle of relativity (how unexpectedly powerful it turns

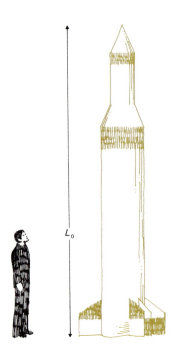

An observer and a spacecraft are at rest on the ground. The observer finds the spacecraft's length to be L_0.

The spacecraft is moving at the velocity v. The observer on the ground finds its length to be $L = L_0 \sqrt{1 - v^2/c^2}$.

The observer is moving at the velocity v and the spacecraft is on the ground. The observer finds its length to be $L = L_0 \sqrt{1 - v^2/c^2}$.

The relativistic shortening of distances is an example of the quite general *Lorentz contraction* of lengths:

$$L = L_0\sqrt{1 - v^2/c^2}.$$ Lorentz contraction

The symbols in this formula have these meanings:

L_0 = length measured when the object is at rest,

L = length measured when the object is in relative motion,

v = velocity of relative motion,

c = velocity of light.

FASTER ALWAYS MEANS SHORTER

The length of an object in motion with respect to an observer is measured by the observer to be shorter than when it is at rest with respect to him. This shortening works both ways; to a man in a spacecraft, measurements indicate that objects on the earth are shorter than they were when he was on the ground, and someone on the ground finds that the spacecraft is shorter in flight than when it was at rest. (To the man in the spacecraft, its length is the same whether on the ground or in flight, since it is always at rest with respect to him.) The length of an object is a maximum when determined in a reference frame in which it is stationary, and its length is less when determined in a reference frame in which it is moving. (4)

Only lengths in the direction of motion undergo Lorentz contractions. Thus a spacecraft is shorter in flight than on the ground, but it is not narrower.

The Lorentz contraction is negligible at ordinary velocities, but is an important effect at velocities close to that of light. A velocity of 10^6 m/s—which is 621 miles per second—seems enormous to us, yet it results in a shortening in the direction of motion to only

$$\frac{L}{L_0} = \sqrt{1 - \frac{v^2}{c^2}}$$

$$= \sqrt{1 - \frac{(10^6 \text{ m/s})^2}{(3 \times 10^8 \text{ m/s})^2}}$$

$$= 0.9999944$$

$$= 99.99944\%$$

of the length at rest. On the other hand, something traveling at 9/10 of the velocity of light is shortened to

$$\frac{L}{L_0} = \sqrt{1 - \frac{(0.9c)^2}{c^2}}$$

$$= 0.436 = 43.6\%$$

of its length at rest, a significant change.

THE ULTIMATE VELOCITY

Nothing material can travel faster than light, or indeed even as fast as light. The principle of relativity together with the constancy of the velocity of light can be used to show that v must always be less than c.

Suppose you are in a spacecraft moving at the velocity of light or faster, and I am watching

5

you from the ground. The lamps in the spacecraft suddenly go out, and you want to use a flashlight to find the fuse box at the front of the craft. You turn the flashlight on and, since the ordinary laws of physics hold within the spacecraft, you locate the fusebox and change the blown fuse. (5)

From the ground, though, I see something quite different. To me, since your speed is equal to or greater than c, the light from your flashlight cannot ever reach the front of the spacecraft. (The velocity of this light is the same to both of us, of course.) In fact, if $v > c$ the light illuminates the *back* of the spacecraft. I am forced to conclude that the laws of physics are different in your frame of reference from what they are in mine—which contradicts the principle of relativity. (6)

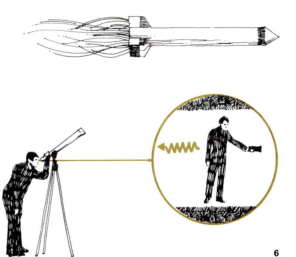

6

The only way to avoid such fundamental difficulties is to assume that nothing can move as fast as or faster than light. This assumption has been tested experimentally many times and has invariably been found correct.

EXERCISES

1. Of the following quantities, which always has the same value to all observers: the length of an object, the speed of an object, the speed of light, or the duration of a time interval?

2. The length of a rod is measured by several observers, one of whom is stationary with respect to the rod. What must be true of the figure obtained by the latter observer?

3. As an object 1 m long when at rest approaches the speed of light, what does its length approach?

4. A spacecraft in flight is observed to be 99% of its length when it was at rest on the earth. What is its speed relative to the earth?

5. An airplane 100 ft long on the ground is flying at 400 mi/hr (180 m/s). How much shorter does it appear to somebody on the ground?

6. An astronaut whose height on the earth is exactly 6 ft is lying parallel to the axis of a spacecraft traveling at 2.7×10^8 m/s relative to the earth. What is his height as measured by another astronaut in the spacecraft? By an observer on the earth?

PART 2

THE LAWS OF MOTION

7

MASS

"Words like *mass* and *force* had, of course, been used earlier . . . In mechanics, however, these words acquire a new precise meaning; they are artificial words, perhaps the first to be coined. Their sound is the same as words of ordinary speech, but their meaning can only be found from a specially formulated definition." Max Born (1882–1970)

The tendency of a body at rest to remain at rest is a property of matter called inertia. In turn, inertia depends upon *mass*: the more mass something has, the harder it is to set in motion. Although the mass of a body refers, in a general way, to the amount of matter it contains, mass is defined in terms of inertia because in this way we can specify exactly how to measure it. The unit of mass is the kilogram.

PHYSICAL QUANTITIES

Until now the only physical quantities we have considered have been the obvious ones of length and time. Many other quantities also turn out to be useful in describing various aspects of the physical universe: mass, linear momentum, angular momentum, force, energy, temperature, electric charge, . . .; the list is long. There is nothing sacred about these quantities—it is entirely possible to dispense with any of them, but only at the expense of making physics a good deal more complicated than it already is. The idea behind the definition of each of the various physical quantities is to single out something that unifies a wide range of observations, to boil down to a brief, clear statement a large number of separate discoveries about nature. Thus the awareness all of us have of the reluctance of objects to be set in motion or, once in motion, to be brought to a stop, a reluctance more or less vaguely referred to in everyday life as inertia, is given a precise description in the concepts of mass and linear momentum together with the principle of conservation of linear momentum.

THE FIRST LAW OF MOTION

In everyday life it is a familiar observation that stationary objects tend to remain station-

Sudden
start

Sudden
stop

1

2

ary and moving objects tend to continue moving. A certain amount of effort is needed to start a cart moving on a level road, and once in motion a certain amount of effort is also required to bring it to a stop. Of course, in the case of a cart friction is a factor in its reluctance to begin to move, but friction is only part of the story. Even without friction, a cart at rest on a level road will remain at rest unless something pushes it. And without friction, a cart moving along a level road at a certain velocity will continue to move at that velocity indefinitely.

Newton's first law of motion is a statement of the above behavior.

A body at rest will remain at rest and a body in motion will continue in motion in a straight line at constant velocity in the absence of any interaction with the rest of the universe.

We might object that, although everyday experience does indicate that bodies in motion *tend* to remain in motion along a straight line at constant velocity, soon or later they invariably come to a stop and, often, deviate from a straight path as well. Even celestial bodies such as the sun, moon, and planets are neither at rest nor pursue straight paths. But these observations do not invalidate the first law of motion, they merely emphasize how difficult it is to avoid interactions between a body and its environment. A cart rolling along a smooth, perfectly level road will not continue forever owing to friction and air resistance, but we are at liberty to imagine what would happen if the air were to disappear and the friction were to vanish.

The resistance a body offers to any change in its state of rest or uniform motion is a property of matter known as *inertia*. When a bus suddenly starts to move, standing passengers seem to be pushed backward. What is actually happening is that inertia tends to keep their bodies in place relative to the earth while the bus carries their feet forward. (**1**)

When a bus suddenly stops, on the other hand, standing passengers seem to be pushed forward. What is actually happening is that inertia tends to keep their bodies moving while the bus has come to a halt. (**2**)

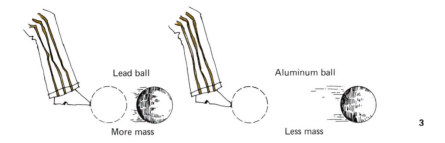

Lead ball

Aluminum ball

More mass

Less mass

3

MASS

A quantitative measure of the inertia of a body at rest is its *mass*. The greater the resistance a body offers to being set in motion, the greater its mass. The inertia of a lead ball exceeds that of an aluminum ball of the same size, as we can tell by kicking them in turn, so the mass of the lead ball exceeds that of the aluminum one. (**3**)

We can arrive at a precise definition of mass by using a simple experiment to compare the inertias of two bodies. What we do is put a small spring between them, push them together so the spring is compressed, and tie a string between them to hold the assembly in place. (**4**)

Now we light a match and burn the string. The compressed spring pushes the bodies apart, and body A flies off to the left at the velocity v_A while body B flies off to the right at the velocity v_B. Body A has a lower velocity than body B, and we interpret this difference to mean that A exhibits more inertia than B. (**5**)

4

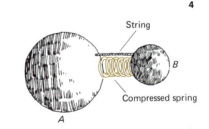

String

B

Compressed spring

A

5

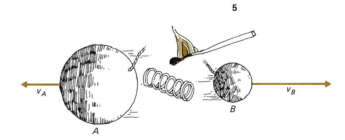

v_A

A

B

v_B

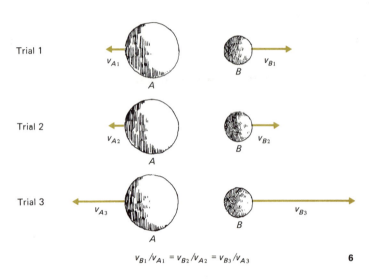

Trial 1 v_{A_1} A B v_{B_1}

Trial 2 v_{A_2} A B v_{B_2}

Trial 3 v_{A_3} A B v_{B_3}

$$v_{B_1}/v_{A_1} = v_{B_2}/v_{A_2} = v_{B_3}/v_{A_3}$$

6

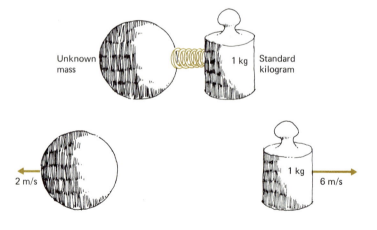

Unknown mass   1 kg Standard kilogram

2 m/s   1 kg   6 m/s

7

We repeat the experiment a number of times using springs of different stiffness, so that the recoil velocities are different in each case. What we find is that body A moves more slowly than body B each time, and that, regardless of the exact values of v_A and v_B, their *ratio*

$$\frac{v_B}{v_A}$$

is always the same. (**6**)

The fact that the velocity ratio v_B/v_A is constant gives us a way to specify what we mean by mass unambiguously. If we denote the masses of bodies A and B by the symbols m_A and m_B, respectively, we define the ratio of these masses to be

$$\frac{m_A}{m_B} = \frac{v_B}{v_A}.$$

The body with the greater mass has the lower velocity, and vice versa. The above procedure provides a definite experimental method for finding the ratio between the masses of any two bodies—it is an *operational definition*.

The next step is to select a certain object to serve as a standard unit of mass. By international agreement this object is a platinum cylinder at Sèvres, France, called the *standard kilogram*. The mass of any other body in the world can be determined by a recoil experiment with the standard kilogram. (There are easier ways to measure mass, needless to say, but we are interested in basic principles for the present.)

Problem. In a recoil experiment with the standard kilogram, a body of unknown mass moves off at a velocity of 2 m/s and the standard kilogram moves off at a velocity of 6 m/s. What is the mass of the body? (**7**)

Solution. If we call the body of unknown mass A and the standard kilogram B, then $m_B = 1$ kg, $v_A = 2$ m/s, and $v_B = 6$ m/s. We find that

$$m_A = 1 \text{ kg} \times \frac{6 \text{ m/s}}{2 \text{ m/s}}$$

$$= 3 \text{ kg}.$$

Why is it necessary to define mass in this seemingly roundabout way? After all, everybody knows that the mass of an object refers to the amount of matter it contains. The trouble is that "amount of matter" is a nebulous concept: it could refer to an object's volume, to the number of atoms it contains, or to yet other properties. It has proved most fruitful to choose the inertia of an object as a measure of the quantity of matter it contains, and to define the object's mass in terms of this inertia as manifested in an appropriate experiment.

DENSITY

A characteristic property of every substance is its *density*, which is its mass per unit volume. When we speak of lead as a "heavy" metal and of aluminum as a "light" one, what we really mean is that lead has a higher density than aluminum: a cubic meter of lead has a mass of 11,300 kg, while a cubic meter of aluminum has a mass of only 2700 kg. In the metric system the proper unit of density is the kg/m^3.

The symbol for density is d. Hence we can write

$$\text{Density} = \frac{\text{mass}}{\text{volume}},$$

$$d = \frac{m}{V}.$$

Frequently densities are given in g/cm^3. To express such densities in kg/m^3 and vice versa we note that

$$1 \frac{g}{cm^3} = 1 \frac{g}{cm^3} \times \left(10^2 \frac{cm}{m}\right)^3 \times \frac{1}{10^3 \text{ g/kg}}$$

$$= 10^3 \frac{kg}{m^3}.$$

A density in g/cm^3 is multiplied by 10^3 to give the equivalent in kg/m^3; a density in kg/m^3 is divided by 10^3 to give the equivalent in g/cm^3.

Densities at atmospheric pressure and room temperature are given in Table 7.1.

Table 7.1

Substance	d (kg/m³)	Substance	d (kg/m³)
Air	1.3	Hydrogen	0.09
Alcohol (ethyl)	7.9×10^2	Ice	9.2×10^2
Aluminum	2.7×10^3	Iron	7.8×10^3
Balsa wood	1.3×10^2	Lead	1.1×10^4
Bromine	3.2×10^3	Mercury	1.4×10^4
Carbon dioxide	2.0	Nitrogen	1.3
Concrete	2.3×10^3	Oak	7.2×10^2
Gasoline	6.8×10^2	Oxygen	1.4
Gold	1.9×10^4	Water, pure	1.00×10^3
Helium	0.18	Water, sea	1.03×10^3

Problem. Fifty cm of snow fall on the horizontal roof of a house. The roof is 15 m long and 10 m wide. If the density of the snow is 130 kg/m^3, find the mass of snow on the roof.

Solution. The volume of the snow is

$$V = \text{length} \times \text{width} \times \text{height}$$
$$= 15 \text{ m} \times 10 \text{ m} \times 0.5 \text{ m}$$
$$= 75 \text{ m}^3.$$

The mass of the snow is therefore

$$m = dV$$

$$= 130 \, \frac{\text{kg}}{\text{m}^3} \times 75 \, \text{m}^3$$

$$= 9750 \, \text{kg}.$$

THE RELATIVITY OF MASS

The mass of a body is not constant but depends upon the body's velocity with respect to the observer who measures its mass. The faster the velocity, the greater the mass. The variation of mass with velocity is not noticeable for ordinary moving bodies, but in the subatomic world the variation is conspicuous because such minute particles as electrons and protons may move at velocities near that of light.

A straightforward though subtle analysis shows that the mass of an object moving at the velocity v relative to an observer will be found by that observer to be equal to

$$m = \frac{m_0}{\sqrt{1 - v^2/c^2}}$$

where the symbols have these meanings:
m_0 = mass measured when object is at rest,
m = mass measured when object is in relative motion,
v = velocity of relative motion,
c = velocity of light.

The quantity m_0 is usually called the *rest mass* of the object.

Since the quantity $\sqrt{1 - v^2/c^2}$ is always less than one, an object will always appear more massive when in relative motion than when at rest. This mass increase is reciprocal: as

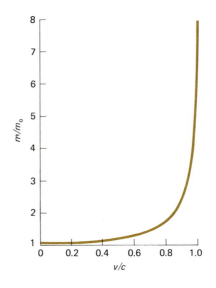

seen from the ground, a spacecraft in flight has a greater mass than when at rest, and to an observer in the spacecraft, all objects on the ground have masses that are greater than their rest masses. **(8)**

EXAMPLES OF RELATIVISTIC MASS INCREASES

Relativistic mass increases are significant only at velocities approaching that of light. The rest mass of the Apollo 11 spacecraft (apart from its launch vehicle, which dropped away after accelerating the spacecraft) was about 63,070 kg (nearly 70 tons). On its way to the moon, the spacecraft's velocity was about 10,840 m/s (6,730 mi/s) and so its mass in flight, as measured by an observer on the earth, increased to

$$m = \frac{m_0}{\sqrt{1 - v^2/c^2}}$$

$$= \frac{63{,}070 \text{ kg}}{\sqrt{1 - \dfrac{(1.084 \times 10^4 \text{ m/s})^2}{(3 \times 10^8 \text{ m/s})^2}}}$$

$$= 63{,}070.00000000065 \text{ kg}.$$

This is not much of a change.

Smaller objects can be given much higher velocities. It is not very difficult to accelerate electrons (rest mass 9.109×10^{-31} kg) to velocities of, say, $0.9999c$. The mass of such an electron is

$$m = \frac{m_0}{\sqrt{1 - \dfrac{v^2}{c^2}}}$$

$$= \frac{9.109 \times 10^{-31} \text{ kg}}{\sqrt{1 - \dfrac{(0.9999c)^2}{c^2}}}$$

$$= 644 \times 10^{-31} \text{ kg}.$$

The electron's mass is 71 times its rest mass! The mass increases predicted by the relativistic mass formula, even such remarkable ones as this, have been experimentally verified without exception.

EXERCISES

1. A jet airplane is descending toward an airport. How can a passenger tell whether the airplane is accelerating, decelerating, or maintaining a constant velocity by watching the curtains of his window?

2. An object moving in a circle at the constant speed v has an average velocity v of zero. Would you expect it to have a relativistic mass increase?

3. A 5-kg rifle recoils with a speed of 4 m/s when it fires a 10-g bullet. What is the bullet's speed?

4. A 30-kg boy and a 40-kg boy face each other on frictionless roller skates. The larger boy pushes the smaller one, who moves away at a speed of 1.2 m/s. With what speed does the larger boy move?

5. How many cubic meters does a metric ton (1000 kg) of pure water occupy? a metric ton of air at sea level?

6. A certain room is 10 m long, 5 m wide, and 3 m high. How many kg of air does it contain at sea level?

7. Which of the following speeds must an object have for its mass to be double its rest mass? $c/2$, $\sqrt{3}c/2$, c, $2c$.

8. Find the mass of an object whose rest mass is 1.000 g when it is traveling at 10, 90, and 99% of the speed of light.

9. A man has a mass of 100 kg on the ground. When he is in a spacecraft in flight, an observer on the earth measures his mass to be 101 kg. How fast is the spacecraft moving?

8

MOMENTUM

"We have not always brought home to ourselves the peculiar character of that Aristotelian universe in which the things that were in motion had to be accompanied by a mover all the time . . . It was a universe in which unseen hands had to be in constant operation, and sublime intelligences had to roll the planetary spheres around. Alternatively, bodies had to be endowed with souls and aspirations, with a disposition to certain kinds of motion . . . The modern law of inertia, the modern theory of motion, is the great factor which in the seventeenth century helped to drive spirits out of the world and opened the way to a universe that ran like a piece of clockwork." Herbert Butterfield (1900–   )

According to the first law of motion, a body at rest tends to remain at rest and a body in motion tends to continue in motion. An alternative way to describe these observations is in terms of inertia: the inertia of a body is a measure of its tendency to remain at rest or to continue in motion. The inertia of a body at rest depends upon its mass, and the inertia of a body in motion depends upon another quantity, its *momentum*.

LINEAR MOMENTUM

We all know that a baseball struck squarely by a bat is harder to stop than the same baseball thrown gently, which suggests that the inertia of a moving body depends in some way upon its velocity. We also know that the heavy iron ball used for the shotput is harder to stop than a baseball whose velocity is the same, which suggests that the inertia of a moving body depends in some way upon its mass.

The quantity which specifies the inertia of a moving body of mass m and velocity v is called its *linear momentum* and is equal to the product of m and v. The symbol for linear momentum is p, and so its definition is written

$$\text{Linear momentum} = \text{mass} \times \text{velocity},$$

$$p = mv.$$

Linear momentum is a vector quantity whose direction is the direction of v. Because p is a measure of the tendency of a moving body to continue in motion at constant velocity along a straight line, it is referred to as *linear momentum*. (**1**) Another quantity, *angular mo-*

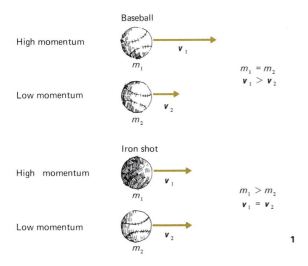

High momentum

Baseball

v_1

m_1

$m_1 = m_2$
$v_1 > v_2$

Low momentum

v_2

m_2

High momentum

Iron shot

v_1

m_1

$m_1 > m_2$
$v_1 = v_2$

Low momentum

v_2

m_2

1

mentum, describes the tendency of a spinning body such as a top to continue to spin.

CONSERVATION PRINCIPLES

One of the chief distinctions between the physical sciences and nearly all other scientific disciplines is that in the former certain very general conservation principles have been found valid. A conservation principle states that no matter what changes a system of some kind that is isolated from the rest of the universe undergoes, a certain quantity keeps the same value it had originally. For example, the law of conservation of mass revolutionized chemistry by holding that the total mass of the products of a chemical reaction is the same as the total mass of the original substances. The increase in mass of a piece of iron when it rusts therefore indicates that the iron has combined with some other material,

rather than having decomposed, as the early chemists believed. In fact, the gas oxygen was discovered in the course of seeking this other material.

Given one or more conservation principles that apply to a given system, we can immediately determine which classes of events can take place in the system and which cannot. Thus when iron rusts, the gain in mass means that it has combined chemically with something else. In physics it is often possible to draw some conclusions about the behavior of the particles that make up a system without a detailed investigation, basing our analysis simply upon the conservation of some particular quantities. The power of this method of approach is exemplified by the great success of physics in understanding natural phenomena, a success largely due to the variety of conservation principles that have been discovered.

CONSERVATION OF LINEAR MOMENTUM

The first conservation principle we shall study is the conservation of linear momentum. This principle states:

The total linear momentum of a system of particles isolated from the rest of the universe remains constant regardless of what events occur within the system.

By "total linear momentum" is meant the vector sum of the individual momenta of the particles that constitute the system.

The first thing we notice about the principle of conservation of linear momentum is that it incorporates both Newton's first law of motion and the definition of mass. If our

A body at rest will remain at rest and a body in motion will continue in motion in a straight line at constant velocity in the absence of any interaction with the rest of the universe.

 t_1

 t_2

$p_1 = 0$

$p_2 = p_1$

p_1

$p_2 = p_1$

2

system of particles is a single body, then its linear momentum $p = mv$ does not change as long as it does not interact with the rest of the universe, which is another way to state the first law of motion. When the body is initially at rest, then $p = 0$ at all times in the future; when it has some initial momentum p, then $p =$ constant at all times in the future, which means that v retains both its original magnitude and direction. **(2)**

Let us apply conservation of linear momentum to the system of two bodies connected by a compressed spring that was used earlier in defining mass. The total linear momentum of this system is initially

$$p_A + p_B = m_A v_A + m_B v_B$$

$$= 0. \textbf{(3)}$$

Now we burn the string, and the bodies fly apart. **(4)**

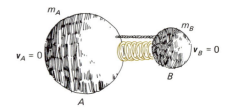

m_A m_B

$v_A = 0$ $v_B = 0$

B

A

3

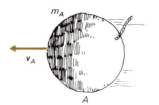

m_A

v_A

A

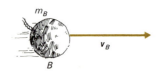

m_B

v_B

B

4

The total momentum of the system must still be zero, since the burning of the string and the subsequent motions of the bodies took place within the system. Hence

$$p_A + p_B = 0,$$
$$p_A = - p_B,$$
$$m_A v_A = - m_B v_B.$$

This is a vector equation, and the minus sign signifies that v_A and v_B are in opposite directions. We can rewrite this equation in scalar form (that is, in terms of the magnitudes of v_A and v_B) as

$$\frac{m_A}{m_B} = \frac{v_B}{v_A},$$

which is the same as the definition of mass that was given earlier.

The principle of conservation of linear momentum would hardly be worth expressing if it only included what is already stated in Newton's first law of motion and in the definition of mass. But the principle goes a good deal further, since it can be applied to systems that consist of any number of bodies that interact with one another in any manner whatsoever, and it holds in three dimensions, not just in one. The conservation of linear momentum is a generalization based upon innumerable experiments and observations, and no exception to it has ever been found.

With the help of advanced mathematics it is possible to show that, if the laws of nature are the same at every point in space, then the principle of conservation of linear momentum must follow as an inevitable consequence. Thus this principle gives us a hint of a profound order underlying the physical universe as well as being a useful relationship for solving practical problems.

APPLICATIONS OF MOMENTUM CONSERVATION

Momentum considerations are helpful in analyzing (in general terms) collisions and explosions. Another conservation principle, the conservation of energy, is also sometimes needed in order to understand such events, but momentum conservation alone is often sufficient.

The recoil of a rifle is easily interpreted in terms of conservation of linear momentum. When a rifle is fired, the forward momentum of the bullet must be balanced by the backward momentum of the rifle in order that the total momentum of the system of rifle + bullet remains zero. Hence

$$m_{\text{rifle}} v_{\text{rifle}} = m_{\text{bullet}} v_{\text{bullet}}.$$

Since the bullet is much lighter than the rifle, the rifle's recoil velocity is much less than the muzzle velocity of the bullet.

Problem. A 4.0-kg rifle fires a 16-g (0.016 kg) bullet at a muzzle velocity of 600 m/s. Find the recoil velocity of the rifle.

Solution. Here we have

$$v_{\text{rifle}} = \frac{m_{\text{bullet}}}{m_{\text{rifle}}} \times v_{\text{bullet}}$$

$$= \frac{0.016 \text{ kg}}{4.0 \text{ kg}} \times 600 \text{ m/s}$$

$$= 2.4 \text{ m/s}.$$

The momentum of the bullet in the forward direction is

$$m_{\text{bullet}} v_{\text{bullet}} = 0.016 \text{ kg} \times 600 \text{ m/s}$$

$$= 9.6 \frac{\text{kg-m}}{\text{s}},$$

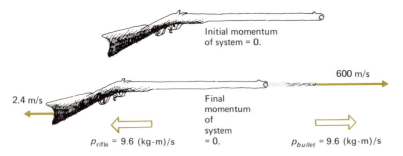

Initial momentum
of system = 0.

600 m/s

2.4 m/s

Final
momentum
of
system
= 0.

p_{rifle} = 9.6 (kg-m)/s

p_{bullet} = 9.6 (kg-m)/s

5

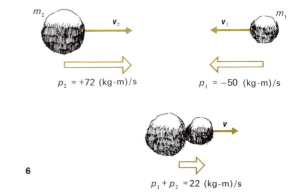

m_2 v_2 v_1 m_1

p_2 = +72 (kg-m)/s p_1 = −50 (kg-m)/s

v

6

$p_1 + p_2$ = 22 (kg-m)/s

and the momentum of the rifle in the back-ward direction is

$$m_{rifle}v_{rifle} = 4.0 \text{ kg} \times 2.4 \text{ m/s}$$

$$= 9.6 \frac{\text{kg-m}}{\text{s}}. \quad \textbf{(5)}$$

In a collision between two or more bodies, the vector sum of their linear momenta before they interact equals the vector sum of their linear momenta afterward. The essential effect of the collision is to redistribute the total momentum present. Only collisions in which the bodies stick together afterward can be completely analyzed without recourse to conservation of energy, however, and we shall postpone discussing other kinds of collision until Section 15.

Problem. A 5-kg lump of clay that is moving at 10 m/s to the left strikes a 6-kg lump of clay moving at 12 m/s to the right. The two lumps stick together after they collide. Find the final velocity of the composite body.

Solution. If we call the mass of the final body M and its velocity V, conservation of linear momentum requires that

Momentum afterward = momentum before,

$$MV = m_1v_1 + m_2v_2.$$

Adopting the convention that motion to the right is + and to the left is −, we have

$$m_1 = 5 \text{ kg}, \qquad\qquad m_2 = 6 \text{ kg},$$

$$v_1 = -10 \frac{m}{s}, \qquad v_2 = +12 \frac{m}{s},$$

$$M = m_1 + m_2 = 11 \, kg, \qquad V = ?$$

Solving for V yields

$$V = \frac{m_1 v_1 + m_2 v_2}{M}$$

$$= \frac{5 \, kg \times (-10 \, m/s) + 6 \, kg \times 12 \, m/s}{11 \, kg}$$

$$= \frac{-50 \, kg\text{-}m/s + 72 \, kg\text{-}m/s}{11 \, kg} = \frac{22 \, kg\text{-}m/s}{11 \, kg}$$

$$= 2 \frac{m}{s}.$$

Since V is positive, the composite body moves off to the right. (**6**)

It is essential to keep in mind the directional character of linear momentum. Sometimes the problem under consideration involves bodies that move along a straight line, as in the preceding example, but in general the bodies may move in two or three dimensions and we must be sure to take this into account by a vector calculation.

ROCKET PROPULSION

The principle underlying rocket flight is conservation of linear momentum. The total linear momentum of a rocket on its launching pad is zero. When it is fired, exhaust gases shoot downward at high velocity, and the rocket moves upward to balance the momentum of the gases. The total momentum of the system of rocket plus exhaust gases remains at its initial value of zero. Rockets do not operate by "pushing" against their launching pads, against the atmosphere, or against anything else; in fact, they perform

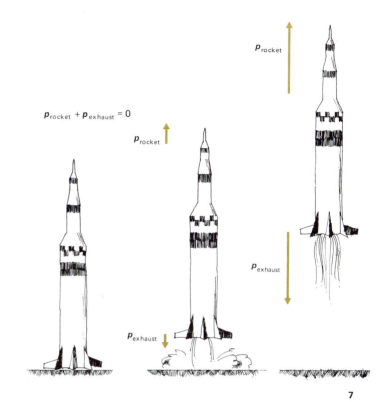

7

best in space, where there is no air to impede their motion. (**7**)

To attain higher velocities than a single rocket is capable of, two or more rocket stages can be used. The first stage is a large rocket whose payload is another, smaller rocket. When the fuel of the first stage has burnt up, its fuel tanks and engine are cast loose. Then the second stage is fired starting from a high initial velocity instead of from rest and without the burden of the fuel tanks and engine of the first stage. This process can be repeated a number of times, depending upon the final velocity required. The Saturn V launch vehicle that propelled the Apollo 11 spacecraft to the moon employed three stages,

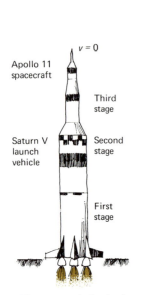

Apollo 11
spacecraft

Third
stage

Saturn V
launch
vehicle

Second
stage

First
stage

$v = 0$

First stage is ignited.

$v = 2750$ m/s

First stage drops off,
second stage is ignited.
$\Delta v = 2750$ m/s.

$v = 6940$ m/s

Second stage drops off,
third stage is ignited.
$\Delta v = 4190$ m/s.

$v = 10,840$ m/s

Third stage drops off,
spacecraft continues to
moon.
$\Delta v = 3900$ m/s.

8

as shown. At ignition, the entire assembly was 111 m (364 ft) long and had a mass of 2.9×10^6 kg (equivalent to 3240 tons). (**8**)

ANGULAR MOMENTUM

The rotational analog of linear momentum is *angular momentum*. The angular momentum *L* of a rotating body depends upon three factors. As we might expect, the greater the mass of the body and the faster it turns, the greater the magnitude of *L*. Less obviously, the farther from the axis of rotation the mass is distributed, the more the angular momentum. A short, fat top has more angular momentum than a tall, thin one of the same mass and speed of rotation. (**9**)

Angular momentum is a vector quantity whose direction is given by the right-hand rule shown here. (**10**)

Like linear momentum, angular momentum is conserved:

The total angular momentum of a system of particles isolated from the rest of the universe remains constant regardless of what events occur within the system.

A skater or ballet dancer doing a spin capitalizes upon conservation of angular momentum. As illustrated here, a skater starts a spin with her arms and one leg outstretched. Then she brings her arms and extended leg inward, which brings their mass closer to the axis of rotation. To maintain *L* constant, the skater spins faster than before. (**11**)

Because angular momentum is a vector quantity, its conservation means that the direction of *L* as well as its magnitude tends to remain unchanged. This is the principle behind the spin stabilization of projectiles,

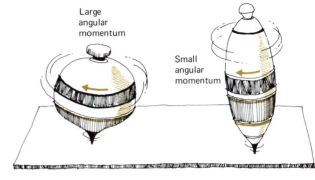

Large angular momentum

Small angular momentum

9

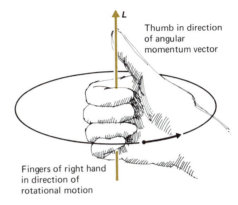

L

Thumb in direction of angular momentum vector

Fingers of right hand in direction of rotational motion

10

11

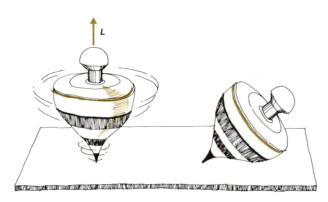

12

such as footballs and rockets. Such projectiles are set spinning about axes in their directions of motion so that they do not tumble and thereby offer excessive air resistance. A top is another illustration of the vector nature of angular momentum. A stationary top set on its tip falls over at once, but a rotating top stays upright until its angular momentum is dissipated by friction between its tip and the ground. (**12**)

Like the conservation principle of linear momentum, the conservation of angular momentum turns out to be a consequence of a symmetry property of the universe. If the laws of nature are independent of direction in space —that is, if the laws of nature are the same regardless of how an observer is oriented with respect to an event of some kind—then angular momentum must be conserved in all interactions.

EXERCISES

1. When an object at rest breaks up into two parts which fly off, must they move in exactly opposite directions?

2. When a moving object strikes a stationary one and the two do not stick together, must they move off in exactly opposite directions?

3. A railway car is at rest on a frictionless track. A man at one end of the car walks to the other end.
   a) Does the car move while he is walking?
   b) If so, in which direction?
   c) What happens when the man comes to a stop?

4. An empty coal car coasts at a certain speed along a level railroad track without friction.
   a) It begins to rain. What happens to the speed of the car?
   b) The rain stops, and the collected water gradually leaks out. What happens to the speed of the car now?

5. Two different rifles, one with a mass of 4 kg and the other with a mass of 5 kg, fire identical bullets with the same muzzle velocities. Do the rifles both have the same recoil momentum? If not, which one has the greater recoil momentum?

6. All helicopters have two propellers; some have both propellers on vertical axes but rotating in opposite directions, and others have one propeller on a vertical axis and the other on a horizontal axis perpendicular to the helicopter body at the tail. Why is a single propeller never used?

7. If the polar icecaps melt, how will the length of the day be affected?

8. A 60-kg woman dives horizontally from a 100-kg boat with a speed of 1.5 m/s. What is the recoil speed of the boat?

9. A 100-kg astronaut outside an orbiting satellite throws away his 0.5-kg camera in disgust when it jams. The speed of the camera is 4 m/s, and

the astronaut begins to move away from the satellite as a result.

a) He fires a burst of gas from a special pistol to bring himself to a stop. If the gas emerges at 50 m/s, how much gas must he discharge?

b) If he did not use his gas pistol, how far away from the satellite would he be in 1 hr?

10. A $\frac{1}{2}$-kg stone moving at 4 m/s collides head-on with a 4-kg lump of clay moving in the opposite direction at 1 m/s. The stone becomes embedded in the clay. What is the speed of the composite body after the collision?

11. A $\frac{1}{2}$-kg stone moving at 4 m/s overtakes a 4-kg lump of clay moving in the same direction at 1 m/s. The stone becomes embedded in the clay. What is the speed of the composite body after the collision?

12. A driverless 2000-kg car is moving along a road at 25 m/s. In order to stop the car, an 8000-kg tank makes a head-on collision with it. What should the tank's speed be in order that both tank and car come to a stop as a result of the collision?

13. A hunter has a rifle that can fire 0.06-kg bullets with a muzzle velocity of 900 m/s. A 40-kg leopard springs at him at a speed of 10 m/s. How many bullets must the hunter fire into the leopard in order to stop him in his tracks?

14. A 12,000-kg freight car rolls at 2 m/s along a horizontal track. It collides with a stationary 18,000-kg freight car and the two cars couple together. What is the speed of the cars afterward?

9

FORCE

"Thus we have two laws of motion, the law of inertia, and that of the force being proportional to the acceleration, which are given from experience. They are the most natural and the most simple which it is possible to imagine, and are without doubt derived from the nature itself of matter; but this nature being unknown, they are with respect to us solely the consequences of observation, and the only ones which the science of mechanics requires from experience." Pierre Laplace (1749–1827)

Left to itself, a body has a constant momentum —if at rest, it remains at rest, and if moving at the velocity v, it continues to do so. A *force* is any influence capable of changing a body's momentum. Thus every acceleration can be traced to the action of a force. There are only four basic interactions between elementary particles that give rise to all of the forces in the natural world.

NEWTON'S LAWS OF MOTION

Almost three centuries ago Isaac Newton (1642–1727) formulated three principles, based upon observations he and others had made, which summarize so much of the behavior of interacting bodies that they have become known as the laws of motion. In the years since Newton's work it has become clear that the principle of conservation of momentum not only incorporates the first and third laws of motion but goes beyond them in scope and usefulness, and that the concept of force, which is defined by the second law of motion, has only a limited range of application; the interactions among elementary particles that lead to the existence of atoms, molecules, and solids are better treated in terms of energy than in terms of force. But Newton's laws of motion are still valuable because force, despite the ability of theoretical physics to dispense with it, is a physical quantity whose meaning all of us find easy to appreciate. And the laws of motion frequently offer the most convenient way to understand phenomena that involve objects whose velocities are small relative to the velocity of light and which are larger in size than atoms and molecules—a rather extensive class of objects.

FORCE

According to the principle of conservation of linear momentum, the linear momentum of a

$p_2 = mv$

$p_1 = 0$

1

body remains constant as long as the body is isolated from the rest of the universe. When the body interacts with something else, its momentum may change. We can interact with a football by kicking it, and the result is a change in the football's momentum from $p_1 = 0$ to some value $p_2 = mv$. (**1**)

Or the interaction can take the form of catching a moving football, in which case again the football's momentum changes. (**2**)

An interaction can lead to a change in the *direction* of p as well as to a change in its magnitude, as for instance when a football bounces off a tree. (**3**)

$p_1 = mv$

$p_2 = 0$

2

p_1

p_2

3

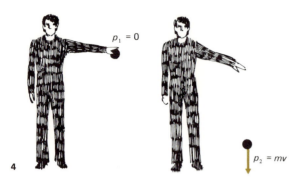

$p_1 = 0$

$p_2 = mv$

4

A downward force exerted by the earth causes dropped objects to fall.

Some interactions are more effective than others in causing momentum changes. A swift kick affects the momentum of a football more than a gentle tap does. The concept of *force* can be used to put the matter on a precise basis. In general:

A force is any influence that can produce a change in the linear momentum of a body.

This definition is in accord with the notion of a force as a "push" or a "pull," but it goes further since no direct contact is necessarily implied. No hand reaches up from the earth to pull a dropped stone downward, yet the increasing downward momentum of the falling stone testifies to the action of a force on it. **(4)**

BALANCED AND UNBALANCED FORCES

It is entirely possible for two or more forces to act upon a body without affecting its momentum; the forces may have such magnitudes and directions as to cancel one another out. What is required for a momentum change is a *net* force, often called an *unbalanced force*. When a body is acted upon by several forces whose vector sum is zero, the forces are said to be *balanced forces* and the body is said to be in *equilibrium*. But each one of the forces acting by itself must, by definition, be capable of producing a momentum change. **(5)**

SECOND LAW OF MOTION

In his *second law of motion* Isaac Newton proposed a quantitative definition of force:

The net force acting upon a body is equal to the rate of change of the body's linear momentum.

In equation form,

$$\text{Force} = \frac{\text{linear momentum change}}{\text{time interval}},$$

$$F = \frac{p_2 - p_1}{\Delta t},$$

where

F = force acting on a body,
p_1 = initial linear momentum of the body,
p_2 = linear momentum of the body after the force has acted for a certain time interval,
Δt = duration of time interval.

Force is a vector quantity since every force is characterized by a direction as well as a magnitude.

Like all definitions, this one is arbitrary, and it has been accepted for three hundred years because it leads to a particularly simple theoretical framework (called "Newtonian" or "classical" mechanics) for analyzing the behavior of moving bodies.

In the above form, the second law of motion holds regardless of any mass changes a body many undergo as its momentum changes under the action of a force. Very often the velocities involved in a particular situation are so small

5

compared with the velocity of light that the masses of the bodies present remain constant for all practical purposes. This is certainly true for the macroscopic objects that populate the world around us: as we found in Section 7, even the swiftest spacecraft experiences a mass change of only about 1 part in 10^{14}. However, it is *not* true for elementary particles such as electrons and protons which may already be traveling nearly as fast as light when a force is applied; such particles increase in mass almost in proportion to their momentum increase, since they cannot go more than a trifle faster regardless of how much force is exerted on them.

ACCELERATION AND FORCE

Under the assumption of constant mass, we can substitute $m\boldsymbol{v}$ for $\boldsymbol{p}$ and express the second law of motion as

$$\boldsymbol{F} = \frac{\boldsymbol{p}_2 - \boldsymbol{p}_1}{\Delta t}$$

$$= \frac{m\boldsymbol{v}_2 - m\boldsymbol{v}_1}{\Delta t}$$

$$= m\frac{\boldsymbol{v}_2 - \boldsymbol{v}_1}{\Delta t}.$$

But the quantity $(\boldsymbol{v}_2 - \boldsymbol{v}_1)/\Delta t$ is what was defined in Section 4 as *acceleration*:

$$\text{Acceleration} = \frac{\text{change in velocity}}{\text{time interval}},$$

$$\boldsymbol{a} = \frac{\boldsymbol{v}_2 - \boldsymbol{v}_1}{\Delta t}.$$

We conclude that

$$\text{Force} = \text{mass} \times \text{acceleration},$$

$$\boldsymbol{F} = m\boldsymbol{a}.$$

This is the form in which the second law of motion is usually expressed in classical mechanics.

In words, the formula $\boldsymbol{F} = m\boldsymbol{a}$ states that the net force $\boldsymbol{F}$ acting upon a body of mass m is equal to the product of m and the acceleration $\boldsymbol{a}$ the body undergoes as a result of the application of the force. The effect of applying a net force to a body is to accelerate it.

The second law provides us with a method for analyzing and comparing forces in terms of the accelerations they produce. Thus a force which causes a body to have twice the acceleration another force produces must be twice as great. A body moving to the right but going

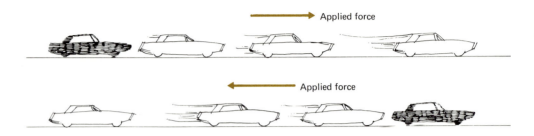

Applied force

Applied force

6

slower and slower is accelerated to the left; hence there is a force toward the left acting on it. The first law of motion is clearly a special case of the second: when the net force on a body is zero, its acceleration is also zero. **(6)**

The acceleration of a body is proportional to the net force applied to it. Successive positions of a ball are shown at 1-s intervals when forces of F, $2F$, and $3F$ are applied. **(7)**

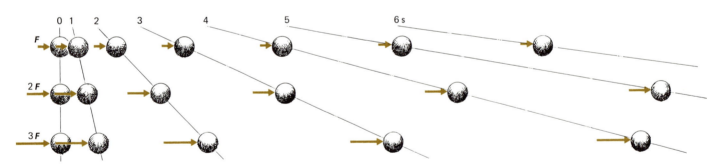

7

When the same force is applied to bodies of different mass, the resulting accelerations are inversely proportional to the masses. Successive positions of balls of mass m, $2m$, and $3m$ are shown at 1-s intervals when identical forces of F are applied. (8)

Let us examine a few problems in order to become familiar with the application of the second law of motion. These problems are, of course, artificial and oversimplified, but they do illustrate the power of the second law in situations involving force and motion.

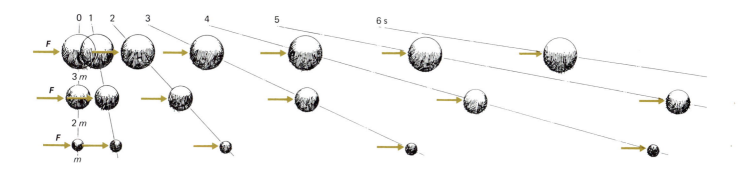

8

THE NEWTON

It is convenient to have a special unit for force. In the system of units we are using, the appropriate unit is the *newton*:

A newton is that force which, when applied to a 1-kg mass, gives it an acceleration of 1 m/s².

The newton is not a fundamental unit like the meter, second, and kilogram, and in some calculations it may have to be replaced by its equivalent in terms of the latter. This equivalent may be found as follows:

$$F = ma,$$

$$1\,\text{N} = 1\,\text{kg} \times 1\,\frac{\text{m}}{\text{s}^2}$$

$$= 1\,\frac{\text{kg-m}}{\text{s}^2}.$$

Problem. A force of 10 N is applied to a 4.0-kg block that is at rest on a perfectly smooth, level surface. Find the velocity of the block and how far it has gone after 6.0 s.

Solution. We start from the second law of motion in scalar form, since only one direction is involved here:

$$F = ma.$$

We know what F and m are, so we find the acceleration a of the block as follows:

$$a = \frac{F}{m}$$

$$= \frac{10\,\text{N}}{4.0\,\text{kg}}$$

$$= 2.5\,\frac{\text{m}}{\text{s}^2}.$$

The direction of the acceleration is the same as that of the force.

To find the velocity of the block after $t = 6.0$ s, we use the formula $v = at$ and obtain

$$v = at$$

$$= 2.5 \, \frac{m}{s^2} \times 6.0 \, s$$

$$= 15 \, \frac{m}{s}.$$

For the distance the block travels in $t = 6.0$ s at an acceleration of $a = 2.5$ m/s² we require the formula

$$s = v_0 t + \frac{1}{2}at^2$$

from Section 4. Here the block started from rest, so $v_0 = 0$ and we have

$$s = \frac{1}{2}at^2$$

$$= \frac{1}{2} \times 2.5 \, \frac{m}{s^2} \times (6.0 \, s)^2$$

$$= 45 \, m.$$

After 6.0 s a 4.0-kg mass acted upon by a 10-N force will have gone 45 m and have a velocity of 15 m/s. **(9)**

Problem. Next let us consider the same block sliding on a level but slightly rough surface without the 10-N force acting. The block starts at a velocity of 15 m/s, and gradually slows down and stops in a distance of 20 m. What was the frictional resistive force that acted?

Solution. We start by determining the acceleration of the block. Since we know that its velocity has gone from 15 m/s to 0 in 20 m, we use the formula

$$v^2 = v_0^2 + 2as$$

from Section 4. Here the final velocity is $v = 0$, the initial velocity is $v_0 = 15$ m/s, and the distance is $s = 20$ m. Hence

$$v^2 = v_0^2 + 2as,$$

$$a = \frac{v^2 - v_0^2}{2s}$$

$$= \frac{0 - (15 \, \text{m/s})^2}{2 \times 20 \, \text{m}}$$

$$= -5.6 \, \frac{m}{s^2}.$$

The minus sign indicates a negative acceleration, that is, an acceleration directed opposite to the velocity of the block. Now that we know

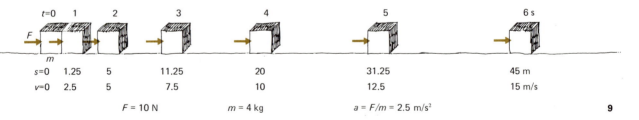

$t=0$	1	2		3	4	5	6 s
m							
$s=0$	1.25	5		11.25	20	31.25	45 m
$v=0$	2.5	5		7.5	10	12.5	15 m/s

$F = 10 \, \text{N} \qquad m = 4 \, \text{kg} \qquad a = F/m = 2.5 \, \text{m/s}^2$ **9**

Successive distances and velocities of a 4-kg mass acted upon by a 10-N force.

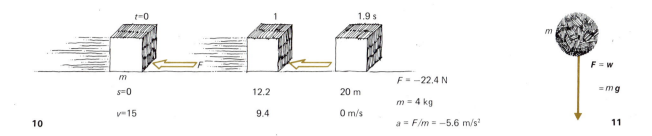

$t=0$ 1   1.9 s

m

$s=0$ 12.2   20 m

$v=15$ 9.4   0 m/s

$F = -22.4$ N

$m = 4$ kg

$a = F/m = -5.6$ m/s^2

m

$F = w$

$= m\mathbf{g}$

A mass of 4 kg with an initial velocity of 15 m/s comes to a stop in a distance of 20 m when a resistive force of 22 N acts on it.

the mass and the acceleration of the block, it is easy to find the resistive force:

$$F = ma$$

$$= 4.0 \text{ kg} \times (-5.6\, \frac{\text{m}}{\text{s}^2})$$

$$= -22 \text{ N}.$$

The minus sign indicates that the direction of the force is opposite to that of the block's velocity. (**10**)

THE FUNDAMENTAL INTERACTIONS

Every force, without exception, arises from one or another of the four fundamental interactions that are possible between the elementary particles of which all matter consists. Two of these interactions are effective only when the particles involved are extremely close together; these are called the "strong" and "weak" nuclear interactions because they are responsible for the ability of protons and neutrons to stick together to form atomic nuclei, and one is more powerful than the other. The others—the gravitational and electromagnetic interactions—are unlimited in range, but their strength decreases with distance. In later sections we shall explore the properties of these four interactions.

WEIGHT IS A FORCE

The gravitational force with which an object is attracted to the earth is called its *weight*. Weight is different from mass, which is a measure of the inertia an object at rest exhibits. Although they are different physical quantities, mass and weight are closely related.

The weight of a stone is the force that causes it to be accelerated when it is dropped. All objects in free fall have a downward acceleration of $g = 9.8$ m/s^2, the acceleration of gravity. If the stone's mass is m, then the downward force on it, which is its weight w, can be found from the second law of motion, $F = ma$, by letting $F = w$ and $a = g$. Evidently

Weight = mass × acceleration of gravity,

$$w = mg.$$

The weight of any object is equal to its mass multiplied by the acceleration of gravity. Since g is a constant near the earth's surface, the weight w of an object is always directly proportional to its mass m: a large mass is heavier than a small one. (**11**)

The mass of an object is a more fundamental property than its weight, because its mass is

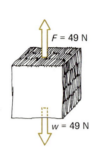

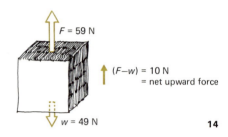

12

13

14

the same everywhere in the universe whereas the gravitational force on it depends upon its position relative to the earth (or relative to some other astronomical body). A 100-kg man weighs 980 N on the earth, but he would weigh 2587 N on Jupiter, 372 N on Mars, 162 N on the moon, and 0 in space far from the sun and other stars. **(12)**

Problem. How much force is needed to lift a 5.0-kg box from the ground? How much force is needed to give it an upward acceleration of 2.0 m/s²?

Solution. The weight of the box is

$$w = mg$$

$$= 5.0 \text{ kg} \times 9.8 \frac{\text{m}}{\text{s}^2}$$

$$= 49 \text{ N}.$$

Therefore an upward force of 49 N is needed to lift the box from the ground. **(13)**

To give the box an upward acceleration, we must apply a force F greater than its weight w

in order that there be a *net* upward force. The net upward force required here is

$$F - w = ma$$

$$= 5.0 \text{ kg} \times 2.0 \frac{\text{m}}{\text{s}^2}$$

$$= 10 \text{ N},$$

and so the applied force must be

$$F = w + ma$$

$$= 49 \text{ N} + 10 \text{ N}$$

$$= 59 \text{ N}. \textbf{(14)}$$

Problem. An elevator whose mass is 1000 kg fully loaded is suspended by a cable whose maximum permissible tension is 25,000 N. What is the greatest upward acceleration possible for the elevator under these circumstances? What is the greatest downward acceleration?

Solution. This problem is similar to the previous one. When the elevator is at rest, or

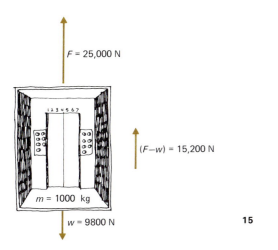

F = 25,000 N

1 2 3 4 5 6 7

(F−w) = 15,200 N

m = 1000 kg

w = 9800 N

15

moving at constant velocity, the tension in the cable is just the elevator's weight of

$$w = mg$$

$$= 1000 \text{ kg} \times 9.8 \frac{\text{m}}{\text{s}^2}$$

$$= 9800 \text{ N.}$$

To accelerate the elevator upward, an additional tension is required in order to provide a net upward force. Here it is given that the maximum tension cannot exceed 25,000 N, so that the greatest net force that can be applied to the elevator is (25,000 − 9800) N or 15,200 N. The acceleration of the elevator when a net force of 15,200 N is applied to it is

$$a = \frac{F}{m}$$

$$= \frac{15,200 \text{ N}}{1000 \text{ kg}}$$

$$= 15.2 \frac{\text{m}}{\text{s}^2}. \quad (15)$$

For the elevator to exceed the downward acceleration of gravity, which is $g = 9.8 \text{ m/s}^2$, a downward force in addition to the weight of the elevator is required. Since this cannot be provided by a supporting cable, the maximum downward acceleration is 9.8 m/s².

LIMITATIONS OF THE SECOND LAW

Every statement in science must be capable of experimental verification, however indirect, if it is to mean anything at all. The second law defines the net force on an object in terms of the rate of change of the object's linear momentum. When we are considering the motions of things we can actually see, there is no real difficulty in measuring momenta and time intervals, and the second law is a useful intellectual tool for analyzing the behavior of balls and spacecraft, specks of dust and planets.

On the level of electrons and protons, atoms and molecules, however, the process of measurement itself becomes a serious problem. As we shall learn in Section 33, the wave character of moving bodies (imperceptible on a macroscopic scale of size, but dominant on a microscopic scale) makes it impossible, even in principle, to perform experiments of unlimited accuracy. The *uncertainty principle* is a quantitative statement of the degree of inaccuracy inherent in measurements of various kinds; not instrumental inaccuracy, which may also be present, but an inaccuracy due to the very nature of what is being measured.

In the atomic world, the second law of motion does not yield consistent results when applied to observations made on individual particles. However, it is still a useful abstraction of experience in another sense, as a statement of *probability* rather than of *truth*. If we examine the responses of electrons (say) to applied forces, we would find that $F = (p_2 - p_1)/\Delta t$ is obeyed *on the average*, but that the individual data show deviations from this formula. The ordinary objects we encounter in our daily lives are aggregates of such vast numbers of particles that their average behavior is all we perceive. Hence the second law of motion is a reliable guide to macroscopic events, but it does not have the—apparently—universal validity of the principles of relativity and of conservation of linear momentum.

EXERCISES

1. When a body is accelerated, a force is invariably acting upon it. Does this mean that, when a force is applied to a body, it is invariably accelerated?

2. The moon revolves around the earth in an approximately circular orbit. Is the moon accelerated in its motion? Does a force act on the moon? If so, in which direction?

3. It is less dangerous to jump from a high wall onto loose earth than onto a concrete pavement. Why?

4. Compare the tension in the coupling between the first two cars with that in the coupling between the last two cars in a train
   a) when the train's velocity is constant, and
   b) when the train is accelerating.

5. When a force equal to its weight is applied to a body free to move, what is its acceleration?

6. A force of 1 N is applied in turn to an object of mass 1 kg and an object of weight 1 N. Which receives the greater acceleration?

7. An empty truck whose mass is 2000 kg has a maximum acceleration of 1 m/s². What is its maximum acceleration when it is carrying a 1000-kg load?

8. A force of 10 N gives an object an acceleration of 5 m/s². What force would be needed to give the same object

a) an acceleration of 1 m/s²?
b) an acceleration of 10 m/s²?

9. A force of 4000 N is applied to a 1400-kg car. What will the car's speed be after 10 s if it started from rest?

10. A 1200-kg car accelerates from 10 m/s to 15 m/s in 5 s. Find the force acting on the car.

11. A 2000-kg truck is braked to a stop in 15 m from an initial speed of 12 m/s. How much force was required?

12. How much force must you supply to give a 1-kg object an upward acceleration of 2g? a downward acceleration of 2g?

13. A car moving at 10 m/s (22.4 mi/hr) strikes a stone wall.
    a) The car is very rigid and the 80-kg driver comes to a stop in a distance of 0.2 m. What is his average acceleration and how does it compare with the acceleration of gravity g? How much force acted upon him? Express this force in both newtons and pounds, where 1 N = 0.225 lb.
    b) The car is so constructed that its front end gradually collapses upon impact, and the driver comes to a stop in a distance of 1 m. Answer the same questions for this situation.

14. A 100-kg man slides down a rope at constant speed.
    a) What is the minimum breaking strength the rope must have?
    b) If the rope has precisely this strength, will it support the man if he tries to climb back up?

15. A 1000-kg elevator has a downward acceleration of 1 m/s². What is the tension in its supporting cable?

16. A 1000-kg elevator has an upward acceleration of 1 m/s². What is the tension in its supporting cable?

17. A 100-kg man stands on a scale in an elevator.
    a) What does the scale read when the elevator is ascending with an acceleration of 1 m/s²?
    b) When it is descending with an acceleration of 1 m/s²?

c) When it is ascending at the constant speed of 3 m/s?

d) When it is descending at the constant speed of 3 m/s²?

e) When the cable has broken and it is descending in free fall?

18. A man stands on a scale in an elevator. When the elevator is at rest, the scale reads 700 N. When the elevator starts to move, the scale reads 600 N.

a) Is the elevator going up or down?

b) Does it have a constant speed? If so, what is this speed?

c) Does it have a constant acceleration? If so, what is this acceleration?

19. A mass of 8 kg and another of 12 kg are suspended by a string on either side of a frictionless pulley. Find the acceleration of each mass.

20. A mass of 0.05 kg and another of 0.01 kg are suspended by a string on either side of a frictionless pulley. Find the acceleration of each mass.

$$\frac{F}{a} = m a$$

10

ACTION AND REACTION

An important consequence of conservation of momentum is that forces always come in pairs. For every force one body exerts on another, there is an equal but opposite reaction force the second body exerts on the first. There is no way for just one force to come into being in the universe, although the action and reaction forces act on different bodies so that each body can be acted upon by a single force.

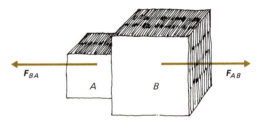

1

"To every action there is always opposed an equal reaction: or, the mutual actions of two bodies upon each other are always equal, and directed to contrary parts. Whatever draws or presses another is as much drawn or pressed by that other. If you press a stone with your finger, the finger is also pressed by the stone." Isaac Newton (1642–1727)

THE THIRD LAW OF MOTION

The third of Newton's laws of motion states:

When a body exerts a force on another body, the second body exerts a force on the first body of the same magnitude but in the opposite direction.

If we call one of the interacting bodies A and the other B, then according to the third law of motion

$$F_{AB} = -F_{BA},$$

where F_{AB} is the force body A exerts on body B and F_{BA} is the force body B exerts on body A. (1)

82

2 3 4

There is no such thing as a single force in the universe. We push against a wall; the wall pushes back on us. (2)

We throw a ball; as we are pushing it into the air, it is pushing back on our hand. (3)

We fire a rifle, and the expanding gases in its barrel push the bullet out; the gases push back on the rifle, which gives rise to the recoil force we feel with our shoulder. (4)

An apple falls because of the downward gravitational pull of the earth; there is an equal upward pull by the apple on the earth which we cannot detect because the earth is so much more massive than the apple, but it is nevertheless there. (5)

5

THE THIRD LAW AND MOMENTUM CONSERVATION

The third law of motion follows from the principle of conservation of linear momentum, just as the first law does. However, the second law, which defines force, is independent of this principle.

Let us consider what happens when an apple falls to the ground. The total linear momentum of the system of the earth plus the apple cannot be affected by the falling of the apple, because the interaction between them does not involve any other body in the universe. Hence it must be true that

Momentum change of earth
+ momentum change of apple = 0,

$$\Delta p_{\text{earth}} + \Delta p_{\text{apple}} = 0,$$

$$\Delta p_{\text{earth}} = - \Delta p_{\text{apple}}.$$

The earth must undergo a momentum change during the time the apple is undergoing a momentum change, but in the opposite direction; as the apple falls down, the earth (so to speak) falls up. If the earth did *not* undergo such a momentum change, then the linear momentum of the system of apple + earth would not be constant, thereby violating the principle of conservation of linear momentum.

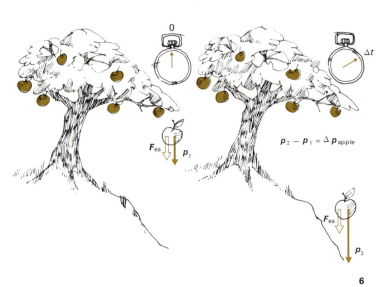

6

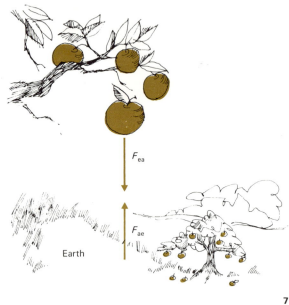

Earth

7

During some time interval Δt while it is falling, the apple gains $\Delta \boldsymbol{p}_{\text{apple}}$ in momentum. The gravitational force the earth exerts on the apple which causes it to fall is $\boldsymbol{F}_{\text{ea}}$, where the "ea" stands for "earth on apple." From the second law of motion,

$$\boldsymbol{F}_{\text{ea}} = \frac{\Delta \boldsymbol{p}_{\text{apple}}}{\Delta t}$$

and so

$$\Delta \boldsymbol{p}_{\text{apple}} = \boldsymbol{F}_{\text{ea}} \Delta t. \quad (6)$$

The momentum change $\Delta \boldsymbol{p}_{\text{earth}}$ of the earth during the same time interval Δt must be the result of a force the apple exerts on the earth, since nothing else in the universe is involved. Hence

$$\Delta \boldsymbol{p}_{\text{earth}} = \boldsymbol{F}_{\text{ae}} \Delta t$$

where "ae" stands for "apple on earth." From the fact that

$$\Delta \boldsymbol{p}_{\text{earth}} = - \Delta \boldsymbol{p}_{\text{apple}},$$

we have

$$\boldsymbol{F}_{\text{ae}} \Delta t = - \boldsymbol{F}_{\text{ea}} \Delta t,$$

or, since the time interval Δt is the same,

$$\boldsymbol{F}_{\text{ae}} = - \boldsymbol{F}_{\text{ea}}.$$

The force the apple exerts on the earth must be equal in magnitude and opposite in direction to the force the earth exerts on the apple. (7)

The above result, generalized to describe the interaction of *any* two bodies A and B, is the third law of motion:

$$\boldsymbol{F}_{AB} = - \boldsymbol{F}_{BA}.$$

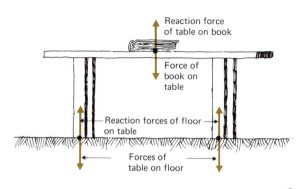

Reaction force
of table on book

Force of
book on
table

Reaction forces of floor
on table

Forces of
table on floor

8

Earth exerts
forward reaction
force on foot

Foot exerts
backward force
on earth

9

ACTION AND REACTION

The third law of motion always applies to two different forces on two different bodies—the *action force* that one body exerts on another, and the equal but opposite *reaction force* that the second exerts on the first.

Let us apply the third law to a few situations in order to appreciate its significance. A 1-kg book lies stationary on a table, pressing down on the table with a force of $mg = 9.8$ N. The table pushes upward on the book with the reaction force of 9.8 N. Why doesn't the book fly upward into the air? The answer is that the upward force of 9.8 N on the book merely balances its weight of 9.8 N, which acts downward. The net force on the book is zero. If the table were not there to cancel out the 9.8-N weight of the book, it would of course be accelerated downward. (**8**)

Another illustration of the third law of motion is the operation of walking. We push backward with one foot, and the earth pushes forward on us. The forward reaction force exerted by the earth causes us to move forward, and at the same time, the backward force of our foot causes the earth to move backward. Owing to the earth's relatively larger mass, its motion cannot be detected practically, but it is there. Why is it that there is no reaction force on us, responding to the earth's push on our foot, to keep us from moving? The explanation is that every aotion–reaction pair of forces acts on *different* bodies. We push on the earth, the earth pushes back on us. If there are no *additional* forces present to impede our motion (for instance, pressure by a wall directly in front of us), we proceed to undergo an acceleration. (**9**)

We might conceivably find ourselves on a frozen lake with a perfectly smooth surface. Now we cannot walk because the absence of friction prevents us from exerting a backward force on the ice which would produce a forward force on us. But what we can do is exert a force on some object we may have with us, say a rock. We throw the rock forward by applying a force to it; at the same time the rock is pressing back on us with the identical force but in the opposite direction, and in consequence we find ourselves moving backward. (**10**)

10

CENTER OF MASS

We have made no distinction thus far between the mechanical behavior of individual particles and that of composite bodies consisting of many particles. The third law of motion is what permits us to proceed in this simple way. All the forces that act upon a system of particles may be divided into two categories:

1. *Internal forces* that the particles within the system exert upon one another;

2. *External forces* that originate outside the system.

According to the third law, each one of the internal forces is matched by an equal and opposite force that acts somewhere else within the system. If particle A exerts the force $\boldsymbol{F}_{AB}$ on particle B, then B exerts the force $\boldsymbol{F}_{BA} = -\boldsymbol{F}_{AB}$ on A. When we add up all of the various forces to determine the net force on the system of particles, the internal forces cancel out in pairs. Only the external forces affect the system as a whole. In considering the motion of any object, then, we need take into account only the external forces that act upon it, and we can ignore the internal forces that are holding it together, pushing it apart, or otherwise affecting its configuration.

Internal and external forces act on this system of three particles. (**11**)

The vector sum of the internal forces is zero because these forces cancel out in pairs, according to the third law of motion. (**12**)

The vector sum of the external forces need not be zero. (**13**)

Only the vector sum of the external forces affects the motion of the system as a whole. (**14**)

A net external force $\boldsymbol{F}$ acts upon a composite body of total mass m, and in response the body is accelerated. As we have found, the body's motion is not affected by internal forces, so its acceleration is given by the second law of motion as

$$\boldsymbol{a} = \frac{1}{m}\boldsymbol{F}.$$

But a body may have a variety of motions, such as expansion, contraction, and rotation, not shared by a particle, and different values for $\boldsymbol{a}$ may be found at the same time depending upon which part of the body is being considered. To what point in the body does the acceleration $\boldsymbol{a}$ refer?

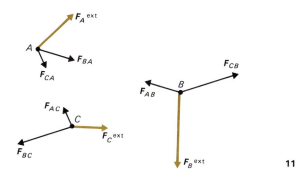

11

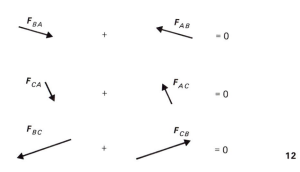

$$F_{BA} + F_{AB} = 0$$

$$F_{CA} + F_{AC} = 0$$

$$F_{BC} + F_{CB} = 0$$

12

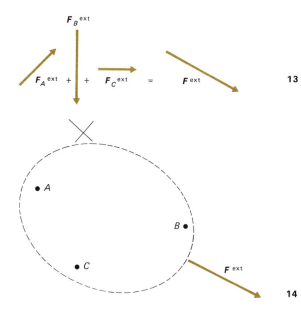

$$F_A{}^{ext} + F_B{}^{ext} + F_C{}^{ext} = F^{ext}$$

13

14

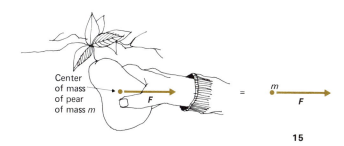

15

The answer is that every body of matter, indeed every system of particles, has a certain *center of mass* which is the effective location of the body's mass insofar as external forces are concerned. The motion of this center of mass obeys the second law of motion. When a net force **F** acts on a body of mass m, then, we can imagine the force as applied to a particle of mass m located at the body's center of mass. (**15**)

The center of mass of a body, which describes a certain aspect of its inertial behavior, is the same as its center of gravity, which describes a corresponding aspect of its gravitational behavior, because inertial and gravitational masses are equal. Hence we have an easy way to locate the center of mass of a body—we merely find that point (the "balance point") from which it can be suspended without tending to rotate. (**16**)

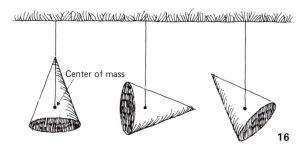

16

Center of mass
of spacecraft

Center of mass
of fragments

The center of mass of a system of particles obeys the first law of motion as well as the second. That is, the center of mass of a system of particles remains at rest or travels at constant velocity along a straight line in the absence of any interaction between the members of the system and the rest of the universe. Thus if an explosion occurs in a spacecraft out in space far from everything else, the center of mass of the fragments continues to travel along the same path as the intact spacecraft followed, even though the fragments themselves may be scattered in all directions. The reason for this is again the third law of motion: during the explosion all the forces exerted on the fragments occurred in equal and opposite pairs. **(17)**

LIMITATIONS OF THE THIRD LAW

When two objects are in contact, the response of one of them to a push by the other is immediate: the reaction force comes into being as soon as the action force is applied. The third law of motion is obeyed by such a pair of objects at all times.

What if the objects are far apart and interact by means of, say, gravitational forces? If object A moves farther away from object B, the forces between them decrease, but they do not decrease at the same moment. The "news" of the motion of A cannot travel faster than the velocity of light, so there must be a time lag between the force change on A and that on B. Eventually, of course, $F_{AB} = - F_{BA}$, but the point is that, owing to the finite velocity with which signals of any kind travel, the third law may be violated at certain times. In everyday life such time lags are not easy to detect, but they exist and introduce an undesirable complication in what we would like to consider as a fundamental law of nature. **(18)**

On the other hand, it is easy to avoid the problem of time lags in the principle of conservation of linear momentum. This is one of the reasons why the latter principle is considered "more fundamental" than the third law of motion. Something (but not necessarily something with rest mass) carries the information from A to B about A's change in position, and all we need do is assume that

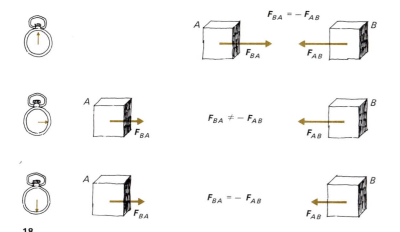

$F_{BA} = -F_{AB}$

$F_{BA} \neq -F_{AB}$

$F_{BA} = -F_{AB}$

18

this something can possess linear momentum. Then it is possible for there to be an exact conservation of linear momentum in the system at all times. This notion has turned out to be a very productive one, as we shall see in later sections when we consider the transport of information between interacting bodies.

The conclusion, then, is that we cannot go wrong when we use conservation of linear momentum to analyze a situation, but that we may have to be careful in dealing with the third law of motion when the interacting objects are distant from one another.

EXERCISES

1. Can we conclude from the third law of motion that a single force cannot act upon a body?

2. Since the opposite forces of the third law of motion are equal in magnitude, how can anything ever be accelerated?

3. A ladder rests against the side of a wall.
   a) What forces are exerted by the foot of the ladder on the ground?

b) What are the reactions to these forces?
   c) What forces are exerted by the top of the ladder on the wall?
   d) What are the reactions to these forces?
   Use a diagram to illustrate your answers.

4. A block is pushed along a rough tabletop.
   a) What forces are exerted by the block on the table?
   b) What are the reaction forces exerted by the table on the block?
   Use a diagram to illustrate your answers.

5. A car that is towing a trailer is accelerating on a level road. The force the car exerts on the trailer is
   a) equal to the force the trailer exerts on the car.
   b) greater than the force the trailer exerts on the car.
   c) less than the force the trailer exerts on the car.
   d) equal to the force the trailer exerts on the road.

6. An engineer designs a propeller-driven spacecraft. Because there is no air in space, he incorporates a supply of oxygen as well as a supply of fuel for the motor. What do you think of the idea?

11

CIRCULAR MOTION

"A centripetal force is that by which bodies are drawn or impelled, or in any way tend, towards a point as to a centre. Of this sort is . . . that force, whatever it is, by which the planets are continually drawn aside from the rectilinear motions, which otherwise they would pursue, and made to revolve in curvilinear orbits." Isaac Newton (1642–1727)

Almost all objects in the physical world travel in curved paths. Often these paths are either circles or are very close to being circles. For example, the orbits of the earth and the other planets about the sun are nearly circular in shape, as is the orbit of the moon about the earth. On a microcosmic scale, a handy way to visualize an atom is to imagine it as having a central nucleus with electrons circling around. And, of course, circular motion is no novelty in daily life.

CENTRIPETAL FORCE

An object traveling in a circle at a velocity whose magnitude is constant is said to undergo *uniform circular motion.*

Although the velocity of such an object has the same *magnitude* all along its path, the *direction* of the velocity changes constantly. A changing velocity means a changing momentum, which in turn signifies that the object must be acted upon by a force. Since the object's path is a circle, the force on it must be directed inside the circle. This force is called *centripetal force*, literally "force seeking the center". Without it, circular motion cannot occur. Thus:

Centripetal force = inward force on an object moving in a circle. (1)

To verify directly the crucial role of centripetal force in circular motion, we can whirl a ball at the end of a string. As the ball swings around, we must continually exert an inward force on it by means of the string. If we let go of the string, the ball flies off tangent to its original circular path. With no centripetal force on it, the ball then proceeds along a straight path at constant velocity as the first law of motion predicts. (Actually, of course, the ball will fall to the ground eventually

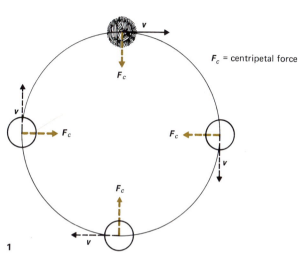

1

String provides
centripetal force

2

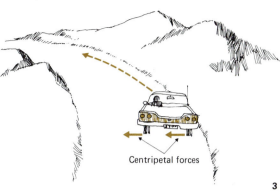

F_c = centripetal force

Centripetal forces

3

because of gravity, but this is irrelevant here.) (2)

A centripetal force is acting whenever rotational motion occurs, since such a force is required to change the direction of motion of a particle from the straight line which it would normally follow to a curved path. Gravitation provides the centripetal forces that keep the planets moving around the sun and the moon around the earth. Friction between its tires and the road provides the centripetal force needed by a car in rounding a curve. If the tires are worn and the road wet or icy, the frictional force is small and may not be enough to permit the car to turn. (3)

CENTRIPETAL ACCELERATION

How large a centripetal force is needed to keep a given object moving in a circle with a certain speed? To find out, we must first compute the acceleration of such an object. In the following derivation we shall consider the circular motion of a particle; the same arguments and conclusions hold for the circular motion of the center of mass of an object of finite size.

A particle is traveling along a circular path of radius r at the constant speed v. At $t = 0$ the

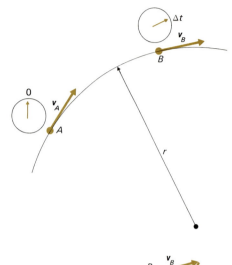

particle is at the point A, where its velocity is v_A, and at $t = \Delta t$ the particle is at the point B, where its velocity is v_B. **(4)**

The change in the particle's velocity in the time interval Δt is

$$\Delta v = v_B - v_A$$

and so its acceleration is

$$\text{Acceleration} = \frac{\text{change in velocity}}{\text{time interval}},$$

$$a = \frac{\Delta v}{\Delta t}. \quad \textbf{(5)}$$

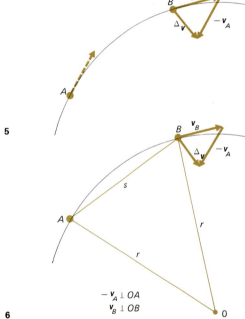

The vector triangle whose sides are $-v_A$, v_B, and Δv is similar to the space triangle whose sides are OA, OB, and s. Since v is the speed of the particle, the magnitudes of $-v_A$ and v_B are both v. Also, OA and OB are radii of the circle, so their lengths are both r. Corresponding sides of similar triangles are proportional, hence

$$\frac{\Delta v}{v} = \frac{s}{r}$$

and

$$\Delta v = \frac{vs}{r}. \quad \textbf{(6)}$$

The distance the particle actually covers in going from A to B is the arc joining these points, the length of which is $v\Delta t$. The distance s, however, is the chord joining A and B. We are finding the *instantaneous* acceleration of the particle, and we are therefore concerned with the case where A and B are very close together, in which case the chord and arc are equal. Hence

$$s = v\Delta t$$

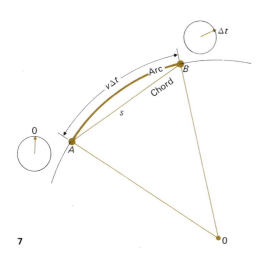

7

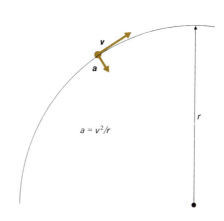

$$a = v^2/r$$

8

and we have

$$\Delta v = \frac{v^2 \Delta t}{r} \quad . \text{(7)}$$

The magnitude of the particle's acceleration is therefore

$$a_c = \frac{\Delta v}{\Delta t} = \frac{v^2}{r}. \quad \text{Centripetal acceleration}$$

This acceleration is called the *centripetal acceleration* of the particle: the centripetal acceleration of a particle in uniform circular motion is proportional to the square of its speed and inversely proportional to the radius of its path. (8)

When the points A and B are a finite distance apart, Δv does not point toward O, the center of the particle's circular path. When A and B are an infinitesimal distance apart, however, Δv *does* point toward O. Because the centripetal acceleration a_c is an instantaneous acceleration, we are solely concerned with the case when Δt and hence s are vanishingly small, and the direction of a is accordingly radially inward. (9)

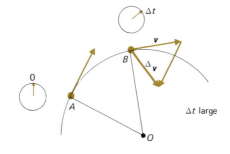

Δt large

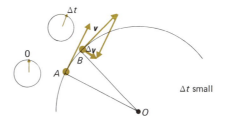

Δt small

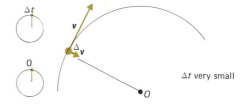

Δt very small

9

Problem. The moon is 3.84×10^8 m from the earth and circles the earth once every 27.3 days. Find its centripetal acceleration.

Solution. The time needed by an object in uniform circular motion to make a complete revolution and return to some starting point is called its *period*, symbol T. The distance the object travels in making a complete circle of radius r is $2\pi r$, the circumference of the circle.

The speed of the body is therefore

$$v = \frac{\text{distance}}{\text{time}}$$

$$= \frac{2\pi r}{T}. \quad (10)$$

In the case of the moon, $r = 3.84 \times 10^8$ m and

$$T = 27.3 \text{ days} \times 24 \frac{\text{hr}}{\text{day}} \times 60 \frac{\text{min}}{\text{hr}} \times 60 \frac{\text{s}}{\text{min}}$$

$$= 2.36 \times 10^6 \text{ s}.$$

Hence the moon's orbital speed is

$$v = \frac{2\pi r}{T}$$

$$= \frac{2\pi \times 3.84 \times 10^8 \text{ m}}{2.36 \times 10^6 \text{ s}}$$

$$= 1.02 \times 10^3 \frac{\text{m}}{\text{s}}.$$

The centripetal acceleration of the moon is therefore

$$a_c = \frac{v^2}{r}$$

$$= \frac{(1.02 \times 10^3 \text{ m/s})^2}{3.84 \times 10^8 \text{ m}}$$

$$= 2.71 \times 10^{-3} \frac{\text{m}}{\text{s}^2}.$$

This acceleration is directed toward the center of the earth, and is only about 1/3600 of the 9.8-m/s^2 gravitational acceleration of an ob-

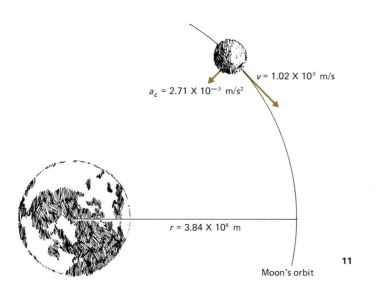

v = 1.02 X 10³ m/s

a_c = 2.71 X 10⁻³ m/s²

r = 3.84 X 10⁸ m

11

Moon's orbit

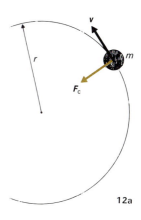

12a

ject near the earth's surface, which is also directed toward the center of the earth. (**11**)

MAGNITUDE OF CENTRIPETAL FORCE

From the second law of motion $F = ma$ we see that the centripetal force F_c which must be acting on an object of mass m that is in uniform circular motion is

$$F_c = ma_c.$$

Since the centripetal acceleration has the magnitude

$$a_c = \frac{v^2}{r},$$

the magnitude of the centripetal force is

$$F_c = \frac{mv^2}{r}. \qquad \text{Centripetal force}$$

Evidently the centripetal force that must be exerted to maintain an object in uniform circular motion increases with increasing mass and with increasing speed, with the force

more sensitive to a change in speed since it is the square of the speed that is involved. An increase in the radius of the path, however, reduces the required centripetal force.

The centripetal force on an object in uniform circular motion is equal in magnitude to mv^2/r. (**12a**)

Doubling the mass doubles the required centripetal force. (**12b**)

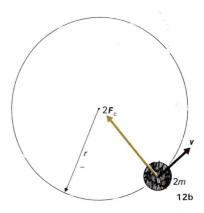

12b

Magnitude of centripetal force   **95**

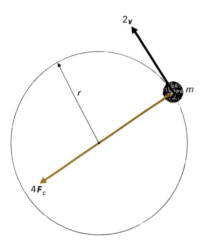

12c

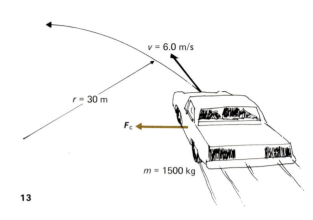

$v = 6.0$ m/s

$r = 30$ m

F_c

$m = 1500$ kg

13

Doubling the speed, however, quadruples the required centripetal force. (**12c**)

Doubling the radius of the circle halves the required centripetal force. (**12d**)

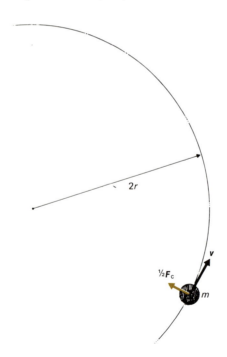

$2r$

v

$\frac{1}{2}F_c$

m

12d

Problem. Find the centripetal force required by a 1500-kg car that makes a turn of radius 30 m at a speed of 6.0 m/s.

Solution. The centripetal force here is

$$F_c = \frac{mv^2}{r}$$

$$= \frac{1.5 \times 10^3 \text{ kg} \times (6.0 \text{ m/s})^2}{30 \text{ m}}$$

$$= 1.8 \times 10^3 \text{ N}.$$

This force (equivalent to about 400 lb) must be provided by the road acting on the car's tires through the agency of friction. (**13**)

BANKED TURNS

Usually the friction between its tires and the road is enough to provide a car with the centripetal force it needs to make a turn. However, if the car's speed is high or the road surface is slippery, the available frictional force may not be enough and the car will skid. To avoid the likelihood of skids, highway curves are usually *banked* so that the roadbed tilts inward. The horizontal component of the reaction force of the road on the car (the action force is the car's weight pressing on the road) then furnishes the required centripetal force.

The proper banking angle θ varies directly with the square of the car's speed and inversely with the radius of the curve. When a car goes around a curve at precisely the design speed, the reaction force of the road provides the centripetal force. If the car goes more slowly than this, friction tends to keep it from sliding down the inclined roadway; if the car goes faster, friction tends to keep it from skidding outward. (14)

CENTRIFUGAL FORCE

When a person sits beside the driver in a car making a left turn, he finds himself pressing against the right-hand door of the car. As this situation appears to him, there is a force—called *centrifugal force*, literally "force fleeing from the center"—which is pushing him outward. (15)

To somebody standing beside the road, however, there is no such outward force at all. What is happening is that, as the car turns to the left, the man's body tends to continue in motion along a straight line. If the car had no door and the seat were very smooth, the car would move away to the left under him as he continued on ahead by virtue of his momentum. This does not occur here because the car's door keeps him inside. The true force on the man, as seen by the outside observer, is the *inward* centripetal force provided by the door.

Centrifugal force, then, is not a real force in the same sense as centripetal force, which is the force that actually causes a moving object to be deflected into a curved path. Centrifugal

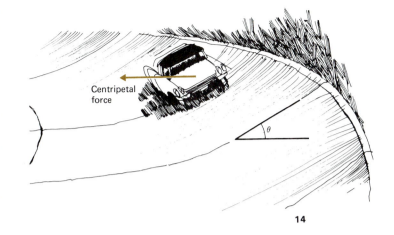

Centripetal force

θ

14

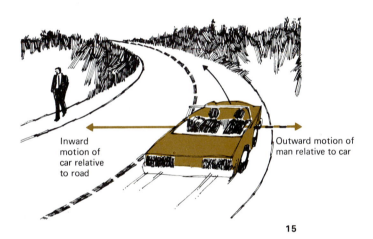

Inward motion of car relative to road

Outward motion of man relative to car

15

force is that force which a person in a frame of reference (here the car) moving in a curved path *invents* in order to account for his outward movement *relative to that frame of reference*. Thus centrifugal force occurs only in an accelerated frame of reference. In a frame of reference which is not accelerated, centrifugal forces do not arise. The man beside the road is in such a frame of reference, and to him the only force present is the centripetal force of the road on the car. It is normally best to use a nonaccelerated frame of reference from which to analyze phenomena that involve moving objects, since the laws of motion hold only in such frames.

To clarify the matter further, let us return to the case of a ball being whirled at the end of a string. If the string breaks, the ball does not fly off because "centrifugal force pushes it out," as we might be tempted to think. When the string breaks there is no longer any centripetal force on the ball, and so it simply proceeds tangentially to its former path in order to conserve the linear momentum it had the instant it was set free.

We shall not refer to centrifugal forces any further in this book. Centripetal forces—the forces that cause circular motion to occur from the point of view of someone in a non-accelerated frame of reference—will figure in a number of important discussions, however.

EXERCISES

1. Under what circumstances (if any) can an object move in a circular path without being accelerated?

2. Where should you stand on the earth's surface to experience the most centripetal acceleration? the least?

3. Can an observer in a windowless laboratory determine whether the laboratory is moving in a circle at a uniform speed?

4. A phonograph record rotates $33\frac{1}{3}$ times per minute. What is the centripetal acceleration of a fly standing 10 cm from the center of the record?

5. The minute hand of a large clock is 0.5 m long. What is the centripetal acceleration of its tip?

6. A boy swings a pail of water in a vertical circle 1 m in radius. What is the minimum speed the pail must have if the water is not to spill?

7. What is the centripetal force needed to keep a 3-kg mass moving in a circle of radius 0.5 m at a speed of 8 m/s? Neglect gravity.

8. A string 0.8 m long is used to whirl a 2-kg stone in a vertical circle. What must be the minimum speed of the stone if the string is to be just taut when the stone is at the top of the circle? How does this speed compare with that required for a 1-kg stone in the same situation?

9. A 2-kg stone at the end of a string 1 m long is whirled in a vertical circle at a constant speed of 4 m/s. What is the tension in the string when the stone is at the top of the circle? at the bottom of the circle?

10. A 2000-kg car is rounding a curve of radius 200 m on a level road. The maximum frictional force the road can exert on the tires of the car is 4000 N. What is the highest speed at which the car can round the curve?

12

GRAVITATION

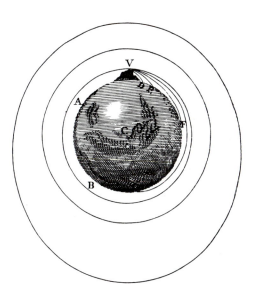

"Thus Sir Isaac Newton discovered and fully described, from undisputed observations and unexceptionable calculations, this simple principle of the gravitation of the particles of matter towards each other; which being extended over the system to all distances, and diffused from the centre of every globe, is the chain that keeps the parts of each together, and preserves them in their regular motions about their proper centres." Colin Maclaurin (1698–1746)

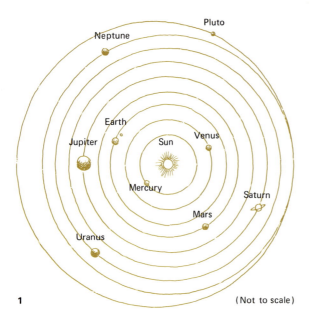

1 (Not to scale)

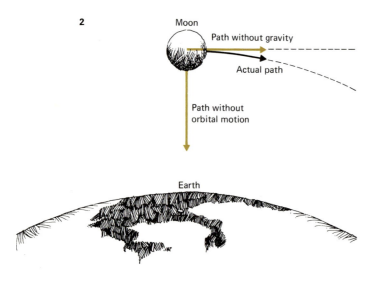

The earth and the other planets pursue approximately circular orbits around the sun. We conclude that the planets are being acted upon by centripetal forces that originate in the sun, since the sun is at the center of all the orbits. This much was generally understood by the middle of the seventeenth century, when Newton turned his mind to the question of exactly what the nature of the centripetal forces was. **(1)**

THE LAW OF GRAVITATION

Newton proposed that the inward force exerted by the sun that is responsible for the planetary orbits is merely one example of a universal interaction, called *gravitation*, that occurs between all bodies in the universe by virtue of their possession of mass.

Another example of gravitation, according to Newton, is the attraction of the earth for nearby bodies. Thus the centripetal acceleration of the moon and the downward acceleration g of objects dropped near the earth's surface have an identical cause, namely the gravitational pull of the earth.

Without gravity, the moon would move off in space along a straight line; without a tangential component of velocity, the moon would fall directly to the earth like a falling stone; the combination of the pull of gravity and the tendency to conserve linear momentum yields a nearly circular path about the earth. The moon is a falling body, as are the planets, even though they get no closer to their parent bodies as time goes on. **(2)**

Newton was able to arrive at the form of the *law of universal gravitation* from an analysis of the motions of the planets about the sun:

Every object in the universe attracts every other object with a force directly proportional to each of their masses and inversely proportional to the square of the distance separating them.

The law of gravitation is expressed in equation form as

$$F_{\text{grav}} = G\,\frac{m_A m_B}{r^2},\qquad \text{Law of gravitation}$$

where

F_{grav} = magnitude of the gravitational force between objects A and B,

$G = 6.67 \times 10^{-11}\ \dfrac{\text{N-m}^2}{\text{kg}^2}$

   = a universal constant,

m_A = mass of object A,

m_B = mass of object B,

r = distance between objects A and B.

The direction of the gravitational force is always along a line joining the two objects A and B. The force on A exerted by B is equal in magnitude to that on B exerted by A, but is in the opposite direction. (3)

The gravitational force between two objects is directly proportional to each of their masses. (4)

Doubling one mass doubles the force on both objects. (5)

Doubling both masses quadruples the force on both objects. (6)

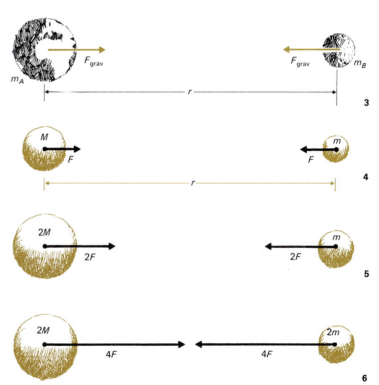

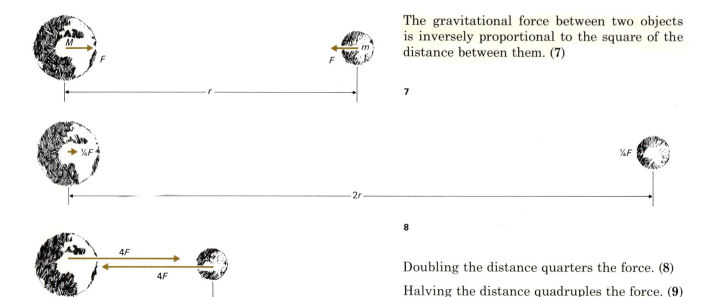

The gravitational force between two objects is inversely proportional to the square of the distance between them. (7)

7

8

Doubling the distance quarters the force. (8)

Halving the distance quadruples the force. (9)

IS GRAVITATION UNIVERSAL?

What is the justification for assuming that the law of gravitation, obtained from data on the solar system, is also valid for the entire universe, describing the gravitational attraction of objects both larger and smaller than the members of the solar system?

There is no simple answer to this legitimate query; instead we must invoke a broad body of knowledge that bears upon the subject. For example, we observe that all the matter on the earth's surface experiences the same acceleration in free fall, which suggests identical gravitational behavior. Careful analysis of the light reaching us from the stars and galaxies throughout the visible universe indicates that the matter of which these bodies are composed behaves identically with matter found on the earth; and so on. Nowhere do we find reason to suspect there should be any objects in the universe that do not obey Newton's law of gravitation, and it is un-

9

reasonable to postulate the existence of such objects with no evidence whatever for the necessity of doing so.

CENTER OF MASS

In applying the law of gravitation, between what points in the respective objects concerned should r be measured? The planets are sufficiently far from the sun for their dimensions to be negligible compared with the appropriate values of r, but this is certainly not true for, say, the moon and the earth. In computing the gravitational force the earth exerts on the moon, should r be the distance between their facing surfaces, between their centers, or between some other points?

Newton was able to solve this problem, but only with the help of his mathematical invention, calculus. The conclusion he reached is that a spherical object behaves gravitationally as though its entire mass were concentrated at its center. (10)

Thus the distance r between the earth and the moon for calculating the gravitational force between them is the distance between their centers. (11)

For objects of other shapes the problem is more difficult, but it turns out that the point from which r is to be taken for the purpose of gravitational calculations is always the *center of mass* of the object in question—the same center of mass used in reckoning the motion of that object. Since the gravitational and inertial masses of an object are equal, it is not surprising that its respective centers of mass should be at the same location.

In the event that the two interacting bodies are so close together that r is small compared with their dimensions, for instance an apple at the earth's surface, local irregularities

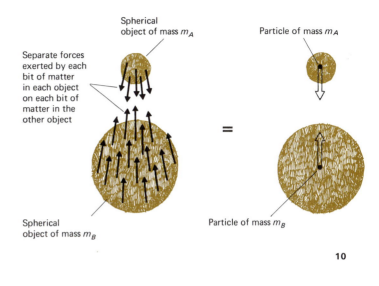

Spherical object of mass m_A

Particle of mass m_A

Separate forces exerted by each bit of matter in each object on each bit of matter in the other object

Spherical object of mass m_B

Particle of mass m_B

10

Earth

Moon

r

11

become detectable. In fact, a common tool of the mineral prospector is an instrument for determining variations in the gravitational force on a standard object at different locations.

The earth's gravitational pull on an object varies inversely with the square of its distance from the center of the earth. Hence a person's weight at a distance r from the center of the earth is $(r_{\text{earth}}/r)^2$ of his weight on the earth's surface, where

$$r_{\text{earth}} = 6.4 \times 10^6 = 6400 \text{ km.}$$

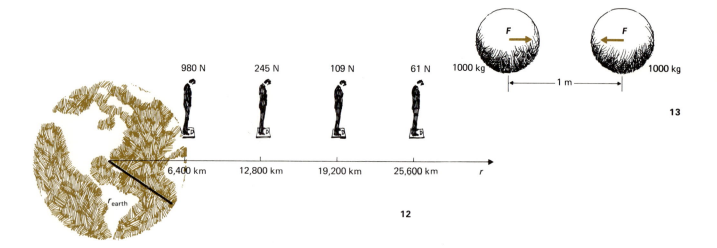

980 N   245 N   109 N   61 N

r_{earth} 6,400 km   12,800 km   19,200 km   25,600 km r

1000 kg   F   1 m   F   1000 kg

13

12

Here we see how the weight of a 100-kg man decreases as his distance from the earth increases. His rest mass, of course, is the same 100 kg everywhere in the universe. (**12**)

The moon is 3.84×10^8 m from the earth, which is 60 times the earth's radius. The gravitational force the earth exerts on the moon is therefore $(1/60)^2 = 1/3600$ as strong as the force the earth would exert on it if the moon were at the earth's surface. The acceleration a of the moon toward the earth in turn ought to be 1/3600 of the acceleration of an object at the earth's surface. Since the latter acceleration is g,

$$a = \frac{g}{3600}$$

$$= \frac{9.8 \text{ m/s}^2}{3600}$$

$$= 2.7 \times 10^{-3} \frac{\text{m}}{\text{s}^2},$$

which is the same as the value of the moon's centripetal acceleration that follows from its period of revolution about the earth (see page

94). The correspondence between the observed centripetal acceleration of the moon and the acceleration inferred from the law of gravitation was used by Newton as evidence for the universal validity of the latter law, which was originally derived from data on planetary motion about the sun.

THE CONSTANT OF GRAVITATION

The constant of gravitation G cannot be determined from astronomical data alone, as Newton realized. A direct measurement of the gravitational force between two known masses a known distance apart is required. The difficulty here is that gravitational forces are minute between objects of laboratory size. For instance, two 1000-kg iron spheres whose centers are 1.00 m apart attract each other with a force of only

$$F = G\frac{m_A m_B}{r^2}$$

$$= 6.67 \times 10^{-11} \frac{\text{N-m}^2}{\text{kg}^2} \times \frac{10^3 \text{ kg} \times 10^3 \text{ kg}}{(1.00 \text{ m})^2}$$

$$= 6.67 \times 10^{-5} \text{ N}.$$

Blowing gently on one of the spheres produces a good deal more force than this. (**13**)

Henry Cavendish finally determined G in 1798, over a century after Newton's work. He used an instrument called a *torsion balance* with which he measured the forces exerted on the small spheres when the large ones were brought close to them in terms of the resulting twist in the fine suspending thread. The torsion balance is the rotational analog of an ordinary spring balance which measures forces in terms of the extension of a spring. (**14**)

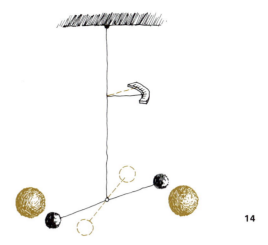

14

MASS OF THE EARTH

A knowledge of the value of G makes it possible to determine the mass of the earth, something we could not hope to accomplish by a direct experiment. Let us consider an apple of mass m at the earth's surface. The gravitational pull of the earth on the apple is the apple's weight of

$$w = mg. \text{ (15)}$$

15

A spherical object behaves gravitationally as though its mass were concentrated at its center. Thus the earth–apple system can be represented by two point masses M and m a distance r_e apart, where M is the earth's mass and r_e is its radius. (**16**)

According to Newton's law of gravitation, the force the earth exerts on the apple is

$$F = G\frac{Mm}{r_e^2}.$$

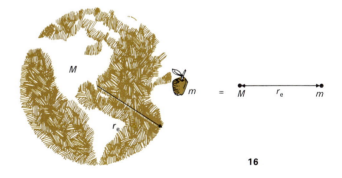

16

This force must equal the apple's weight w, and so

$$F = w,$$

$$G\frac{Mm}{r_e^2} = mg.$$

When we solve this equation for M we see that the apple's mass m drops out. The mass of the earth is

$$M = \frac{gr_e^2}{G}$$

$$= \frac{9.8 \text{ m/s}^2 \times (6.4 \times 10^6 \text{ m})^2}{6.7 \times 10^{-11} \text{ N-m}^2/\text{kg}}$$

$$= 6.0 \times 10^{24} \text{ kg}.$$

SATELLITE ORBITS

Let us use what we know about gravitation and circular motion to investigate the orbits of earth satellites. In the following discussion the frictional resistance of the atmosphere, which ultimately brings down all artificial satellites, will be neglected.

Near the earth's surface the gravitational force on an object of mass m is its weight

$$w = mg$$

where g is the acceleration of gravity. For uniform circular motion about the earth this force must provide the object with the centripetal force

$$F_c = \frac{mv^2}{r}.$$

Hence the condition for a stable orbit is

$$w = F_c,$$

$$mg = \frac{mv^2}{r},$$

$$v = \sqrt{rg}.$$

Here v is the satellite speed, r is the radius of its orbit, and g is the acceleration of gravity at the orbit.

For an orbit just above the earth's surface,

$$v = \sqrt{rg}$$

$$= \sqrt{6.4 \times 10^6 \text{ m} \times 9.8 \text{ m/s}^2}$$

$$= 7.9 \times 10^3 \text{ m/s},$$

which is nearly 18,000 mi/hr. Any object sent off parallel to the earth's surface with this speed will become a satellite of the earth. If it is sent off at a higher speed, its orbit will be elliptical rather than circular. If the object's speed is 11.2×10^3 m/s, 41% more than the minimum orbital speed, it is able to escape from the earth permanently. This escape velocity is about 25,000 mi/hr. (17)

Table 12.1

Body	Escape velocity
Sun	618×10^3 m/s
Moon	2.4
Mercury	4.3
Venus	10.3
Earth	11.2
Mars	5.1
Jupiter	57.5
Saturn	35.4
Uranus	21.9
Neptune	24.4
Pluto	?

The velocity an object needs to escape from an astronomical body depends only upon the mass m and radius r of that body. The formula for escape velocity is

$$v_{escape} = \sqrt{\frac{2Gm}{r}}.$$

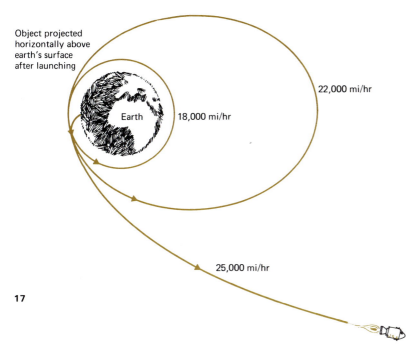

Object projected
horizontally above
earth's surface
after launching

22,000 mi/hr

Earth   18,000 mi/hr

25,000 mi/hr

17

THE GRAVITATIONAL FIELD

In everyday life, forces seem to be transmitted by what seems to be "direct contact": something pushes or pulls something else. Gravitational forces, however, act in the absence of direct contact and are able to produce their effects through millions of miles of empty space. In the modern view of gravitational phenomena, the interaction between two bodies is considered as involving the region in which the bodies are located. In this view, the presence of a single body alters the properties of the region around it, setting up a *gravitational field* which stands ready to interact with any other bodies brought into it. Thus the sun is surrounded by a gravitational field, and it is the forces exerted by this field on the planets which are the centripetal forces that hold them in their orbits.

As we shall learn, *all* forces can be interpreted as arising through the intermediacy of a force field; in the case of "direct contact" forces between solid objects, this field is the electromagnetic field.

WHAT IS A FIELD?

A field—in the sense that physicists use the word—is a region of space in which a certain quantity has a definite value at every point. Thus it is appropriate to speak of the temperature field in a room, of the velocity field of the water in a river, and of the gravitational field around the earth. But it is not appropriate to speak of the "chair field" in a room, or of the "rowboat field" in a river, or of the "airplane field" around the earth because

18

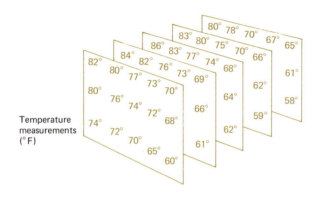

Temperature
measurements
(°F)

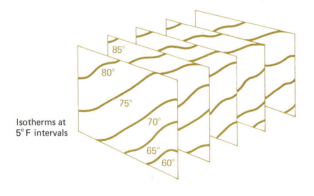

Isotherms at
5°F intervals

19

chairs, rowboats, and airplanes are definite objects; it is easier to just specify their individual positions and motions because their presence somewhere is not a property of a region of space that varies throughout that region in the way that temperature, water speed, and gravitational force vary.

To determine the temperature field in a room, we must measure the temperature at a great many points with a thermometer. The results might be displayed by a series of cross-sectional maps of the room with temperature values written in at each point of measurement. (**18**)

A better way to picture the temperature field is to draw a series of lines on each map that connect points having the same temperature. These lines are called *isotherms*, and might be drawn for temperatures that are 5°F apart. Naturally the actual measurements are not necessarily 60°, 65°, 70°, and so on; it is necessary to interpolate between the actual measurements to find the contours of the various isotherms. Temperature is a scalar quantity because it involves a magnitude only. To say that the temperature somewhere is 70°F describes it completely. The field of a scalar quantity is a *scalar field*, and it can always be pictured by a plot of isolines (lines joining sets of identical values) as in the case of a temperature field. (**19**)

A vector quantity has direction as well as magnitude, so picturing a *vector field* is not a simple matter. One method is to draw lines on a map of the region occupied by the field so that the lines always point in the direction of the field quantity at each point. For example, the velocity field in a river can be shown with the help of lines called *streamlines* that represent the paths of successive particles of water. Several rubber balls thrown into a river and photographed with a movie camera would permit us to make a plot of the velocity field of the river since each ball would follow a streamline as it moves with the current. (20)

LINES OF FORCE

A *force field* is a region of space at every point of which an appropriate test object would experience a force. Thus a *gravitational field* is a region of space in which an object by virtue of its mass is acted on by a force. To visualize a gravitational field, we may use *lines of force*. These are constructed as follows: at several points in space we draw arrows in the direction a test mass would go if released there. Then these arrows are connected to form smooth curves (the lines of force) whose concentration near any point is proportional to the magnitude of the force on the test mass at that point. If we place a mass in the field, it will experience a force in the direction of the line of force it is on; the magnitude of the force will depend upon how close together the lines of force in its vicinity are. (21)

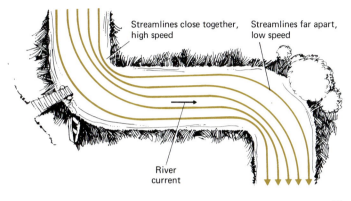

Streamlines close together, high speed

Streamlines far apart, low speed

River current

20

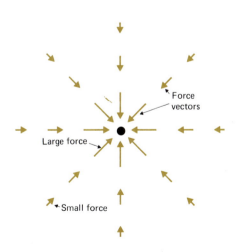

Force vectors

Large force

Small force

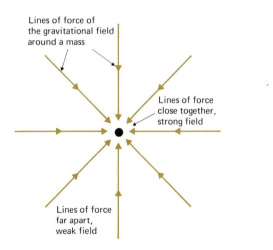

21

Lines of force of the gravitational field around a mass

Lines of force close together, strong field

Lines of force far apart, weak field

22

Lines of force near two identical bodies. A mass in the center cannot reach either body if released, since the net force on it there is zero. At distances large relative to the dimensions of the bodies and their separation, the gravitational field becomes that of a single point mass. **(22)**

Lines of force near two bodies of different mass (for instance, the earth and the moon). Again, at distances large relative to the dimensions of the bodies and their separation, the gravitational field becomes that of a single point mass. **(23)**

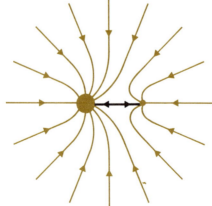

23

It is important to keep in mind that lines of force do not actually exist threading space; they are simply a device for giving intuitive form to our thinking about force fields. The use of lines of force is not limited to gravitational fields. Electric and magnetic fields, for instance, are often described in this manner.

EXERCISES

1. A hole is drilled to the center of the earth and a stone dropped into it. Compare the mass and weight of the stone at the center of the earth with their values at the earth's surface.

2. A woman takes her bathroom scale with her on a vacation trip to a mountain resort. If her mass does not change, will her weight appear more, less, or the same as at sea level?

3. The value of g decreases with increasing distance from the earth's surface. What happens to the value of G?

4. A man has a mass of 60 kg at the earth's surface. What would his mass be at a height above the earth's surface of one earth's radius?

5. Compare the magnitude of the gravitational force the sun exerts on the earth with the magnitude of the centripetal force involved in the earth's orbital motion.

6. The moon's mass is 1.2% of the earth's mass. Is the gravitational force exerted by the moon on the earth less than, the same as, or greater than the gravitational force exerted by the earth on the moon?

7. The earth's equatorial radius is about 21 km greater than its polar radius, a phenomenon known as the "equatorial bulge."
   a) How is this fact related to the daily rotation of the earth on its axis?
   b) If the earth were to rotate twice as fast as it now does, would you expect the equatorial bulge to be more than, the same as, or less than it is now?

8. If a planet existed whose mass and radius were both twice those of the earth, what would the

acceleration of gravity at its surface be in terms of g?

9. Two satellites are launched from a certain station with the same initial speeds relative to the earth's surface. One is launched toward the west, the other toward the east. Will there be any difference in their orbits? If so, what will it be and why?

10. An artificial earth satellite is placed in an orbit whose radius is half that of the moon's orbit. Is its time of revolution longer or shorter than that of the moon?

11. What can you tell about the force an object would experience at a given point in a gravitational field by looking at a sketch of the lines of force of the field?

12. Can lines of force ever intersect in space? Explain.

13. What is the acceleration of a meteor when it is one earth's radius above the surface of the earth?

14. An object dropped near the earth's surface falls 4.9 m in the first second. How far does the moon fall toward the earth in each second? Why doesn't the moon ever reach the earth?

15. The moon's mass is 7.3×10^{22} and the average radius of its orbit is 3.8×10^8 m. At what point could an object be placed between the earth and the moon where it would experience no resultant force? (Neglect the gravitational attractions of the sun and the other planets.)

16. Most of the stars in the galaxy of which the sun is a member (the Milky Way) are concentrated in an assembly about 100,000 light-years across, whose shape is roughly that of a fried egg. The sun is about 30,000 light-years from the center of the galaxy, and revolves around it with a period of about 2×10^8 yr. A reasonable estimate for the mass of the galaxy may be obtained by considering this mass to be concentrated at the galactic center with the sun revolving around it like a planet around the sun. On this basis, calculate the mass of the galaxy. How many stars having the mass of the sun is this equivalent to? (1 yr $= 3.16 \times 10^7$ s, 1 light-year $= 9.46 \times 10^{15}$ m, and $m_{sun} = 2.0 \times 10^{30}$ kg).

17. In *Alice in Wonderland* this statement appears: "Now, here, you see, it takes all the running you can do, to stay in the same place. If you want to get somewhere else, you must run at least twice as fast as that!" Check the correctness of this statement by comparing the velocity needed for a stable satellite orbit near the surface of a planet with the escape velocity for that planet. If the ratio between them is not exactly 2, what is it?

18. The moon's radius is 27% of the earth's radius and its mass is 1.2% of the earth's mass.
   a) What is the acceleration of gravity in terms of g on the surface of the moon?
   b) How much would a boy weigh there whose weight on the earth is 500 N?

19. The mass of the planet Jupiter is 1.9×10^{27} kg, its radius is 7.0×10^7 m, its orbital radius is an average of 7.8×10^{11} m, and the mass of the sun is 2.0×10^{30} kg.
   b) Assuming that Jupiter has a circular orbit, what is its orbital speed?
   c) What is the acceleration of gravity on the surface of Jupiter?

20. Find the radius of a satellite orbit whose period is exactly 1 day. (Be sure to take into account the variation of g with r.) Such a satellite will remain indefinitely over a particular location on the earth; most communications relay satellites are placed in orbits of this kind.

PART 3

ENERGY AND MATTER

13

WORK

"The idea of work for machines, or natural processes, is taken from comparison with the working power of man . . . But one thing is common to all [machines]; they all need a *moving force*, which sets and keeps them in motion, just as the works of the human hand all need the moving force of the muscles." Hermann von Helmholtz (1821—1894)

All changes in the physical universe are the result of forces. Forces set objects in motion, change their paths, and bring them to a stop; forces pull things together and push them apart. The quantity called *work* is a measure of the amount of change (in a general sense) a force gives rise to when it acts upon something.

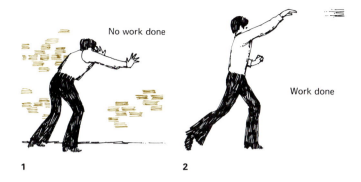

No work done

Work done

1   2

ENERGY AND CHANGE

We all use the word *energy*, but how many of us know exactly what it means? We speak of the energy of a lightning bolt or of an ocean wave; we say that an active person is energetic; we hear a candy bar described as being full of energy; we read that most of the world's electricity will come from nuclear energy in the years to come. What do a lightning bolt, an ocean wave, an active person, a candy bar, and an atomic nucleus have in common?

In general terms, energy refers to *an ability to accomplish change*. All changes in the physical world involve energy, usually with energy being transformed from one sort into another. But "change" is not a very precise concept, and we must clarify our ideas before going further. What we shall do is first define a quantity called *work*, and then see how it permits us to discuss energy and its relation to change in the orderly manner of science.

WORK

When we push against a brick wall, nothing happens. We have applied a force, but the wall has not yielded and shows no effects. (1)

However, when we apply exactly the same force to a ball, the ball flies through the air for some distance. Now something has been

accomplished because of our push, whereas in the former case there was no result. (2)

What is the essential difference between the two situations? In the first case, where we pushed against a wall, the wall did not move. But in the second case, where we threw the ball, the ball *did* move while the force was being applied and before it left our hand. The displacement of the object while the force acted on it was what made the difference.

DEFINITION OF WORK

If we think carefully along these lines, we will see that, whenever a force acts so as to produce motion in an object, the force acts during a displacement of the object.

In order to make this notion definite, a physical quantity called *work* is defined as follows:

The work done by a force acting on an object is equal to the magnitude of the force multiplied by the distance through which the force acts.

In equation form,

$$\text{Work} = \text{force} \times \text{distance}$$

$$W = Fs.$$

Work is a scalar quantity, with no direction associated with it.

$W=Fs$

Work = force × distance through which force acts.

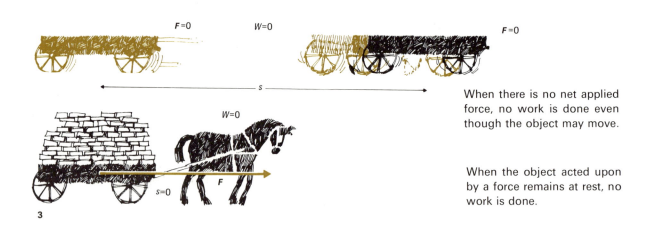

$F=0$ $W=0$ $F=0$

When there is no net applied force, no work is done even though the object may move.

When the object acted upon by a force remains at rest, no work is done.

$W=0$

$s=0$ F

3

This definition is a great help in clarifying the effects of forces. Unless a force acts through a displacement of the object it acts upon, no work is done, no matter how great the force. And even if an object moves, no work is done unless a force is acting upon it or it exerts a force on something else. (3)

Our intuitive concept of work is in accord with the above definition: when something happens because a person applied a force of some kind, we say that he has performed work. Here we have simply broadened the concept to include inanimate forces. (We still must be careful, though; while we may become tired after pushing against a brick wall for a long time, we still have done no work on the wall if it remains in place.)

The definition of work has an important qualification: the force F must be in the same direction as the vector displacement s. If F

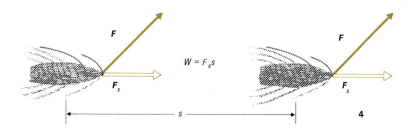

$W = F_s s$

4

and s are not parallel, we must replace F in the formula $W = Fs$ by the magnitude of its component F_s in the direction of the displacement s. (**4**)

When the force and the displacement are perpendicular, $F_s = 0$ and no work is done. In order for work to be done, the force must have a component in the direction of the displacement. (**5**)

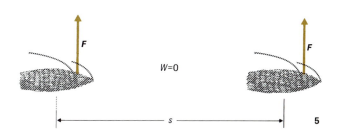

$W = 0$

5

THE JOULE

Work is given a special unit, the *joule* (J).

One joule is equal to the work done by a force of 1 newton acting through a distance of 1 meter.

That is,

$$1 \text{ J} = 1 \text{ N-m. } (\mathbf{6})$$

1 N   1 N

1 m

6

Problem. A man pulls an 80-kg crate for 20 m across a level floor using a rope that is held horizontal. The man exerts a force of 150 N on the rope. How much work does he perform? (**7**)

Solution. The work done by the man is

$$W = Fs$$
$$= 150 \text{ N} \times 20 \text{ m}$$
$$= 3.0 \times 10^3 \text{ J.}$$

We note that the 80-kg mass of the crate has no significance here—it is the force exerted by the man that determines how much work he does.

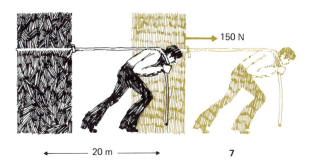

150 N

20 m   **7**

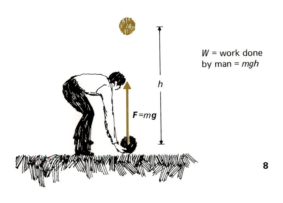

W = work done
by man = mgh

h

$F = mg$

8

mg

h

W = work done
by gravity = mgh

9

s

mg

W=0

10

WORK DONE AGAINST GRAVITY

It is easy to compute the work done in lifting an object against gravity. The force of gravity on an object of mass m is the same as its weight mg. Hence in order to raise the object to a height h above its original position, a force of mg must be exerted on it. Since

$$F = mg,$$

$$s = h$$

here, the work done is

$$W = Fs$$

$$= mgh.$$

To lift an object of mass m to a height h requires the performance of the amount of work mgh. **(8)**

When an object of mass m falls from a height h, the force of gravity does the work mgh on it. **(9)**

The force of gravity does no work on objects that move parallel to the earth's surface, however. **(10)**

Problem. How much work is done in lifting a 30-kg load of bricks to a height of 20 m on a building under construction?

Solution. Here $m = 30$ kg and $h = 20$ m, so that

$$W = mgh$$

$$= 30 \text{ kg} \times 9.8 \frac{\text{m}}{\text{s}^2} \times 20 \text{ m}$$

$$= 5.9 \times 10^3 \text{ J.}$$

It is important to note that only the height h is involved in work done against the force of gravity. The particular route taken by an object being raised is not significant; excluding any frictional effects, exactly as much work must be expended to climb a flight of stairs as to go up in an elevator to the same floor (though not by the person involved!). **(11)**

POWER

Often the time needed to perform a task is just as significant as the actual amount of work required. Given enough time, even the feeblest motor can raise the Sphinx. However, if we want to carry out a certain operation quickly, we try to obtain a motor whose output of work is rapid in terms of the total required. The rate at which work is done is therefore an important engineering quantity. This rate is called *power*: the faster some agency can do work, the more *powerful* it is.

If an amount of work W is performed in a time interval Δt, the power involved is

$$\text{Power} = \frac{\text{work done}}{\text{time interval}},$$

$$P = \frac{W}{\Delta t}.$$

The unit of power is the *watt* (W), where

$$1\ \text{W} = 1\ \frac{\text{J}}{\text{s}}.$$

Thus a 1-W motor is capable of performing work at the rate of 1 J per second.

A common unit of power in engineering is the *horsepower* (hp), which was introduced two centuries ago by James Watt to compare the power output of the steam engine he had perfected with a more familiar source of work. Its equivalent in watts is

$$1\ \text{hp} = 746\ \text{W}.$$

Problem. A 24-hp electric motor provides power for the elevator of a 6-story building.

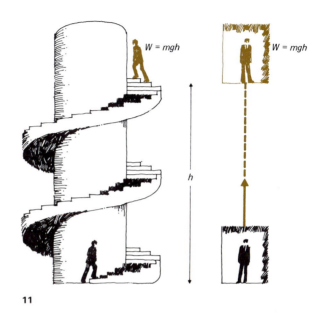

$W = mgh$ $W = mgh$

h

11

If the total mass of the loaded elevator is 1000 kg, how long does it take to rise the 30 m from the ground floor to the top floor?

Solution. The work done in raising the elevator through a height h is

$$W = mgh.$$

According to the definition of power, $P = W/\Delta t$, a motor whose power output is P performs the amount of work W in the time interval

$$\Delta t = \frac{W}{P}.$$

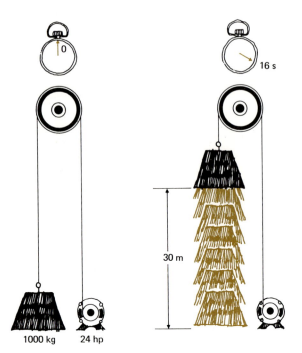

1000 kg   24 hp

30 m

Here

$$P = 24 \text{ hp} \times 746 \frac{\text{W}}{\text{hp}}$$

$$= 1.8 \times 10^4 \text{ W},$$

and so

$$t = \frac{W}{P}$$

$$= \frac{mgh}{P}$$

$$= \frac{10^3 \text{ kg} \times 9.8 \text{ m/s}^2 \times 30 \text{ m}}{1.8 \times 10^4 \text{ W}}$$

$$= 16 \text{ s. } (12)$$

EXERCISES

1. A man holds a 10-kg package 1.2 m above the ground for 1 min. How much work does he perform?

2. The sun exerts a force of 4×10^{28} N on the earth, and the earth travels 9.4×10^{11} m in its annual orbit of the sun. How much work is done by the sun on the earth in the course of a year?

3. Electrical energy is usually reckoned by utility companies in kilowatt-hours (kWh). How many joules are there in a kWh?

4. Two men set out to climb to the summit of a 3000-m mountain starting from sea level. One of them sets out along a slope that averages $30°$ above the horizontal, the other along a slope that averages $40°$ above the horizontal. Each man has a mass of 80 kg and carries a 10-kg knapsack. Find the work done by each of them.

5. a) A force of 130 N is used to lift a 12-kg mass to a height of 8 m. How much work is done by the force?
   b) A force of 130 N is used to push a 12-kg mass on a horizontal, frictionless surface for a distance of 8 m. How much work is done by the force?

6. Four thousand joules are used to lift a 30-kg mass. If the mass is at rest before and after its elevation, how high does it go?

7. A horse is towing a barge with a rope that makes an angle of $20°$ with the canal. If the horse exerts a force of 80 N, how much work does it do in moving the barge 1 km (1 km = 10^3m)? Use a vector diagram.

8. A 50-kg boy runs up a staircase to a floor 10 m higher in 9.8 s. What is his power output in watts? in horsepower?

9. In 1932 five members of the Polish Olympic ski team climbed from the 5th to the 102nd floor of the Empire State Building, a distance of approximately 300 m, in 21 min. One of these men had a mass of 75 kg. How many horsepower did he develop in the ascent?

10. An escalator carries passengers from one floor of a building to another 10 m higher. It is designed to have a capacity of 200 passengers per minute, assuming an average mass per passenger of 70 kg. Find the required horsepower of the motor if half the work it does is dissipated as heat.

11. In 1970 approximately 2×10^{20} J of work were performed throughout the world by inanimate devices of all kinds, ≈ 15 times as much as muscle power provided in that year. The work was used for heat, light, transport, manufacturing, and so forth. About 98% of the work was ultimately derived from the fossil fuels coal, natural gas, and oil, the rest mainly from water power with a small (0.25% of the total) contribution from nuclear power stations.
    a) Express the power consumption in 1970 in watts.

    b) Find the average power consumption per person in watts and in horsepower on the assumption that the world's population in 1970 was 3.5×10^9.

12. Sunlight reaches the earth at the rate of 1400 W per m^2 of surface perpendicular to the direction from the sun.
    a) How much power does the entire earth ($r = 6.4 \times 10^6$ m) receive from the sun?
    b) How much is this per person? Compare these results with the answers to the previous problem.

13. A waterfall is 30 m high and 10^4 kg of water flows over it per second.
    a) How much power does this flow represent?
    b) If all this power could be converted to electricity, how many 100-W light bulbs could be supplied?

14

ENERGY

"Energy . . . we know only as that which in all natural phenomena is continually passing from one portion of matter to another." James Clerk Maxwell (1831–1879)

Energy is that property whose possession enables something to perform work. Two familiar kinds of energy are the kinetic energy which a moving object has by virtue of its motion and the potential energy which an object acted upon by a force has by virtue of its position.

ENERGY

From the straightforward notion of work we proceed to the complicated and many-sided concept of *energy*:

Energy is that property whose possession enables something to perform work.

When we say that something has energy, we mean it is capable (directly or indirectly) of exerting a force on something else and doing work on it. On the other hand, when we do work on something, we have added to it an amount of energy equal to the work done. Like linear momentum, energy is conserved in all known interactions on every scale, from the interior of the atom to the universe as a whole. The unit of energy is the same as the unit of work, the joule.

There are three broad categories of energy:

1. *Kinetic energy*, which is the energy something possesses by virtue of its motion;
2. *Potential energy*, which is the energy something possesses by virtue of its position in a force field;
3. *Rest energy*, which is the energy something possesses by virtue of its mass.

In the above descriptions the word "something" was used instead of "object" because, as we shall see later, such nonmaterial entities as force fields and massless particles may also possess energy.

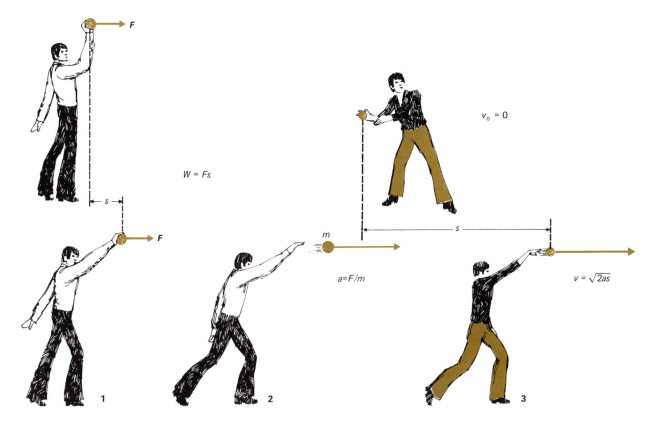

W = Fs

F

s

$v_0 = 0$

s

m

a=F/m

$v = \sqrt{2as}$

1   2   3

All modes of energy possession fit into one or another of these three categories. For instance, it is convenient for many purposes to think of heat as a separate form of energy, but what this term actually refers to is the sum of the kinetic energies of the randomly moving atoms and molecules in a body of matter.

KINETIC ENERGY

When we perform work on a ball by throwing it, what becomes of this work?

Let us suppose we apply the uniform force F to the ball for a distance s before it leaves our hand. The work done on the ball is therefore Fs. (1)

The mass of the ball is m. As we throw it, its acceleration has the magnitude

$$a = \frac{F}{m}$$

according to the second law of motion, $F = ma$. (2)

We know from the formula (see page 29),

$$v^2 = v_0^2 + 2as,$$

that when an object starting from rest ($v_0 = 0$) undergoes an acceleration of magnitude a through a distance s, its final speed v is related to a and s by

$$v^2 = 2as. (3)$$

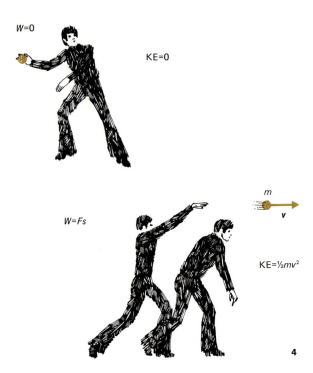

$W=0$ KE=0

$W=Fs$ m v KE=½mv^2

4

If we substitute F/m for a in the latter formula we find that

$$v^2 = 2as$$
$$= 2\frac{F}{m}s,$$

which we can rewrite as

$$Fs = \frac{1}{2}mv^2.$$

The quantity on the left-hand side, Fs, is the work our hand has done in throwing the ball. The quantity on the right-hand side, $\frac{1}{2}mv^2$, must therefore be the energy acquired by the ball as a result of the work we did on it. This energy is *kinetic energy*, energy of motion.

That is, we interpret the preceding equation as follows:

Work done on ball = kinetic energy of ball,

$$Fs = \frac{1}{2}mv^2.$$

The symbol for kinetic energy is KE. The kinetic energy of an object of mass m and speed v is therefore

$$\text{KE} = \frac{1}{2}mv^2. \qquad \text{Kinetic energy}$$

A moving object is able to perform an amount of work equal to $\frac{1}{2}mv^2$ in the course of being stopped. (**4** and **5**)

Some kinetic energies

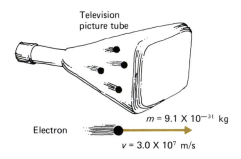

Television picture tube

Electron

$m = 9.1 \times 10^{-31}$ kg

$v = 3.0 \times 10^7$ m/s

$KE = \frac{1}{2}mv^2$
$= \frac{1}{2} \times 9.1 \times 10^{-31}$ kg $\times$ $(3.0 \times 10^7$ m/s$)^2$
$= 4.1 \times 10^{-16}$ J

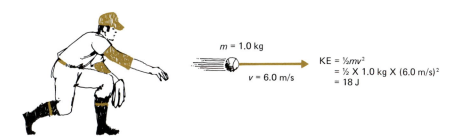

$m = 1.0$ kg

$v = 6.0$ m/s

$KE = \frac{1}{2}mv^2$
$= \frac{1}{2} \times 1.0$ kg $\times (6.0$ m/s$)^2$
$= 18$ J

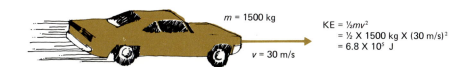

$m = 1500$ kg

$v = 30$ m/s

$KE = \frac{1}{2}mv^2$
$= \frac{1}{2} \times 1500$ kg $\times (30$ m/s$)^2$
$= 6.8 \times 10^5$ J

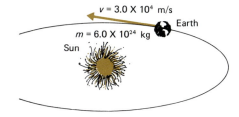

$v = 3.0 \times 10^4$ m/s

Earth

$m = 6.0 \times 10^{24}$ kg

Sun

$KE = \frac{1}{2}mv^2$
$= \frac{1}{2} \times 6.0 \times 10^{24}$ kg $\times$ $(3.0 \times 10^4$ m/s$)^2$
$= 2.7 \times 10^{33}$ J

5

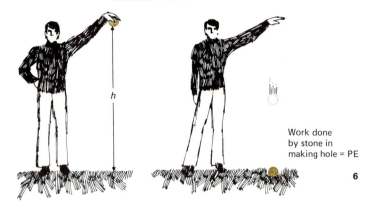

Raised stone has PE

Work done by stone in making hole = PE

6

$F = mg$

m

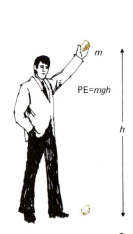

m

PE = mgh

h

8

POTENTIAL ENERGY

When we drop a stone from a height h, it falls faster and faster and finally strikes the ground. In striking the ground the stone does work; if it is sufficiently heavy and has fallen from a great enough height, the work done by the stone is manifest as a hole. Evidently the stone at its original location h above the ground had a capacity to do work, even though it was stationary at the time. The work the stone can perform in falling to the ground is called its *potential energy*, symbol PE. (**6**)

To lift a stone of mass m requires an upward force equal in magnitude to the magnitude mg of its weight. (**7**)

The work done in raising the stone to the height h above the ground is

$$\text{Work} = \text{force} \times \text{distance},$$

$$W = mg \times h.$$

Hence the potential energy of the stone is

$$\text{PE} = mgh \qquad \text{Gravitational potential energy}$$

relative to the ground. (**8**)

The gravitational potential energy of an object depends upon the reference level from which its height h is measured. For example, the potential energy of a 1.0-kg book held 10 cm above a desk is

$$\text{PE} = mgh = 1.0\,\text{kg} \times 9.8\,\frac{\text{m}}{\text{s}^2} \times 0.10\,\text{m} = 0.98\,\text{J}$$

with respect to the desk. However, if the book is 1.0 m above the floor of the room, its potential energy is

$$\text{PE} = mgh = 1.0\,\text{kg} \times 9.8\,\frac{\text{m}}{\text{s}^2} \times 1.0\,\text{m} = 9.8\,\text{J}$$

with respect to the floor. And the book may conceivably be 100 m above the ground, so its potential energy is

$$PE = mgh = 1.0 \text{ kg} \times 9.8 \frac{\text{m}}{\text{s}^2} \times 100 \text{ m} = 980 \text{ J}$$

with respect to the ground. The height h in the formula $PE = mgh$ means nothing unless the base height $h = 0$ is specified. (**9**)

In general, potential energy is a relative quantity. Just as the KE of a moving object depends upon the frame of reference in which its mass and velocity are measured, so the PE of an object subject to a force depends upon the reference position chosen.

We have spoken of only one type of potential energy, namely that possessed by an object by virtue of being raised above some reference level in the earth's gravitational field. The concept of potential energy is a much more general one, however, for it refers to the energy something has as a consequence of its position regardless of the nature of the force acting on it. The earth itself, for instance, has potential energy with respect to the sun, since if its orbital motion were to cease it would fall toward the sun. An iron nail has potential energy with respect to a nearby magnet, since it will fly to the magnet if released. An object at the end of a stretched spring has potential energy with respect to its position when the spring has its normal extension, since if let go the object will move as the spring contracts. In each of these cases the object in question has the potentiality of doing work in its original position. (**10 and 11**)

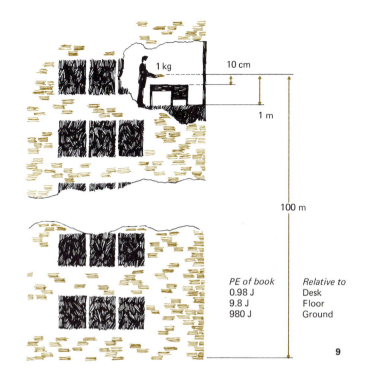

PE of book	Relative to
0.98 J	Desk
9.8 J	Floor
980 J	Ground

9

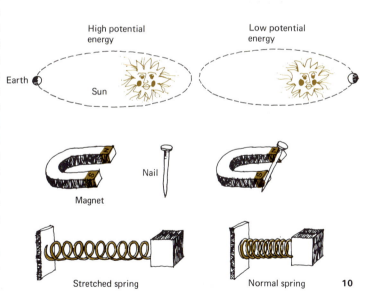

High potential energy   Low potential energy

Earth   Sun

Magnet   Nail

Stretched spring   Normal spring   10

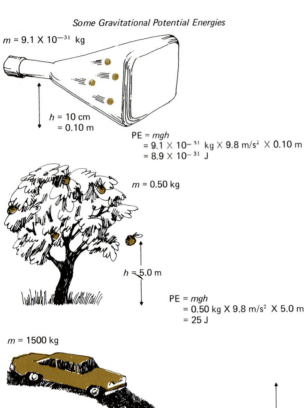

Some Gravitational Potential Energies

$m = 9.1 \times 10^{-31}$ kg

$h = 10$ cm
$= 0.10$ m

$PE = mgh$
$= 9.1 \times 10^{-31}$ kg $\times$ 9.8 m/s^2 $\times$ 0.10 m
$= 8.9 \times 10^{-31}$ J

$m = 0.50$ kg

$h = 5.0$ m

$PE = mgh$
$= 0.50$ kg $\times$ 9.8 m/s^2 $\times$ 5.0 m
$= 25$ J

$m = 1500$ kg

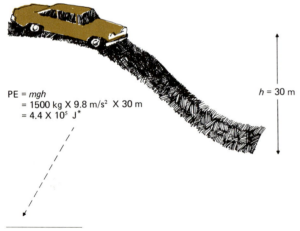

$PE = mgh$
$= 1500$ kg $\times$ 9.8 m/s^2 $\times$ 30 m
$= 4.4 \times 10^5$ J*

$h = 30$ m

---

* This is *less* than the KE of the same car when its speed is 30 m/s. Thus a crash at 30 m/s (67 mi/hr) into a stationary obstacle will yield more work (that is, damage) than dropping the car 30 m (98 ft).

EXERCISES

1. Is kinetic energy a scalar or a vector quantity? Is potential energy a scalar or a vector quantity?

2. Does every moving body possess kinetic energy? Does every stationary body possess potential energy? Can something possess both kinetic and potential energy?

3. At what point in its motion is the kinetic energy of a pendulum bob a maximum? At what point is its potential energy a maximum?

4. An iron ball and a wooden ball of the same size are dropped simultaneously from a tower. When they are 1 m above the ground, the two balls have the same
   a) linear momentum.
   b) kinetic energy.
   c) potential energy.
   d) acceleration.

5. The potential energy of a golf ball in a hole is negative relative to the ground. Under what circumstances (if any) is its kinetic energy negative?

6. A bomb dropped from an airplane explodes in midair. What happens to its total kinetic energy? to its total linear momentum?

7. Is it possible for an object to have more kinetic energy but less linear momentum than another object? Is the converse possible?

8. When the momentum of an object is doubled, what happens to its kinetic energy?

9. When the kinetic energy of an object is doubled, what happens to its momentum?

10. Car A has a mass of 1000 kg and a speed of 20 m/s; car B has a mass of 2000 kg and a speed of 10 m/s. What is the kinetic energy of car A relative to that of car B?

11. Is more work needed to bring a car's speed from 10 mi/hr to 20 mi/hr or from 50 mi/hr to 60 mi/hr?

12. What speed must a 1-kg object have in order to have 1 J of KE?

13. An object of mass 1 kg and another of weight 1 N both have potential energies of 1 J relative to the ground. What are their respective heights above the ground?

14. A 0.02-kg bullet has a speed of 500 m/s. What is its kinetic energy?

15. A 2-kg ball is at rest when a horizontal force of 5 N is applied. In the absence of friction, what is the speed of the ball after it has gone 10 m?

16. A 3-kg stone is lifted to a height of 100 m and then dropped. What is its kinetic energy when it is 50 m from the ground?

17. a) A force of 8 N is used to push a 0.5-kg ball over a horizontal, frictionless table a distance of 3 m. If the ball starts from rest, what is its final kinetic energy?

    b) The same force is used to lift the same ball a height of 3 m. If the ball starts from rest what is its final kinetic energy?

15

CONSERVATION OF ENERGY

"According to the principle that masses remain the same under any physical or chemical changes, mass appeared to be the essential (because unvarying) quality of matter . . . Physicists accepted this principle up to a few decades ago, but it proved inadequate in the face of the special theory of relativity. It was therefore merged with the energy principle —just as, about sixty years before, the principle of the conservation of mechanical energy had been combined with the principle of the conservation of heat." Albert Einstein (1879–1955)

Every object of rest mass m_0 has the rest energy $m_0 c^2$, where c is the velocity of light. In a system of any kind that does not interact with the outside world, the total amount of energy present always remains the same no matter what happens within the system, provided that rest energy is taken into account. This statement constitutes the principle of conservation of energy, the most powerful and wide-ranging principle in all of science.

KINETIC ENERGY AT HIGH VELOCITY

In deriving the formula $\text{KE} = \frac{1}{2}mv^2$ for the kinetic energy of a moving object, we ignored the relativistic variation of its mass with its velocity. At velocities small compared with the velocity of light, this is an acceptable procedure, and the m in the formula then refers to the rest mass m_0 of the object. Thus the proper way to write the above formula is

$$\text{KE} = \frac{1}{2}m_0 v^2, \quad v \ll c.$$

How do we go about finding an expression for the kinetic energy of a moving object that is correct at all velocities? Essentially we follow the same procedure as before and consider the work needed to accelerate an object from rest to some final velocity v, but now we take into account the increase in its mass m as v increases. Although this calculation is quite straightforward, it requires the use of integral calculus and must therefore be omitted here. The result turns out to be

$$\text{KE} = \frac{m_0 c^2}{\sqrt{1 - v^2/c^2}} - m_0 c^2,$$

which is rather different from the $\frac{1}{2}m_0 v^2$ we found for low velocities.

It is not hard to verify that the relativistic formula for KE reduces to $\frac{1}{2}m_0 v^2$ when $v \ll c$.

With the help of the approximation (see page 44)

$$\frac{1}{\sqrt{1 - v^2/c^2}} \approx 1 + \frac{1}{2}\frac{v^2}{c^2} \qquad v \ll c,$$

we find that

$$\mathrm{KE} = \frac{m_0 c^2}{\sqrt{1 - v^2/c^2}} - m_0 c^2$$

$$\approx \left[1 + \frac{1}{2}\frac{v^2}{c^2}\right] m_0 c^2 - m_0 c^2$$

$$\approx \frac{1}{2} m_0 v^2 \qquad v \ll c.$$

The relativistic formula is correct for all velocities, and the formula $\mathrm{KE} = \frac{1}{2}m_0 v^2$ is actually the low-velocity approximation to it. The degree of accuracy required is what determines whether the approximation is valid. For instance, when $v = 10^7$ m/s ($0.033c$), the formula $\frac{1}{2}m_0 v^2$ understates the true kinetic energy by only 0.08%; when $v = 3 \times 10^7$ m/s ($0.1c$) it understates the true kinetic energy by 0.8%; but when $v = 1.5 \times 10^8$ m/s ($0.5c$) the understatement is a significant 19%, and when $v = 0.999c$ the understatement is a whopping 4300%. Since 10^7 m/s is about 6210 mi/s, the nonrelativistic formula $\frac{1}{2}m_0 v^2$ is entirely satisfactory for finding the kinetic energies of ordinary objects, and it fails only at the extremely high velocities reached by elementary particles under certain circumstances.

REST ENERGY

The relativistic kinetic-energy formula does far more than give us a way to calculate the kinetic energies of objects of any velocity, important though that is. Since the mass m of something whose rest mass is m_0 is

$$m = \frac{m_0}{\sqrt{1 - v^2/c^2}}$$

when its velocity is v, we can express the relativistic kinetic-energy formula as

$$\mathrm{KE} = \frac{m_0 c^2}{\sqrt{1 - v^2/c^2}} - m_0 c_2$$

$$= mc^2 - m_0 c^2$$

$$= (m - m_0)c^2.$$

The latter equation states that the kinetic energy of a moving object is equal to its increase in mass $m - m_0$ multiplied by c^2, the square of the velocity of light.

Evidently there is a connection of some kind between mass and energy. To pinpoint this connection, we rewrite the above equation in still another way:

$$mc^2 = m_0 c^2 + \mathrm{KE}.$$

This equation has the following interpretation, which has been confirmed by experiment. The quantity mc^2 is the total energy of an object whose mass is m. If the object is at rest, $m = m_0$ and the object has a *rest energy* of

$$E_0 = m_0 c^2. \qquad \text{Rest energy}$$

If the object is in motion, it has both the rest energy $m_0 c^2$ and the kinetic energy KE.

Every body of matter thus possesses a certain inherent amount of energy even if it is not moving (so that $\mathrm{KE} = 0$) and is not being acted upon by a force field (so that $\mathrm{PE} = 0$).

THE LIBERATION OF REST ENERGY

Why are we not aware of rest energy as we are aware of kinetic and potential energies? After all, a 1-kg object—such as this book—contains the rest energy

$$m_0 c^2 = 1 \text{ kg} \times (3 \times 10^8 \text{ m/s})^2$$

$$= 9 \times 10^{16} \text{ J}$$

which is enough energy to send a payload of perhaps a million tons to the moon. How can so much energy be bottled up without revealing itself in some manner?

In fact, all of us *are* familiar with processes in which rest energy is liberated, only we do not usually think of them in these terms. In every chemical reaction in which energy is given off, for instance a fire, a certain amount of matter is being converted into energy in the form of heat, which is molecular kinetic energy. But the amount of matter that vanishes in such reactions is so small that it escapes our notice. When 1 kg of dynamite explodes, 6×10^{-11} kg of matter is transformed into energy. The lost mass is so minute a fraction of the total mass involved as to be impossible to detect directly (hence the "law" of conservation of mass in chemistry), but it results in the evolution of

$$m_0 c^2 = 6 \times 10^{-11} \text{ kg} \times (3 \times 10^8 \text{ m/s})^2$$

$$= 5.4 \times 10^6 \text{ J}$$

of energy, which is hard to avoid detecting.

The conversion of matter into energy occurs on an especially grand scale in the sun. The physical process involved is the fusion of hydrogen atomic nuclei to form helium atomic nuclei (see page 360). In each reaction sequence four hydrogen nuclei (which are protons) each of mass 1.673×10^{-27} kg join together in a series of steps to yield a helium nucleus of mass 6.646×10^{-27} kg. But the total mass of four hydrogen nuclei is $4 \times 1.673 \times 10^{-27}$ kg $= 6.692 \times 10^{-27}$ kg, which is 4.6×10^{-29} kg more than the mass of a helium nucleus. Thus each time the reaction sequence occurs

$$m_0 c^2 = 4.6 \times 10^{-29} \text{ kg} \times (3 \times 10^8 \text{ m/s})^2$$

$$= 4.1 \times 10^{-12} \text{ J}$$

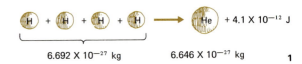

6.692 X 10^{-27} kg   6.646 X 10^{-27} kg   **1**

of energy is liberated. About 4×10^9 kg of matter is converted into energy in the sun per second which yields a power output of 3.6×10^{26} W. Rest energy is just as conspicuous as any other form of energy, provided it is recognized as such. (**1**)

CONSERVATION OF ENERGY

The principle of conservation of energy states:

The total amount of energy in a system isolated from the rest of the universe always remains constant, although energy transformations from one form to another, including rest energy, may occur within the system.

This principle is perhaps the most fundamental generalization in all of science, and no violation of it has ever been found.

It was mentioned earlier that, if space is homogeneous and isotropic (that is, if the laws of nature are the same everywhere and independent of direction), then linear momentum and angular momentum must be conserved in all interactions. Thus both kinds of momentum conservation are not chance relationships, but arise from underlying symmetries in nature. Again through the use of advanced mathematics, it is possible to show that if *time* is homogeneous (that is, if the laws of nature were always the same as they are now and will always remain the same), then energy must be conserved in all interactions. It is deeply satisfying to the physicist to find connections between principles he has come upon through experiment and such indications of inherent order in the universe.

CONSERVATION OF MECHANICAL ENERGY

In a great many physical processes the rest masses, and hence the rest energies, of the participating objects do not change. In such processes mechanical energy is conserved: the sum of the kinetic and potential energies of the objects involved is constant. An increase in potential energy means a decrease in kinetic energy, and vice-versa.

A falling stone provides a simple example of conservation of mechanical energy. As it falls, its initial potential energy is converted into kinetic energy, so that the total energy of the stone remains the same. The potential energy of a 1-kg stone 50 m above the ground is $mgh = 490$ J and its total mechanical energy is 490 J until it interacts with the ground and transfers energy to it.

Table 15.1

Height		PE = mgh	KE = $\frac{1}{2}mv^2$	PE + KE
50 m		490 J	0 J	490 J
40		392	98	490
30		294	196	490
20		196	294	490
10		98	392	490
0		0	490	490

Another example is the motion of a planet about the sun. Planetary orbits are elliptical, so that at different points in its orbit the planet is at different distances from the sun.

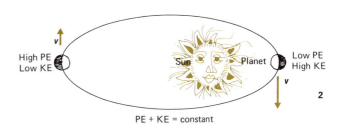

High PE
Low KE

Sun   Planet

Low PE
High KE

2

PE + KE = constant

When the planet is close to the sun, it has a low potential energy, just as a stone near the ground has a low potential energy; when the planet is far from the sun, it has a high potential energy. Since the sum of the planet's PE and KE must be constant, we conclude (correctly) that the kinetic energy of the planet is a maximum when it is nearest the sun and a minimum when it is farthest from the sun. (**2**)

Newton's laws of motion enable us—in theory—to solve all mechanical problems, that is, problems that involve forces and moving objects. However, these laws are actually useful only in the simplest cases, because in order to apply them we must take into detailed account all the various forces acting on each object at every point in its path, which is usually a difficult and complicated procedure. The great advantage of the principle of conservation of mechanical energy is that it permits us to draw definite conclusions about the relationship between the initial and final states of motion of some object or system of objects without having to investigate exactly what happens in between.

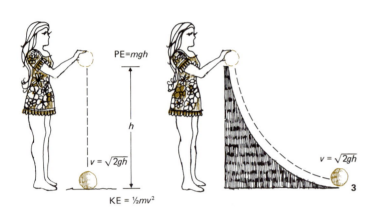

PE=mgh

h

$v = \sqrt{2gh}$

$v = \sqrt{2gh}$

KE = ½mv²

3

A ball slides down a smooth, curved track so that it is moving horizontally when it reaches the bottom. What is its final velocity? Because the path of the ball is curved, to apply the laws of motion directly means an involved calculation, but conservation of mechanical energy makes the problem ridiculously easy. When it is let go, the ball has a potential energy relative to the bottom of its path of

$$\text{PE}_{\text{top}} = mgh.$$

At the bottom the kinetic energy of the ball is

$$\text{KE}_{\text{bottom}} = \frac{1}{2}mv^2.$$

Conservation of mechanical energy requires that

$$\text{KE}_{\text{bottom}} = \text{PE}_{\text{top}},$$

$$\frac{1}{2}mv^2 = mgh,$$

$$v = \sqrt{2gh}.$$

This is the same velocity the ball would have if it were simply dropped. (3)

Some problems require more than one conservation principle for their solution. An example is a collision between a moving billiard ball and a stationary one on a level table. The potential energies of the balls remain the same, so the sum of their kinetic energies before the collision must equal the sum of their kinetic energies afterward. By itself this fact does not tell us what will happen after the collision, although it does tell us what *won't* happen—the velocities of the balls afterward cannot be such that the total kinetic energy of the system is either more or less than the initial kinetic energy of the first ball. To find the answer we must also make use of the conservation of linear momentum.

Let us call the initial and final speeds of the balls v_1, v_2 and v'_1, v'_2. Conservation of momentum and of energy require that

	Before		After
Momentum:	$m_1v_1 + m_2v_2$	$=$	$m_1v'_1 + m_2v'_2,$
Energy:	$\frac{1}{2}m_1v_1^2 + \frac{1}{2}m_2v_2^2$	$=$	$\frac{1}{2}m_1v_1'^2 + \frac{1}{2}m_2v_2'^2.$

Since we have said that the balls are identical, $m_1 = m_2$, and since the second ball was originally at rest, $v_2 = 0$; hence

$$v_1 = v'_1 + v'_2, \qquad v_1^2 = v_1'^2 + v_2'^2.$$

The *only* way of solving these equations is to have either v'_1 or v'_2 equal zero. If v'_2 were zero, it would mean that the first ball traveled completely *through* the second ball. Because this is impossible, we must have as the solution

$$v'_1 = 0, \qquad v'_2 = v_1;$$

the first ball stops, and the second begins to move with the original speed of the first ball. (4)

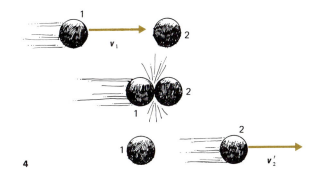

4

HEAT AND ENERGY

In the operation of a pile driver, a motor performs an amount of work mgh in raising a hammer of mass m to a height h above the top of the pile. Then the hammer, with a potential energy of mgh, is released, and as it drops, its potential energy becomes kinetic energy. When the pile is struck by the hammer, the kinetic energy of the latter is converted into work as the pile is driven into the ground. If it requires a force F to drive the pile downward, the depth d to which the pile will go is given by

$$d = \frac{mg}{F}\, h,$$

a formula we obtain simply by equating the initial potential energy mgh of the hammer with the work Fd that it does. (5)

What has happened to the work done in pushing the pile further into the ground? In its new position, the pile is stationary and so has no KE, and its PE is actually less than before because the pile is now closer to the center of the earth.

To find out where the work has gone, all we need do is touch the pile after it has been given a few blows by the hammer. It is warmer than

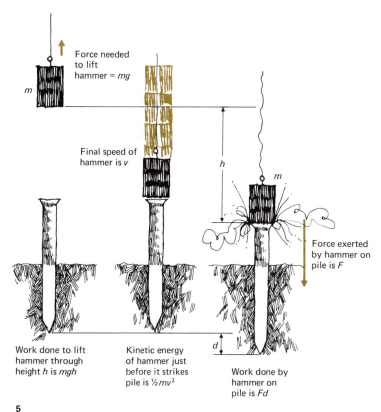

Force needed to lift hammer = mg

m

Final speed of hammer is v

h

m

Force exerted by hammer on pile is F

Work done to lift hammer through height h is mgh

Kinetic energy of hammer just before it strikes pile is $\frac{1}{2}mv^2$

d

Work done by hammer on pile is Fd

5

before, just as two sticks rubbed together by a Boy Scout become hot. This is a quite general observation: work done against frictional forces produces a rise in the temperature of the objects involved. What is happening is that the energy that has disappeared on a macroscopic level reappears on a microscopic level as additional molecular kinetic energy, which is manifested in a rise in temperature. Temperature and heat are examined in detail in later sections.

EXERCISES

1. Can the potential energy of an object be negative? Can its kinetic energy? its mass energy? its linear momentum?

2. Two identical watches, one wound and the other unwound, are dropped into beakers of acid and completely dissolved. Is there any difference between the two reactions?

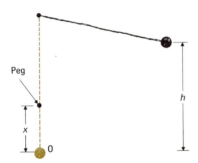

3. A ball on a string is swung to one side and released a height h above its lowest point, as in the figure. What is the maximum value of x in order that the string wind around the peg?

4. Certain particles are known which travel at the speed of light. What must the rest mass of such a particle be? Must it possess energy?

5. A man skis down a slope 100 m high. His speed at the foot of the slope is 20 m/s. What per-

centage of his initial potential energy was dissipated?

6. A boy slides down a sliding pond from a starting point 3 m above the ground. His speed at the bottom is 4 m/s. What percentage of his initial potential energy was dissipated?

7. At her highest point, a 40-kg girl on a swing is 2 m from the ground while at her lowest point she is 0.8 m from the ground. What is her maximum speed? On another swing a 50-kg boy undergoes exactly the same motion. What is his maximum speed?

8. A force of 500 N is used to lift a 20-kg object to a height of 10 m. There is no friction present.
   a) How much work is done by the force?
   b) What is the change in the potential energy of the object?
   c) What is the change in the kinetic energy of the object?

9. A man uses a rope and system of pulleys to lift an 80-kg object to a height of 2 m. He exerts a force of 220 N on the rope and pulls a total of 8 m of rope through the pulleys in the course of raising the object, which is at rest afterward.
   a) How much work does the man do?
   b) What is the change in the potential energy of the weight?
   c) If these answers are different, explain.

10. A 0.5-kg ball leaves the hand of a pitcher at 30 m/s. If his hand moves through 1 m in the act of throwing the ball, what is the average force he exerts? What is this force in lb, where 1 N = 0.225 lb?

11. In the operation of a certain pile driver, a 1000-kg hammer is dropped from a height of 5 m above the head of a pile. If the pile is driven 25 cm into the ground with each impact of the hammer, how does the average force on the pile when it is struck compare with the weight of the hammer?

12. A sledge hammer whose head has a mass of 5 kg is used to drive a spike into a wooden beam. The workman is tired and merely allows the hammer to drop on the spike from a height

0.4 m above it. If the spike is driven 1 cm at each blow, what is the average force on it when struck?

13. Dynamite liberates about 5.4×10^6 J/kg when it explodes. What proportion of its total energy content is this?

14. A ball is dropped from a height of 1 m and loses 10% of its kinetic energy when it bounces on the ground. To what height does it rise?

15. Is the relativistic formula for <u>kinetic</u> energy equal to $\frac{1}{2}mv^2$, where $m = m_0/\sqrt{1 - v^2/c^2}$ is the mass of an object whose rest mass is m_0 and whose speed is v?

16. What is the speed of a particle whose kinetic energy is equal to its rest energy?

17. A ballistic pendulum consists of a wooden block of mass M suspended by long cords from the ceiling. A bullet of mass m and speed v is fired horizontally into the block, which swings away until its height is the amount h above its original height. Find a formula that gives v in terms of g and the readily measurable quantities m, M, and h.

18. A 0.5-kg stone moving at 4 m/s overtakes a 4-kg lump of clay moving at 1 m/s. The stone becomes embedded in the clay.

a) What is the speed of the composite body after the collision?

b) How much kinetic energy is lost?

19. A body of mass m_1 and speed v_1 collides with a stationary body of mass m_2. The ratio between the original kinetic energy of m_1 and the kinetic energy imparted to m_2 in the collision is $4(m_2/m_1)/(1+m_2/m_1)^2$.

a) Verify that this formula is consistent with the conservation of both kinetic energy and momentum.

b) In a nuclear reactor the fast neutrons produced during fission are slowed down by elastic collisions with the atomic nuclei of the "moderator." Two substances often used as moderators are deuterium, whose nuclei have masses approximately double that of the neutron, and carbon, whose nuclei have masses approximately twelve times that of the neutron. (These substances are used because they have little tendency to absorb neutrons.) With the help of the above formula, find the percentage of the initial energy lost by a neutron colliding head-on with (1) a stationary deuterium nucleus, and (2) a stationary carbon nucleus.

16

HEAT

"It appears to me to be extremely difficult, if not quite impossible, to form any distinct idea of anything capable of being excited and communicated in the manner in which heat was excited and communicated in these experiments, except it be *motion*." Benjamin Thompson, Count Rumford (1753–1814)

Every body of matter contains internal energy in the form of the kinetic energies of its atoms or molecules. The higher the temperature of a body, the faster its constituent particles move, and the more internal energy it contains. Heat is, in a sense, internal energy in transit: when heat is added to a body, its temperature increases, and when heat is removed, its temperature decreases.

TEMPERATURE

Temperature, like force, is a key concept in physics which, while we have a clear idea of its meaning in terms of our sense impressions, requires a roundabout definition in order to be specified precisely. We shall attain such precision later, but for the moment we shall dodge the issue and accept temperature merely as that which is responsible for sensations of hot and cold.

There are a number of properties of matter that vary with temperature, and these can be used to construct *thermometers*, devices for measuring temperature. For example, when an object is heated sufficiently, it glows, at first a dull red, then bright red, and finally, at a high enough temperature, it becomes "white hot." By measuring the color of the light it gives off, we can accurately determine the temperature of an object. This method can only be used at rather high temperatures, however.

Of wider application is the fact that matter usually expands when its temperature is increased and contracts when its temperature is decreased. Railroad tracks must be laid with gaps between successive rails to allow for expansion in the summer; heated air above a radiator rises as it expands and becomes lighter than the surrounding air; a column of mercury in a glass tube changes length with a change in temperature. All three of these observations have resulted in practical thermometers.

Two strips of different metals that are joined together bend to one side with a change in temperature owing to different rates of expansion in the two metals, a fact employed in constructing household oven thermometers. The higher the temperature, the greater the deflection. When cooled, such a bimetallic strip bends in the opposite direction. (1)

In a constant-volume gas thermometer, which is a very sensitive laboratory instrument, the height of the mercury column at the left is adjusted until the mercury column at the right just touches the gas bulb. The difference in heights of the two mercury columns is a measure of the pressure needed to maintain the gas in a fixed volume, and hence a measure of the temperature. (2)

Mercury (or colored alcohol) expands more when heated than glass does, and so the length of the liquid column in a liquid-in-glass thermometer is a measure of the temperature of the thermometer bulb. (3)

THE CELSIUS SCALE

Before we can use any of these or other thermal properties of matter to construct a practical thermometer, we must begin by specifying a temperature scale and the method by which we shall calibrate the thermometer in terms of this scale. Water is a readily available liquid which freezes into a solid, ice, and vaporizes into a gas, steam, at definite temperatures at sea level atmospheric pressure. We can establish a temperature scale by defining the freezing point of water (or, more exactly, the point at which a mixture of ice

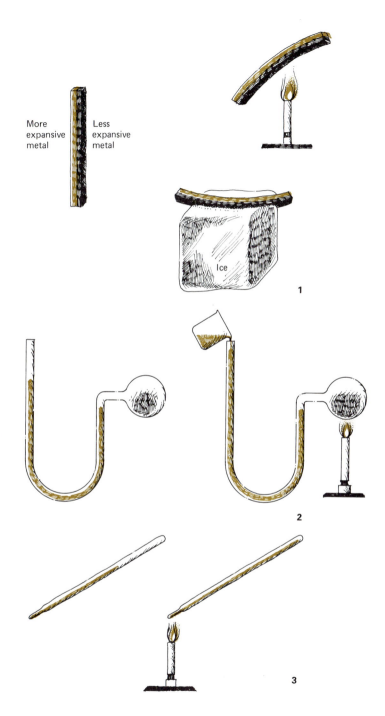

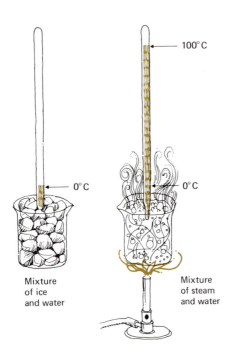

Mixture
of ice
and water

Mixture
of steam
and water

and water is in equilibrium, with exactly as much ice melting as water freezing) as 0° and the boiling point of water (or, more exactly, the point at which a mixture of steam and water is in equilibrium) as 100°. This scale is called the *celsius* scale, and temperatures measured in it are written, for example, "40°C." In the United States, the celsius scale is sometimes called the *centigrade* scale.

To calibrate a thermometer, say an ordinary mercury thermometer, we first plunge it into a mixture of ice and water. When the mercury column has come to rest we mark the position of its top 0°C on the glass. Then we plunge it into a mixture of steam and water, and when the mercury column has again come to rest, we mark the new position of the top of the mercury column 100°C. Finally we divide the interval between the 0°C and 100°C markings into 100 equal parts, each representing a change in temperature of 1°C, and extend the scale with divisions of the same length beyond 0°C and 100°C as far as is convenient. In doing this we have, of course, assumed that changes in the length of the mercury column are always directly proportional to the changes in temperature that brought them about. **(4)**

THE FAHRENHEIT SCALE

Although the celsius scale is used in most of the world, a different temperature scale called the *fahrenheit* scale is commonly used for nonscientific purposes in English-speaking countries.

In the fahrenheit scale the freezing point of water is 32°F and the boiling point of water is 212°F. This means that 180°F separates the freezing and boiling points of water, whereas 100°C separates them in the celsius scale. Therefore fahrenheit degrees are 100/180 or

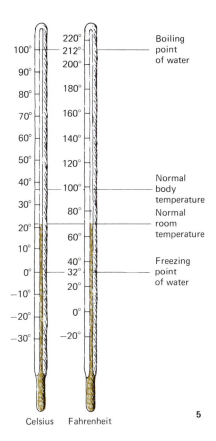

100°	220°	
	212°	Boiling point of water
90°	200°	
80°	180°	
70°	160°	
60°	140°	
50°	120°	
40°	100°	Normal body temperature
30°	80°	Normal room temperature
20°	60°	
10°		
0°	40°	Freezing point of water
	32°	
−10°	20°	
−20°	0°	
−30°	−20°	

Celsius   Fahrenheit   5

5/9 as large as celsius degrees. We can convert temperatures from one scale to the other with the help of the formulas

$$°F = \tfrac{9}{5}°C + 32°,$$

$$°C = \tfrac{5}{9}(°F − 32°). \quad (5)$$

CALORIC

Because we are so familiar with the idea that heat is a form of energy, it may not be easy for us to sympathize with the struggles early scientists had in trying to understand its nature. In particular, the idea that heat is an actual substance called *caloric* is hard for us to take seriously, although it was not a bad notion when first proposed.

According to the caloric theory of heat, an object gets hot when it absorbs caloric and gets cold when caloric escapes from it. Because nobody could see or smell or weigh caloric, it was supposed to be an invisible, odorless, massless fluid. The caloric theory had its downfall in the observation that rubbing two pieces of metal (or anything else) together produces heat as long as the rubbing continues, which means that there must be an unlimited amount of caloric in them. But according to the theory, since the pieces of metal were at room temperature to begin with, they must have had only a certain definite amount of caloric to begin with—which disagrees with the experimental findings.

Today we not only know that heat is a form of energy, but it has been possible to trace the energy to the motions of the atoms and molecules of which every substance is composed. It is not necessary to assume the existence of any mysterious fluids or any new physical principles to explain heat; all we have to do is apply the same laws of mechanics to molecules that we apply to balls, cars, and planets. In this section we shall consider various aspects of heat and related phenomena in terms of everyday experience, leaving their detailed explanation until later.

HEAT AND INTERNAL ENERGY

Every body of matter contains a certain amount of *internal energy* in addition to any kinetic or potential energy it may possess by virtue of its motion or position. This internal energy resides in the random motions of the

High temperature but small mass, hence little internal energy

Low temperature but large mass, hence much internal energy

Internal energy always flows from hot body to cold body, regardless of internal energy content

6

atoms or molecules of which the body is composed. The total amount of internal energy a body contains depends upon its temperature, upon its composition, upon its mass, and upon its physical state (solid, liquid, or gas). However, the temperature of the body alone is what determines whether internal energy will be transferred from it to another body with which it is in contact, or vice-versa. A large block of ice at 0°C has far more internal energy than a cup of hot water, yet when the water is poured on the ice some of the ice melts and the water becomes cooler, which signifies that energy has passed from the water to the ice. **(6)**

When the temperature of a body increases, it is customary to say that *heat* has been added to it; when the temperature of a body decreases, it is customary to say that heat has been removed from it. Thus we can think of heat as internal energy in transit. In fact, we need not even know that heat is a form of energy in order to make an adequate working definition of heat:

Heat is a quantity that causes an increase in the temperature of a body of matter to which it is added and a decrease in the temperature of a body of matter from which it is removed, provided that the matter does not change state during the process.

The latter part of the definition is required because changes of state (for instance, from ice to water or water to steam) involve the transfer of heat to or from a body without any change in temperature.

The term heat remains in the vocabulary of physics partly because of convenience and partly because of tradition. Temperature, on the other hand, is a unique concept both in a macroscopic sense as an indicator of the direction of internal energy flow and (as we shall see later) in a microscopic sense as a measure of average molecular kinetic energy.

THE KILOCALORIE

In the metric system the unit of heat, called the *kilocalorie* (abbreviated kcal), is that amount of heat required to raise the temperature of 1 kg of water through 1°C. Similarly, 1 kcal of heat must be removed from 1 kg of water to reduce its temperature by 1°C. Because this amount of heat actually varies slightly with temperature, the kilocalorie is formally defined as the amount of heat involved in changing the temperature of 1 kg of water from 14.5°C to 15.5°C; the difference is insignificant for most purposes, however.

It is entirely possible to use the joule as the unit of heat instead of the kcal. However, despite its inconvenience, the kcal is widely employed, and for the time being at least it is necessary to be familiar with it.

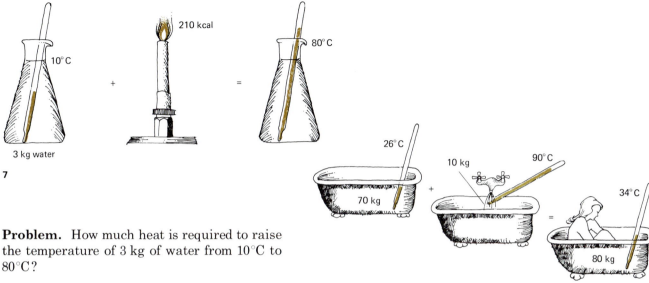

Problem. How much heat is required to raise the temperature of 3 kg of water from 10°C to 80°C?

Solution. The temperature of the water must be raised by 70°C. Since 1 kcal of heat raises the temperature of 1 kg of water by 1°C, 70 kcal is required for each kg of water here. There are 3 kg of water in all, and so 3×70 kcal = 210 kcal of heat are required. (7)

Problem. A bathtub contains 70 kg of water at 26°C. Ten kg of water at 90°C is poured in. What is the final temperature of the mixture?

Solution. The final temperature T of the mixture will be more than 26°C and less than 90°C. The 70 kg of cold water will have gained heat, and the 10 kg of hot water will have lost heat. The heat gained by the cold water is

Heat gained = mass of water × change in temperature

$$= 70 \text{ kg} \times (T - 26°C)$$
$$= (70\,T - 1820) \text{ kcal}.$$

The heat lost by the hot water is

Heat lost = mass of water × change in temperature

$$= 10 \text{ kg} \times (90°C - T)$$
$$= (900 - 10\,T) \text{ kcal}.$$

If we neglect heat losses to the air and to the bathtub itself, the heat gained by the cold water must equal the heat lost by the hot water in order that energy be conserved. Hence

$$\text{Heat gained} = \text{heat lost,}$$
$$70\,T - 1820 = 900 - 10\,T,$$
$$80\,T = 2720,$$
$$T = 34°C. \text{ (8)}$$

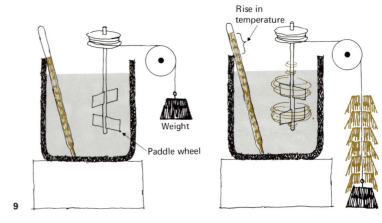

Rise in temperature

Weight

Paddle wheel

9

MECHANICAL EQUIVALENT OF HEAT

If we push a box across a level floor a distance of 10 m with a force of 200 N, we perform 2000 J of work, and yet the box is at rest afterward. Where has the 2000 J of energy gone?

Because the box and the floor are warmer afterward, we might reasonably conclude that the work done in overcoming friction has gone into internal energy, but this observation by itself is not enough to establish that energy is being conserved in the process. What we must also know is exactly how much heat is produced when a certain amount of energy is dissipated. Is it always the same amount of heat? Or does the amount of heat depend upon the kinds of material in contact, or on their initial temperatures, or on the time of day the experiment is carried out? This fundamental question did not receive an adequate answer until a little more than a century ago.

The first definite evidence that the ratio between the energy lost in some mechanical process and the heat that appears as a result of the process is a constant came from the experiments of the English brewer James Prescott Joule (1818–1889). Such constancy means that heat is a form of energy; the ratio between energy lost and heat gained is just the conversion factor between the units of each quantity.

Joule used an apparatus the same in principle as that shown here. As the weight descends it turns paddle wheels inside an insulated container of water, thereby stirring the water. The stirring must be done against the frictional resistance of the water, and the heat produced can be determined by measuring the increase in the water's temperature. The results are always the same: 1 kcal of heat appears for every 4185 J of work done. The quantity

$$J = 4185 \ \frac{\text{J}}{\text{kcal}}$$

is accordingly known as the *mechanical equivalent of heat.* (**9**)

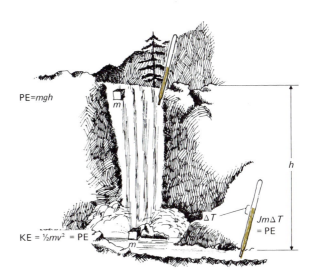

PE=*mgh*

m

h

ΔT

JmΔT = PE

KE = ½*mv²* = PE

m

10

Problem. What is the difference in temperature between the water at the top and at the bottom of a waterfall?

Solution. In the course of dropping through a height h, a mass m of water loses the potential energy PE = mgh. This potential energy appears as kinetic energy in the falling water, and we shall assume it is all dissipated as heat at the bottom of the waterfall. If we call the rise in temperature of the water ΔT, the heat gained by the mass m of water is $m\,\Delta T$. Hence

Potential energy loss = internal energy gain,

$$mgh = Jm\,\Delta T.$$

The mechanical equivalent of heat J is needed in order that both sides of the equation be expressed in the same units. Solving for the temperature rise ΔT yields

$$\Delta T = \frac{gh}{J}$$

$$= 0.0023\ h\ \frac{^\circ\mathrm{C}}{\mathrm{m}}.$$

We can apply this general formula to any waterfall whose height h we know. For example, the Victoria Falls on the Zambesi River are 108 m high, so that the water at the foot of the falls is

$$\Delta T = 0.0023\ \frac{^\circ\mathrm{C}}{\mathrm{m}} \times 108\ \mathrm{m}$$

$$= 0.25\,^\circ\mathrm{C}$$

warmer than at the top. (**10**)

EXERCISES

1. A jar of water is shaken vigorously. What becomes of the work that is done?

2. The normal temperature of the human body is 98.6°F. What is this temperature on the celsius scale?

3. The melting point of lead is 330°C and its boiling point is 1170°C. Express these temperatures on the fahrenheit scale.

4. At what temperature would celsius and fahrenheit thermometers give the same reading?

5. Dry ice (solid carbon dioxide) vaporizes at $-112°F$. What is this temperature on the celsius scale?

6. To make a cup of tea, 200 g of water is to be heated from 20°C to 100°C. How much heat is needed?

7. A sedentary person requires about 30 kcal of energy in his diet per day per kg of body mass. If this energy were used to raise a 1-kg mass above the ground, find its height. [The calorie that dieticians use is the same as the kilocalorie. Thus the energy content of a 130-calorie cupcake is really 130 kcal. A heat unit once widely used is also called the calorie. This is equal to the heat needed to raise the temperature of 1 g of water by 1°C, so that 1000 cal = 1 kcal. The dietician's calorie is sometimes written Calorie to distinguish it from the ordinary calorie.]

8. A 50-kg woman is on a diet that provides her with 2500 kcal daily.
    a) If this amount of energy were used to raise her above the ground, how high would she go?
    b) If 2500 kcal were added to 50 kg of water at 37°C (normal body temperature), what would its final temperature be?

9. How many kcal are evolved per hour by a 60-W light bulb?

10. Radiant energy from the sun arrives at the earth at the rate of about 1400 watts per m^2 of surface perpendicular to the sun's rays. If a reflector 1 m^2 in area is used to concentrate sunlight on a cup containing 0.1 kg of water initially at 20°C, how long will it take for the water to reach 100°C? Assume the water absorbs all the incident energy.

17

HEAT AND MATTER

A body of matter may change its state when heat is added to or removed from it. Thus water becomes ice when cooled at 0°C, and it becomes steam when heated at 100°C. If there is no change of state, the body's temperature changes. The specific heat of a substance is a measure of how great a temperature change accompanies a change in internal energy. Water has the highest specific heat, because it changes temperature least readily of any substance.

"As we ascend from the solid to the fluid and gaseous states, physical properties diminish in number and variety, each state losing some of those which belonged to the preceding state. When solids are converted into fluids, all the varieties of hardness and softness are necessarily lost. Crystalline and other shapes are destroyed. Opacity and color frequently give way to a colorless transparency, and a general mobility of particles is conferred. Passing onward to the gaseous state, still more of the evident characters of bodies are annihilated . . . They now form but one set of substances, and the varieties of density, hardness, opacity, color, elasticity, and form, which render the number of solids and fluids almost infinite, are now supplied by a few slight variations in weight, and some unimportant shades of color." Michael Faraday (1791–1867)

SPECIFIC HEAT

Samples of other substances respond to the addition or removal of a given amount of heat with temperature changes greater than that of an equal mass of water. One kg of water increases in temperature by 1°C when 1 kcal of heat is added to it, but 1 kcal of heat increases the temperature of 1 kg of helium (its volume held constant) by 1.3°C, of 1 kg of ice by 2°C, and of 1 kg of gold by 33°C. (**1**)

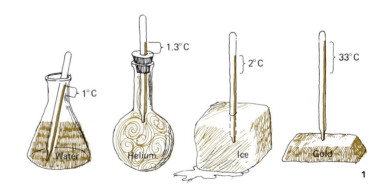

1.3°C — 1°C — 2°C — 33°C

Water — Helium — Ice — Gold

1

The *specific heat* (symbol *c*) of a substance refers to what we might think of as its thermal inertia:

147

The specific heat of a substance is the amount of heat that must be added or removed from 1 kg of it to change its temperature by 1°C.

A high specific heat means a relatively small change in temperature for a given change in internal energy content, just as a large inertial mass means a relatively small acceleration when a given force is applied.

Specific heats of various substances are given in Table 17.1. The values given are averages, since the actual values vary somewhat with temperature.

Table 17.1

Substance	Specific heat, kcal/kg-°C	Substance	Specific heat, kcal/kg-°C
Alcohol (ethyl)	0.58	Marble	0.21
Aluminum	0.22	Mercury	0.033
Copper	0.093	Silver	0.056
Glass	0.20	Steam	0.48
Gold	0.030	Sulfuric acid	0.27
Granite	0.19	Turpentine	0.42
Ice	0.50	Water	1.00
Iron	0.11	Wood	0.42
Lead	0.030	Zinc	0.092

With the help of specific heat we can write a formula for the quantity of heat Q involved when a quantity m of a substance undergoes a change in temperature of ΔT. This formula is simply

$$Q = mc\Delta T,$$

Heat change = mass × specific heat × temperature change.

Problem. How much heat must be removed from 14 kg of aluminum in order to cool it from 80°C to 15°C?

Solution. From the table the specific heat of aluminum is $c = 0.22$ kcal/kg-°C, and so

$$Q = mc\Delta T$$

$$= 14 \text{ kg} \times 0.22 \frac{\text{kcal}}{\text{kg-}°\text{C}} \times (-65°\text{C})$$

$$= -200 \text{ kcal}.$$

The minus sign means that this quantity of heat is to be extracted to achieve the temperature change of $-65°$C.

Problem. 0.20 kg of coffee at 90°C is poured into a 0.50-kg cup at 20°C. Assuming that no heat is transferred to or from the outside, what is the final temperature of the coffee?

Solution. We shall take the specific heat of coffee to be that of water and the specific heat of the cup to be that of glass. To solve the problem, we begin by noting that

Heat gained by cup = heat lost by coffee.

If the final temperature of both coffee and cup is T, then

Heated gained by cup $= m_{cup}c_{cup}(T - 20°\text{C})$

$$= 0.50 \text{ kg} \times 0.20 \frac{\text{kcal}}{\text{kg-}°\text{C}} \times (T - 20°\text{C})$$

$$= (0.10\ T - 2.0) \text{ kcal},$$

and

Heat lost by coffee $= m_{coffee}c_{coffee}(90°\text{C} - T)$

$$= 0.20 \text{ kg} \times 1.0 \frac{\text{kcal}}{\text{kg-}°\text{C}} \times (90°\text{C} - T)$$

$$= (18 - 0.20\ T) \text{ kcal}.$$

Now we set the heat gained by the cup equal

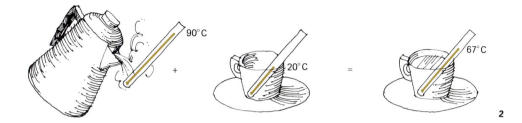

2

to the heat lost by the coffee and solve for T:

$$0.10\,T - 2.0 = 18 - 0.20\,T,$$

$$0.30\,T = 20,$$

$$T = 67\,°C.$$

The temperature of the coffee drops by $23\,°C$ as it warms the cup. Evidently it is necessary to preheat the cup if one wants really hot coffee. (2)

CHANGE OF STATE

Not always does the addition or removal of heat from a sample of matter lead to a change in its temperature. Instead the sample may change its state from solid to liquid or from liquid to gas when heat is added, or it may change from gas to liquid or from liquid to solid when heat is taken away. Such changes of state take place at definite temperatures for most substances, but for a few (glass and wax, for instance) there is only a gradual softening or hardening over a range of temperatures. Substances of the latter kind are not true solids, however; their structures are really those of liquids, and their hardness at room temperature is really a kind of exaggerated viscosity.

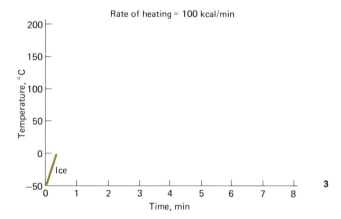

3

Let us examine what happens when we add heat at a constant rate to 1 kg of ice that is initially at $-50\,°C$.

This is a graph of temperature versus time for 1 kg of water, initially at $-50\,°C$, to which heat is added at the constant rate of 100 kcal/min. The specific heat of ice is $0.5\,\text{kcal/kg-}°C$, and so 25 kcal is needed to bring the ice to $0\,°C$. (3)

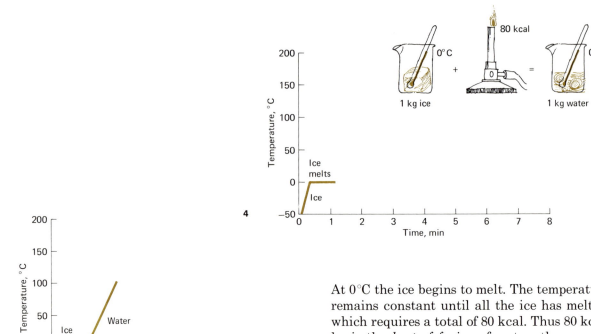

4

At 0°C the ice begins to melt. The temperature remains constant until all the ice has melted, which requires a total of 80 kcal. Thus 80 kcal/ kg is the *heat of fusion* of water: the amount of heat needed to convert 1 kg of ice into 1 kg of water at its melting point of 0°C. (**4**)

When all the ice has turned to water, the temperature goes up once more as further heat is supplied. Since the specific heat of water is 1 kcal/kg-°C, there is now a rise of 1°C per kcal of heat. This rate of change is less than that of ice, since the specific heat of water is greater than that of ice, and so the slope of the graph is smaller. (**5**)

When 100°C is reached, the water begins to turn into steam. The temperature stays constant until a total of 540 kcal is added, at which time all the water has become steam. Thus 540 kcal/kg is the *heat of vaporization* of water: the amount of heat needed to convert 1 kg of water into 1 kg of steam at its boiling point of 100°C. (**6**)

5

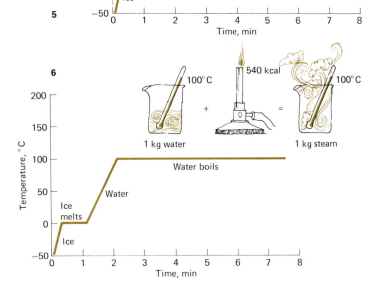

6

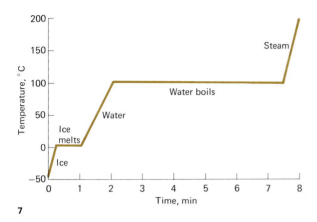

7

After the water has become steam, its temperature rises again. The specific heat of steam is 0.48 kcal/kg-°C, so the temperature increase is 2.1°C per kcal of heat, and the slope of the graph is therefore steeper than it was for ice or water. (**7**)

HEATS OF FUSION AND VAPORIZATION

Substances other than water also have heats of fusion and vaporization which, like their melting and boiling points, are in general different from those of water. Here are broader definitions of these quantities together with a table of some representative values.

The *heat of fusion* of a substance is the amount of heat that must be supplied to change 1 kg of the substance at its melting point from the solid to the liquid state; the same amount of heat must be removed from 1 kg of the substance in the liquid state at its melting point to change it to a solid. The usual symbol for heat of fusion is L_f.

The *heat of vaporization* of a substance is the amount of heat that must be supplied to change 1 kg of the substance at its boiling point from the liquid to the gaseous (or vapor) state; the same amount of heat must be removed from 1 kg of the substance in the gaseous state at its boiling point to change it into a liquid. The usual symbol for heat of vaporization is L_v.

Table 17.2
Heats of fusion and vaporization and melting and boiling points of various substances

Substance	Melting point, °C	L_f, kcal/kg	Boiling point, °C	L_v, kcal/kg
Alcohol (ethyl)	−114	25	78	204
Bismuth	271	12.5	920	190
Bromine	−7	16	60	43
Lead	330	5.9	1170	175
Lithium	186	160	1336	511
Mercury	−39	2.8	358	71
Nitrogen	−210	6.1	−196	48
Oxygen	−219	3.3	−183	51
Sulfuric acid	8.6	39	326	122
Water	0	80	100	540
Zinc	420	24	918	475

Under certain circumstances most substances can change directly from the solid to the vapor state, or vice versa. Both processes are called *sublimation*. For example, "dry ice" (solid carbon dioxide) evaporates directly to gaseous carbon dioxide at temperatures above −78.5°C, and does not pass through the liquid state. With the exception of carbon dioxide and a few other substances, however, sublimation does not occur except at pressures well below that of the atmosphere.

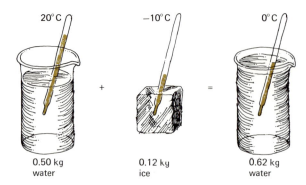

20°C   −10°C   0°C

+   =

8

0.50 kg   0.12 ky   0.62 kg
water   ice   water

Problem. What is the minimum amount of ice at $-10°C$ that must be added to 0.50 kg of water at 20°C in order to bring the temperature of the water down to 0°C?

Solution. We begin, as before, with the statement of energy conservation,

Heat absorbed by ice = heat lost by water.

The heat Q_1 absorbed by the unknown mass of ice in going from $-10°C$ to its melting point of 0°C is

$$Q_1 = m_{ice}c_{ice}\Delta T_{ice}$$

$$= m_{ice} \times 0.50\,\frac{kcal}{kg\text{-}°C} \times 10°C,$$

and the heat Q_2 absorbed by the ice in melting at 0°C is

$$Q_2 = m_{ice}L_{f\ ice}$$

$$= m_{ice} \times 80\,\frac{kcal}{kg}.$$

Hence

Heat absorbed by ice = $Q_1 + Q_2$

$$= (5 + 80)\,m_{ice}\,\frac{kcal}{kg}.$$

The heat lost by the water in cooling to 0°C from 20°C is

Heat lost by water = $m_{water}c_{water}\Delta T_{water}$

$$= 0.50\ kg \times 1.0\,\frac{kcal}{kg\text{-}°C} \times 20°C$$

$$= 10\ kcal.$$

Equating the heat absorbed with the heat lost and then solving for m_{ice} yields

$$85\,m_{ice}\,\frac{kcal}{kg} = 10\ kcal,$$

$$m_{ice} = 0.12\ kg.\ \text{(8)}$$

EXERCISES

1. A cup of hot coffee can be cooled by placing a cold spoon in it. A spoon made of which of the following metals would be most effective for this purpose: aluminum, copper, iron, or silver?

2. Why will the engine of a car whose cooling system is filled with an alcohol antifreeze be more likely to overheat in summer than one whose cooling system is filled with water?

3. Which is more effective in cooling a drink, 10 g of water at 0°C or 10 g of ice at 0°C?

4. How does perspiration give the body a means of cooling itself?

5. Why does turning the flame higher under a pan of boiling water not reduce the time needed to cook an egg in the water?

6. A person decides to lose weight by eating only cold food. A 100-gram piece of apple pie yields about 350 kcal of energy when eaten. If its heat capacity is 0.4 kcal/kg-°C, how much greater is its energy content at 50°C than at 20°C? What percentage difference is this?

7. If all the heat lost by 1 kg of water at 0°C when it turns into ice at 0°C could be turned into kinetic energy, what would the speed of the ice be?

8. How much mass is lost by 1 kg of water at 0°C when it turns into ice at 0°C?

9. When a certain quantity of a vapor condenses into a liquid, what happens to its internal energy content and to its temperature?

10. How much heat must be added to 1 kg of copper to raise its temperature from 20°C to 100°C?

11. How much more heat must be added to 1 kg of ice at 0°C to convert it to steam at 100°C than is required to raise the temperature of 1 kg of water from 0°C to 100°C?

12. Six kilograms of ice at −10°C are added to 6 kg of water at +10°C. Find the temperature of the resulting mixture.

13. How much steam at 120°C is required to melt 0.5 kg of ice at 0°C?

14. A 0.1-kg piece of silver is taken from a bath of hot oil and placed in a 0.08-kg glass jar containing 0.2 kg of water at 15°C. The temperature of the water increases by 8°C. What was the temperature of the oil?

15. A 0.6-kg copper container holds 1.5 kg of water at 20°C. A 0.1-kg iron ball at 120°C is dropped into the water. What is the final temperature of the water?

16. A 5-kg iron bar is taken from a forge at a temperature of 1000°C and plunged into a pail containing 10 kg of water at 60°C. How much steam is produced?

17. By mistake, 0.2 kg of water at 0°C is poured into a vessel containing liquid nitrogen at −196°C. How much nitrogen vaporizes?

18. How much ice at −10°C is required to cool a mixture of 0.1 kg ethyl alcohol and 0.1 kg water from 20°C to 5°C?

19. A 20-kg storage battery has an average specific heat of 0.2 kcal/kg-°C. When fully charged the battery contains 10^6 J of electrical energy. If all this energy were dissipated within the battery, find the increase in its temperature.

20. A 1-kg block of ice at 0°C falls into a lake whose water is also at 0°C, and 0.01 kg of ice melts. What was the minimum altitude from which the ice fell?

21. A lead bullet at 100°C strikes a steel plate and melts. What was its minimum speed?

22. The minimum speed an artificial earth satellite can have is 7.9×10^3 m/s. Aluminum melts at 660°C. If an aluminum satellite reenters the earth's atmosphere when its temperature is 0°C, can it be brought to rest by air resistance without melting? If not, find out how actual spacecraft avoid this dilemma.

18

THE IDEAL GAS

"It is not the abhorrence of the vacuum that causes the quicksilver to stand suspended in the usual experiment, but really the weight and pressure of the air, which balances the weight of the quicksilver." Blaise Pascal (1623–1662)

All gases respond to changes in volume, temperature, and pressure in very nearly the same way, which is not true for the other states of matter. Hence it is possible to define an "ideal gas" which all real gases closely resemble in behavior. This ideal gas serves as a target for theories of the microscopic structure of gases to aim at.

PRESSURE

When a force F acts perpendicular to a surface whose area is A, the *pressure P* exerted on the surface is defined as the ratio between the magnitude F of the force and the area:

$$\text{Pressure} = \frac{\text{force}}{\text{area}},$$

$$P = \frac{F}{A}.$$

Pressure is a scalar quantity. Often a perpendicular force is described as *normal*, which allows us to say that *pressure is the magnitude of normal force per unit area.*

Problem. Find the pressure exerted on the contents of a cylinder by a close-fitting 50-kg piston whose area is 200 cm².

Solution. The force exerted by the piston is equal to its weight *mg*. Hence

$$F = mg$$

$$= 50 \text{ kg} \times 9.8 \, \frac{\text{m}}{\text{s}^2}$$

$$= 490 \text{ N}.$$

Since $1 \text{ m} = 10^2 \text{ cm}$, $1 \text{ m}^2 = 10^4 \text{ cm}^2$, and

$$A = \frac{200 \text{ cm}^2}{10^4 \text{ cm}^2/\text{m}^2}$$

$$= 2.0 \times 10^{-2} \text{ m}^2.$$

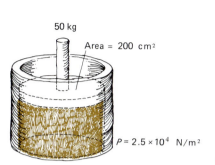

50 kg

Area = 200 cm²

$P = 2.5 \times 10^4$ N/m²

1

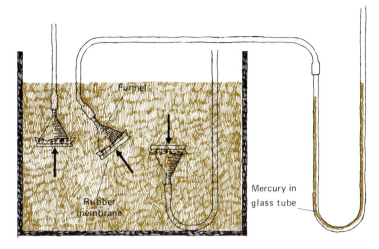

Funnel

Mercury in glass tube

Rubber membrane

2

The pressure exerted on the contents of the cylinder is therefore

$$P = \frac{F}{A}$$

$$= \frac{490 \text{ N}}{2.0 \times 10^{-2} \text{ m}^2}$$

$$= 2.5 \times 10^4 \frac{\text{N}}{\text{m}^2}. \textbf{(1)}$$

PRESSURE ON A FLUID

Pressure is a useful quantity because fluids flow under stress instead of being deformed elastically as solids are. The characteristic lack of rigidity exhibited by fluids has three significant consequences:

1. *The forces a fluid exerts on the walls of its container, and vice versa, always act perpendicular to the walls.*

If this were not so, any sideways force by a fluid on a wall would be met, according to the third law of motion, by a sideways force back on the fluid, which would cause the fluid to move constantly parallel to the wall. But fluids may be at rest in containers of any shape, and so the sole forces they can exert on their containers must be perpendicular to the walls of the latter.

2. *An external pressure exerted on a fluid is transmitted uniformly throughout the volume of the fluid.*

If this were not so, the fluid would flow from a region of high pressure to one of low pressure, thereby equalizing the pressure. We must keep in mind, however, that the above statement refers to a pressure imposed from outside the fluid. The fluid at the bottom of a container is always under greater pressure than that at the top owing to the weight of the overlying fluid. A notable example is the earth's atmosphere, although such pressure differences are ordinarily significant only for liquids.

3. *At any depth in a fluid the pressure on a small surface is the same regardless of the orientation of the surface.*

If this were not so, again, the fluid would flow in such a way as to equalize the pressure. **(2)**

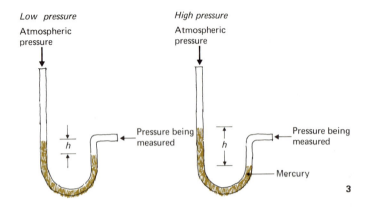

3

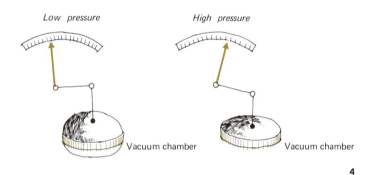

4

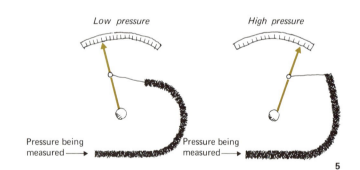

5

Pressures may be measured in a number of ways, three of which are illustrated here. Usually what is directly determined is the difference between the unknown pressure and atmospheric pressure. This difference is called the *gauge pressure*, whereas the true pressure is called the *absolute pressure*. That is,

$$\text{Absolute pressure} = \text{gauge pressure} + \text{atmospheric pressure.}$$

Thus a tire inflated to a gauge pressure of 1.5 atm (1 atm = atmospheric pressure) contains air at an absolute pressure of 2.5 atm.

The metric unit of pressure is the N/m^2, sometimes called the *pascal*. Unfortunately a number of other units of pressure are also in common use, notably the *atmosphere*, which represents the average pressure exerted by the earth's atmosphere at sea level and is equal to 1.013×10^5 N/m^2; the *millibar* (mb), equal to 100 N/m^2; and the millimeter of mercury (mm Hg) or *torr*, which represents the pressure exerted by a column of mercury 1 mm high at sea level and is equal to 133 N/m^2.

A *manometer* measures pressure in terms of the difference h in height of two mercury columns, one open to the atmosphere and the other connected to the source of unknown pressure. (**3**)

An *aneroid* measures pressure in terms of the amount by which the thin, flexible ends of an evacuated metal chamber are pushed in or out by the external pressure. (**4**)

A *Bourdon tube* straightens out when the pressure inside it increases. (**5**)

ATMOSPHERIC PRESSURE

The total mass of the earth's atmosphere is 5.3×10^{18} kg. Although this is only about one millionth of the mass of the earth itself, the atmosphere still consists of an enormous amount of matter, and it exerts a pressure of $1.013 \times 10^5 \text{ N/m}^2$ at sea level. We are not aware of this pressure because our bodies are not airtight, so to speak—the pressure inside our bodies is also $1.013 \times 10^5 \text{ N/m}^2$, which exactly balances the external pressure. If we remove the air from a weak-walled container such as a tin can, however, there is nothing inside it to balance atmospheric pressure and the container collapses. (**6**)

The composition of dry air at sea level is given in Table 18.1. The proportion of water vapor is variable and ranges up to 4%.

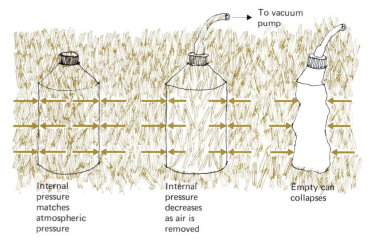

Internal pressure matches atmospheric pressure

Internal pressure decreases as air is removed

Empty can collapses

6

Table 18.1

Gas	Percentage by volume
Nitrogen	78.08
Oxygen	20.95
Argon	0.93
Carbon dioxide	0.03
Neon	0.0018
Helium	0.00052
Methane	0.00015
Krypton	0.00011
Hydrogen, carbon monoxide, xenon, ozone, radon	<0.0001

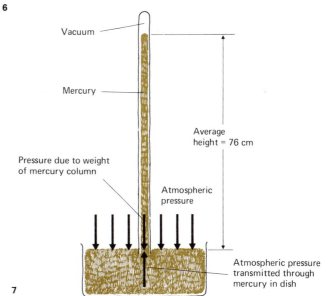

Vacuum

Mercury

Pressure due to weight of mercury column

Average height = 76 cm

Atmospheric pressure

Atmospheric pressure transmitted through mercury in dish

7

A *barometer* is a pressure gauge used to measure atmospheric pressure. Ordinary barometers are of the aneroid type, but precision barometers make use of a mercury column as shown. The weight of the column is balanced by atmospheric pressure transmitted through the mercury in the dish at the bottom. Atmospheric pressure varies with weather conditions over a range of about 15% of its average value. The unit of pressure used in meteorology (the science of weather) is the millibar (mb), equal to 100 N/m^2. At sea level the average atmospheric pressure is 1013 mb. (**7**)

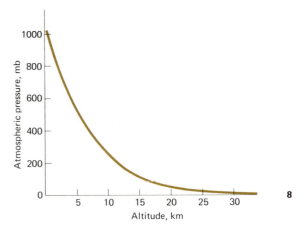

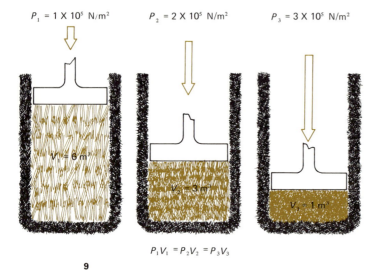

$$P_1V_1 = P_2V_2 = P_3V_3$$

9

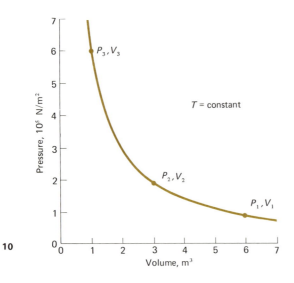

10

Atmospheric pressure decreases with altitude. The pressure is only half its sea level value at an altitude of about 5 km ($\approx$ 3 mi) and a fourth of its sea level value at an altitude of about 10 km ($\approx$ 6 mi). (**8**)

BOYLE'S LAW

When the temperature of a sample of a gas is held constant, the pressure it exerts on its container is found to be inversely proportional to the volume of the container. Expanding the container lowers the pressure; shrinking the container raises the pressure. Conversely, increasing the pressure on a gas sample reduces its volume; decreasing the pressure increases its volume. This relationship is called *Boyle's law* after its discoverer, Robert Boyle (1627–1691). (**9**)

Boyle's law can be expressed in the form

$$PV = \text{constant}, \qquad (T = \text{constant})$$

or alternatively as

$$P_1V_1 = P_2V_2. \qquad (T = \text{constant})$$

In the latter case P_1 is the absolute pressure of the gas when its volume is V_1, and P_2 is the absolute pressure of the gas when its volume is V_2. (**10**)

Problem. How much air at atmospheric pressure can be stored in the 1.2-m³ tank of an air compressor which can withstand an absolute pressure of 8 atm?

Solution. Letting state 1 represent the air in the compressor tank and state 2 the air at atmospheric pressure, we have

$$V_2 = \frac{P_1}{P_2}V_1$$
$$= \frac{8 \text{ atm}}{1 \text{ atm}} \times 1.2 \text{ m}^3$$
$$= 9.6 \text{ m}^3.$$

A total of 9.6 m³ of air at atmospheric pressure can be stored in the 1.2-m³ compressor tank at the high pressure.

ABSOLUTE TEMPERATURE SCALE

Now let us see what happens to a gas sample when its temperature is changed.

When a solid or a liquid is heated or cooled, its volume changes by an amount proportional to the temperature change. When we come to gases, though, a problem arises. Unlike solids and liquids, gases do not have specific volumes at a particular temperature, but expand to fill their containers. The only way to change the volume of a gas is to change the capacity of its container. However, even though its volume may remain the same, another property of a confined gas varies with its temperature, namely the pressure it exerts on the container walls. The air pressure in an automobile tire drops in cold weather and increases in warm weather, an illustration of this property.

Since the pressure a gas sample exerts on the walls of a container of fixed size varies with temperature, then if we hold the gas pressure constant, the sample's volume should vary with temperature. When this prediction is experimentally tested, which was first done

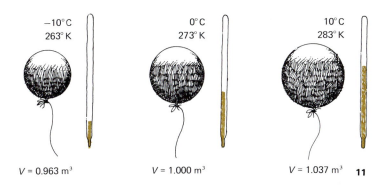

−10°C 263°K   0°C 273°K   10°C 283°K

V = 0.963 m³   V = 1.000 m³   V = 1.037 m³   **11**

over 150 years ago by Charles and Gay-Lussac, it is found that the change in volume of a gas sample at constant pressure when its temperature is changed by ΔT is proportional to both the original volume of the sample and to ΔT. Thus gases at constant pressure behave just as solids and liquids do when they are heated or cooled.

There is one significant difference between the thermal expansion of gases and that of solids and liquids: the percentage change in volume per °C is almost exactly the same for *all* gases, although this percentage change may be very different for different solids and liquids.

Suppose the volume of a certain sample is V_0 at 0°C and some fixed pressure. If the sample is cooled from some initial temperature (not necessarily 0°C), its volume decreases by 1/273 of V_0 for each 1°C the temperature falls. If the gas sample is heated, its volume increases by 1/273 of V_0 for each 1°C the temperature rises. Since 1/273 = 0.0037, a balloon filled with air whose volume is 1.000 m³ at 0°C has a volume of 1.037 m³ at 10°C and a volume of 0.963 m³ at − 10°C. **(11)**

What happens when the balloon is cooled to − 273°C? At that temperature the air in the balloon should have lost 273/273 of its volume

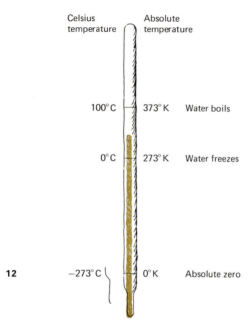

Celsius temperature	Absolute temperature	
100°C	373°K	Water boils
0°C	273°K	Water freezes
−273°C	0°K	Absolute zero

12

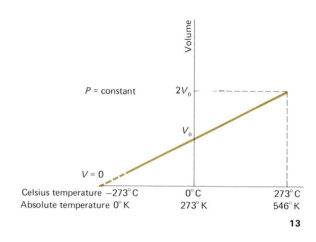

P = constant

Volume

$2V_0$

V_0

$V = 0$

Celsius temperature −273°C	0°C	273°C
Absolute temperature 0°K	273°K	546°K

13

at 0°C, and therefore should have vanished entirely! Actually, all gases condense into liquids at temperatures above − 273°C, so the question has no physical meaning. But − 273°C is still a significant temperature. Let us set up a new temperature scale, the *absolute temperature scale*, and designate − 273°C as the zero point. Temperatures in the absolute scale are denoted °K, after Lord Kelvin (1824–1907), a noted British physicist. To convert temperatures from one scale to the other we note that

$$°K = °C + 273° \quad \textbf{(12)}$$

Absolute temperature

CHARLES'S LAW

The reason for setting up the absolute temperature scale is that, provided the pressure is constant, *the volume of a gas sample is directly proportional to its absolute temperature*. This relationship is called *Charles's law*.

We can express Charles's law in the form

$$\frac{V}{T} = \text{constant}, \qquad (P = \text{constant})$$

or alternatively as

$$\frac{V_1}{T_1} = \frac{V_2}{T_2}. \qquad (P = \text{constant})$$

In the latter equation, V_1 is the volume of a gas sample at the absolute temperature T_1 and V_2 is its volume at the absolute temperature T_2.

If there were a gas that did not liquify before reaching 0°K, then at 0°K its volume would shrink to zero. Since a negative volume has no meaning, it is natural to think of 0°K as *absolute zero*. Actually, 0°K is indeed the lower limit to temperatures capable of being attained, but on the basis of a stronger argument than one based on imaginary gases. This argument is discussed in the next section. **(13)**

160   The ideal gas

IDEAL GAS LAW

Boyle's law and Charles's law can be combined in a single formula called the *ideal gas law*:

$$\frac{P_1 V_1}{T_1} = \frac{P_2 V_2}{T_2}. \quad \text{Ideal gas law}$$

When $T_1 = T_2$, the ideal gas law becomes Boyle's law,

$$P_1 V_1 = P_2 V_2, \quad (T = \text{constant})$$

and when $P_1 = P_2$ it becomes Charles's law,

$$\frac{V_1}{T_1} = \frac{V_2}{T_2}. \quad (P = \text{constant})$$

Another way to express the ideal gas law is

$$\frac{PV}{T} = \text{constant}.$$

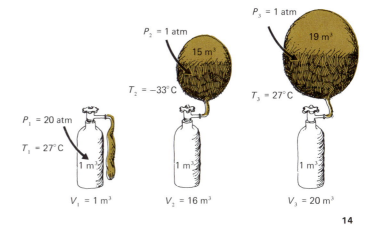

14

The ideal gas law is obeyed approximately by all gases. The significant thing is not that the agreement with experiment is never perfect, but that *all* gases, of whatever kind, behave almost identically. An *ideal gas* is defined as one that obeys $PV/T = \text{constant}$ exactly. While no ideal gases actually exist, they do provide a target for theories of the gaseous state to aim at. It is reasonable to suppose that the ideal gas law is a consequence of the essential nature of gases. Hence the next step is to account for this law and only afterward to seek reasons for its failure to be completely correct.

Problem. (a) A tank with a capacity of 1 m³ contains helium gas at 27°C under a pressure of 20 atm. The helium is used to fill a balloon. When the balloon is filled, the gas pressure inside it is 1 atm, and its temperature has dropped to − 33°C. (The gas has done work in expanding at the expense of its internal energy, and the cooling reflects this loss of internal energy.) What is the volume of the balloon at this time? (b) After a while the helium in the balloon absorbs heat from the atmosphere and returns to its original temperature of 27°C, and it expands further to maintain its pressure at 1 atm. What is the final volume of the balloon?

Solution. (a) The equivalents of 27°C and − 33°C on the absolute scale are 300°K and 240°K respectively. Applying the ideal gas law to the initial expansion, we obtain

$$V_2 = \frac{T_2}{T_1} \frac{P_1}{P_2} V_1$$

$$= \frac{240°\text{K}}{300°\text{K}} \times \frac{20 \text{ atm}}{1 \text{ atm}} \times 1 \text{ m}^3$$

$$= 16 \text{ m}^3.$$

Because the tank's capacity is 1 m², the balloon's volume after the initial expansion is 15 m³. **(14)**

(b) When the helium has reached the outside air temperature of 27°C, which we shall call state 3, then $T_1 = T_3$. Hence we need only apply Boyle's law to states 1 and 3 to obtain the eventual volume of the helium:

$$V_3 = \frac{P_1}{P_3} V_1$$

$$= \frac{20 \text{ atm}}{1 \text{ atm}} \times 1 \text{ m}^3$$

$$= 20 \text{ m}^3.$$

Again we subtract the 1 m³ volume of the tank to find the volume of the balloon itself, which is 19 m³.

EXERCISES

1. A little water is boiled for a few minutes in a tin can, and the can is sealed while it is still hot. Why does the can collapse as it cools?

2. A 50-kg woman balances on the heel of her left shoe, which is 1 cm in diameter. What pressure (in atm) does she exert on the ground?

3. The force on a phonograph needle whose point is 0.1 mm in radius is 0.2 N. What is the pressure it exerts on the record (in atm)?

4. A cork 2 cm in radius is used to close one end of a tube whose other end is connected to a vacuum pump. The pump removes virtually all the air from the tube. How much force would be needed to pull the cork out? [Note: The area of a circle of radius r is πr^2.]

5. Starting from the ideal gas law, obtain an equation relating the pressure and temperature of a gas at constant volume.

6. A sample of gas occupies 100 cm³ at 0°C and 1 atm pressure. What is its volume
   a) at 50°C and 1 atm pressure?
   b) at 0°C and 2.2 atm pressure?
   c) at 50°C and 2.2 atm pressure?

7. A sample of gas occupies 2m³ at 300°K and an absolute pressure of 2×10^5 N/m².
   a) What is its pressure at the same temperature when it has been compressed to a volume of 1 m³?
   b) What is its volume at the same temperature when its pressure has been decreased to 1.5×10^5 N/m²?
   c) What is its volume at a temperature of 400°K and a pressure of 2×10^5 N/m²?

8. A sample of gas occupies a volume of 0.9 m³ at a temperature of 300°K and 1 atm pressure. What is its volume at 400°K and 1 atm pressure?

9. The tires of a parked car on a winter day contain air at a gauge pressure of 1.5 atm at 0°C. After the car has been driven for a while, the air in the tires reaches a temperature of 27°C. If the tires remain unchanged in volume, what is the new pressure of the air they contain?

10. A diver blows an air bubble 1 cm in diameter at a depth of 10 m in a fresh-water lake where the temperature is 5°C. What is the diameter of the bubble when it reaches the surface of the lake where the temperature is 20°C? [Note: The volume of a sphere of radius r is $\frac{4}{3}\pi r^3$, but for this problem it is not necessary to actually calculate the volume of the bubble.]

19

KINETIC THEORY OF MATTER

According to the kinetic theory of gases, a gas consists of a great many randomly moving molecules that interact with one another only during collisions. The impacts of these molecules are responsible for the pressure exerted by a gas. Extensions of this theory to liquids and solids offer valuable insights into how their properties originate.

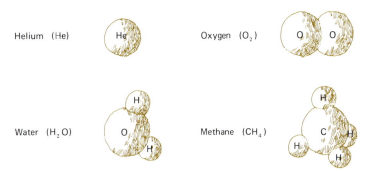

Helium (He) Oxygen (O_2)

Water (H_2O) Methane (CH_4)

1

ATOMS AND MOLECULES

The idea that matter is not infinitely divisible, that all substances are composed of characteristic individual particles, is an ancient one. The ultimate particles of many substances are *molecules*. Although molecules may be further broken down, when this happens they no longer are representative of the original substance. The molecules of a compound consist of the *atoms* of its constituent elements joined together in a definite ratio. Thus each molecule of water contains two hydrogen atoms and one oxygen atom. While the ultimate particles of elements are atoms, many elemental gases consist of molecules rather than atoms. Oxygen molecules, for instance, contain two oxygen atoms each. The molecules of other gases, such as helium and argon, are single atoms. (**1**)

"By convention sweet is sweet, by convention bitter is bitter, by convention hot is hot, by convention cold is cold, by convention color is color. But in reality there are atoms and the void. That is, the objects of sense are supposed to be real and it is customary to regard them as such, but in truth they are not. Only the atoms and the void are real." Democritus (Fifth Century B.C.)

The masses of atoms and molecules are usually expressed in *atomic mass units* (amu) whose magnitude is such that the most abundant type of carbon atom has, by definition, a mass of precisely 12.00000 amu. (The existence of atoms of the same element with different masses is discussed in Section 38; a method for measuring atomic masses is described in Section 26.) The actual mass of the reference carbon atom is 1.992×10^{-26} kg, and so

$$1 \text{ amu} = \frac{1.992 \times 10^{-26} \text{ kg}}{12.00}$$

$$= 1.660 \times 10^{-27} \text{ kg}.$$

The table on page 400 contains a list of the average atomic masses of the elements in amu. We note that ordinary carbon, some of whose atoms are heavier than the most abundant kind, has an average atomic mass of 12.01.

The mass of a molecule is the total of the masses of the atoms it contains. Thus the molecular mass of gaseous oxygen is 32.00 amu, since its formula is O_2 and the atomic mass of oxygen is 16.00 amu.

KINETIC THEORY OF MATTER

According to the assumptions of the kinetic theory of gases, which is part of the kinetic theory of matter, a gas consists of a great many tiny individual molecules that do not interact with one another except when collisions occur. The molecules are supposed to be far apart compared with their dimensions and to be in constant motion, incessantly hurtling to and fro and being kept from escaping by the solid walls of a container or, in the case of the earth's atmosphere, by the pull of gravity.

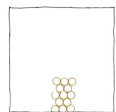

2

A natural consequence of the random motion and large molecular separation is the tendency of a gas to completely fill its container and to be readily compressed or expanded. (**2a**)

In a solid, on the other hand, the molecules are close together and mutual attractive and repulsive forces hold them in place to provide the solid with its characteristic rigidity. (**2b**)

In a liquid the intermolecular forces are sufficient to keep the volume of the liquid constant. However, they are not strong enough to prevent adjacent molecules from sliding past one another, which results in the ability of liquids to flow. (**3**)

3

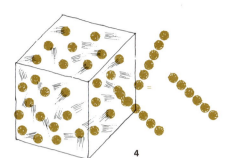

4

5

KINETIC THEORY
OF GASES

Let us see how the kinetic theory of gases accounts for Boyle's and Charles's laws. At first glance it may not seem that there is any straightforward connection between them. But it is also clear that if the kinetic theory is to have any meaning at all, it must yield the same behavior we obtain from experiment. The laws of mechanics bridge the gap between the microscopic picture of a gas as an aggregate of molecules in random motion and the macroscopic picture of a gas as a continuous fluid with certain physical properties. In this way a theory based upon a model may be compared with the data obtained by direct observation.

The pressure a gas exerts originates in the impacts of its molecules, which are in constant, random motion. The vast number of molecules in even a tiny gas sample means that the separate blows of the molecules appear as a continuous force to our senses and measuring instruments. (4)

6

Here is a simplified model of a gas confined to a box. Although the molecules are actually traveling about in all directions, the effects of their collisions with the walls of the box are the same as if one-third of them were moving back and forth between each pair of opposite walls. (5)

When a box containing a gas is doubled in volume, those molecules moving up and down have twice as far to go between impacts. Since their speed is unchanged, the time between impacts is also doubled, and the pressure they exert on the top and bottom of the box falls to half its original value. (6)

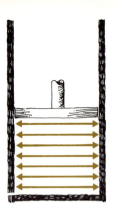
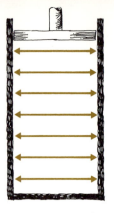

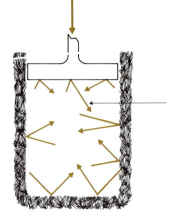

7

8

The expansion of the box means that the molecules moving horizontally are now spread over twice their former area, and the pressure on the sides of the box accordingly falls to half its original value also. Thus doubling the volume means halving the pressure, which is Boyle's law. Like reasoning accounts for a rise in pressure when the volume is reduced. (**7**)

Charles's law follows from the kinetic theory of gases if it is assumed that:

The average kinetic energy of the molecules of a gas is proportional to the absolute temperature of the gas.

This assumption is reasonable since we observe that compressing a gas quickly (so no heat can enter or leave the container) raises its temperature, and such a compression must increase the average energy of the molecules because they bounce off the inward-moving piston more rapidly than they approach it. A familiar example of the latter effect is a baseball rebounding with greater speed when struck by a bat. (**8**)

On the other hand, expanding a gas lowers its temperature, and such an expansion reduces molecular energies since molecules lose speed in bouncing off an outward-moving piston. The association between molecular energy and temperature is thus in accord with experience. (**9**)

Molecules rebounding from inward-moving piston move faster than before.

Molecules rebounding from fixed wall have unchanged speeds.

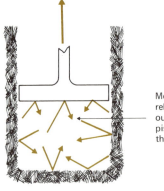

9

Molecules rebounding from outward-moving piston move slower than before.

Molecules rebounding from fixed wall have unchanged speeds.

The interpretation of absolute zero in terms of the elementary kinetic theory of gases is a simple one: it is that temperature at which all molecular translational movement ceases. (**10**)

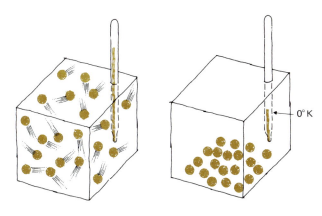

MOLECULAR VELOCITIES

The relationship between the average molecular kinetic energy $\overline{KE}$ in a gas and the absolute temperature T is found to be

Average molecular energy $= \overline{KE} = \frac{3}{2}kT$,

where k, known as Boltzmann's constant, has the value

$$k = 1.38 \times 10^{-23} \; \frac{J}{°K}.$$

This formula holds for all gases regardless of the masses of their molecules.

Different molecules have different kinetic energies, and the $\overline{KE}$ of any specific molecule changes frequently owing to collisions with other molecules. However, the *average* $\overline{KE}$ of the molecules remains the same at a given temperature.

Since the kinetic energy of an object of mass m and velocity v is $\frac{1}{2}mv^2$, in the case of a gas molecule we have for its average velocity at the absolute temperature T

$$\overline{\tfrac{1}{2}mv^2} = \tfrac{3}{2}kT,$$

$$\bar{v} = \sqrt{\frac{3kT}{m}}.$$

Problem. What is the average velocity of oxygen molecules at 0°C?

Solution. As we know, the molecular mass of O_2 is 32.00 amu. Hence an oxygen molecule has a mass in kg of

$$m = 32.00 \text{ amu} \times 1.660 \times 10^{-27} \; \frac{\text{kg}}{\text{amu}}$$

$$= 5.31 \times 10^{-26} \text{ kg}.$$

At an absolute temperature of 273°K (corresponding to 0°C), the average speed of oxygen molecules is therefore

$$v = \sqrt{\frac{3 \times 1.38 \times 10^{-23} \text{ J/°K} \times 273°K}{5.31 \times 10^{-26} \text{ kg}}}$$

$$= 4.61 \times 10^2 \text{ m/s}.$$

which is a little over 1000 mi/hr! Evidently molecular speeds are very large compared with those of the macroscopic bodies familiar to us.

It is important to keep in mind that actual molecular velocities vary considerably on either side of $\bar{v}$. The graph, page 168, shows the distribution of molecular velocities in oxygen at 273°K and in hydrogen at 273°K. The mass of an O_2 molecule is 16 times that of an H_2 molecule. Average molecular velocity de-

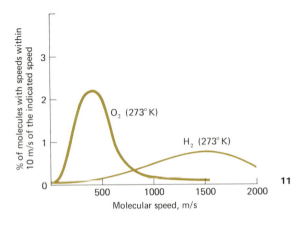

11

creases with molecular mass, hence at the same temperature molecular velocities in hydrogen are on the average greater than in oxygen. At the same temperature the average molecular *energy* is the same for all gases, however. (**11**)

Here we see the distributions of molecular velocities in oxygen at 73°K and at 273°K. The average molecular velocity increases with temperature, as predicted. (**12**)

12

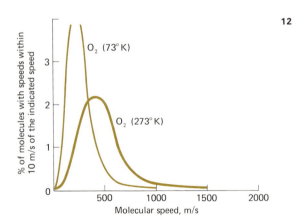

INTERNAL ENERGY OF A GAS

When a certain amount of energy is added to a gas whose molecules are single atoms, all the energy goes into increasing the average velocities of the molecules and is reflected in a rise in the temperature of the gas. (**13**)

However, when this same amount of energy is added to a gas whose molecules are more complex, with two or more atoms per molecule, the temperature rise is *smaller* (for the same number of gas molecules) than it is for the monatomic gas. The reason is that some of the energy input goes into vibrations and rotations of the molecules, whereas only the kinetic energy of molecular translational motion affects the temperature of a gas. Thus the temperature of a gas does not tell us how much internal energy the gas possesses, but only how much kinetic energy of translation its molecules have on the average. (**14**)

MOLECULAR MOTION IN LIQUIDS

The elementary kinetic theory of matter is not as successful when applied to the liquid and solid states as it is when applied to gases; the classical mechanics that gas molecules obey in their translational motion is not adequate to describe the behavior of the molecules in liquids and solids. However, the concept that the internal energy of a substance resides at least in part in the kinetic energies of its molecules helps in understanding a variety of phenomena characteristic of liquids and solids.

The random motion of water molecules led to an important event in the history of science. In 1827 the British botanist Robert Brown noticed that pollen grains in water are in continual, agitated movement. Similar *Brownian motion* is apparent whenever very

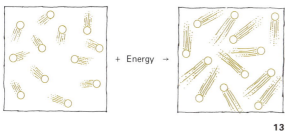

+ Energy →

13

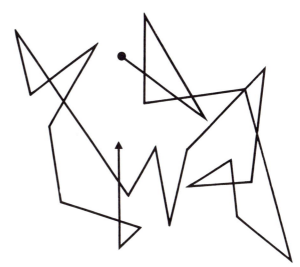

15

Translational motion of diatomic molecule

Vibrational motion of diatomic molecule

Rotational motion of diatomic molecule

14

small particles are suspended in a fluid medium, for example smoke particles in air. (**15**)

According to kinetic theory, Brownian motion originates in the bombardment of the particles by molecules of the fluid. This bombardment is completely random, with successive molecular impacts coming from different directions and contributing different impulses to the particles. Albert Einstein, in 1905, found that he could account for Brownian motion quantitatively by assuming that, as a result of continual collisions with fluid molecules, the particles themselves have the same average kinetic energy as the molecules. Surprising as it may seem, this was the first direct verification of the reality of molecules, and it convinced many distinguished scientists who had previously been reluctant to believe that such things actually exist.

A further kinetic–molecular phenomenon characteristic of the liquid state is evaporation. A dish of water well below its boiling point of 100 °C will nevertheless gradually turn into vapor, growing colder as it does so. The faster the evaporation, the more pronounced the cooling effect; alcohol and ether

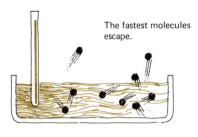

The fastest molecules escape.

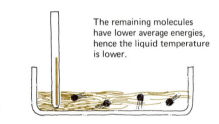

16

The remaining molecules have lower average energies, hence the liquid temperature is lower.

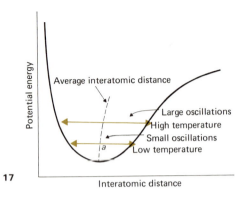

17

chill the skin upon contact because of their extreme volatility. This behavior follows from the variation of molecular speeds in a liquid. The fastest molecules have enough energy to escape through the liquid surface despite the attractive forces of the other molecules. The molecules left behind redistribute the available energy in collisions among themselves, but, because the most energetic ones escape, the average energy that remains is less than before and the liquid is now at a lower temperature. (16)

THERMAL EXPANSION IN SOLIDS

Thermal expansion in a solid also has a straightforward explanation in terms of kinetic theory. Most solids are crystalline in nature, which means that the various atoms that compose them form a regular arrangement in space. (In some crystalline solids the basic constituents are whole molecules, rather than individual atoms, but we shall refer to them as atoms in general here for convenience.) The atoms behave as though they are joined together by tiny springs and constantly oscillate about their equilibrium positions.

The graph shows how the atomic potential energy of a solid varies with interatomic spacing. The normal spacing a corresponds to the lower portion of the curve, where the energy per atom is least. The extent of the vibrations is determined by the width of the curve: when the atomic separation is a minimum or a maximum, the energy of a pair of adjacent atoms is wholly potential, as in the case of a pendulum at each end of its path, while in the middle their energy is wholly kinetic. The average atomic energy is what determines a and hence the dimensions of the solid. (17)

When additional energy of thermal origin is added to a solid, the atomic spacing alternates through a wider range than before. If the potential energy curve were symmetrical about the distance a, no change in the dimensions of the solid would occur, since a would always be halfway between the two parts of the curve. However, the attractive and repulsive forces between atoms vary with distance in different ways, with the repulsive force increasing more rapidly as the atoms move closer together than the attractive force increases as the atoms move farther apart. Consequently the average interatomic spacing becomes larger as the thermal energy of the solid increases and the ranges of the atomic vibrations increase.

EXERCISES

1. At absolute zero, an ideal gas sample would occupy zero volume. Why would an actual gas not occupy zero volume at absolute zero?

2. The temperature of a gas sample is held constant while its volume is decreased. Does the pressure the gas exerts on the walls of its container increase because the molecules strike each square meter of the walls more often, at higher speeds, or with greater force?

3. According to the kinetic theory of matter, molecular motion ceases only at absolute zero. How can this be reconciled with the definite shape and volume of a solid at temperatures well above absolute zero?

4. A sample of nitrogen is compressed to half its original volume while its temperature is kept constant. What happens to the speeds of the nitrogen molecules?

5. Molecular speeds are comparable with those of rifle bullets, yet a gas with a strong odor, such as ammonia, takes a few minutes to diffuse through a room. Why?

6. A bottle contains 1 kg of chlorine gas at $0°C$ and a pressure of 1 atm. When another kg of chlorine is added to the bottle at the same temperature, the pressure rises to 2 atm. Why?

7. One of the assumptions of the kinetic theory is that the average distance between molecules is much greater than the dimensions of the molecules themselves. Oxygen and nitrogen molecules are roughly 4×10^{-10} m in diameter, and there are 2.7×10^{25} molecules in a cubic meter of air at room temperature and atmospheric pressure.

a) On the average, how far apart are the molecules in air?
b) How many molecular diameters is their average separation?

8. The average speed of air molecules is roughly 4×10^2 m/s, and the average distance an air molecule goes between collisions with other molecules is about 10^{-7} m. What is the average number of collisions an air molecule makes per second?

9. Actual molecules attract one another slightly. Does this tend to increase or decrease gas pressures from values calculated from the kinetic theory of gases?

10. The average speed of a hydrogen molecule at room temperature and atmospheric pressure is 1 mi/s. What happens to the average speed of such a molecule if the temperature remains constant but the pressure is doubled?

11. The mass of a nitrogen molecule is seven times greater than that of a hydrogen molecule. What is the temperature of a sample of hydrogen whose average molecular speed is equal to that in a sample of nitrogen at $300°K$?

12. The minimum speed a body must have if it is to leave the earth permanently is 11,200 m/s. Why does hydrogen escape from the atmosphere more readily than oxygen?

13. Which of these gases—CO_2, UF_6, H_2, He, Xe, NH_3—has
a) the highest average molecular speed at a given temperature?
b) the lowest average molecular speed?

14. Two vessels of the same size are at the same temperature. One of them holds 1 kg of H_2 gas and the other holds 1 kg of N_2 gas.
a) Which vessel contains more molecules?
b) Which vessel is under the greater pressure?
c) In which vessel is the average molecular speed greater?

15. To what temperature must a gas sample initially at $27°C$ be raised in order for the average energy of its molecules to double? for their average speed to double?

20

THERMODYNAMICS

"I perceived that, to make the best use of steam, it was necessary, first, that the cylinder should be maintained always as hot as the steam which entered it; and secondly, that when the steam was condensed, the water of which it was composed, and the injection itself, should be cooled down to 100°, or lower where that was possible." James Watt (1736–1819)

The province of thermodynamics is the conversion of heat into mechanical energy. The first law of thermodynamics is simply conservation of energy: the output of energy cannot exceed the input of heat. The second law expresses the fact that the conversion cannot ever be completely efficient—some heat must be wasted in the process. This law is a consequence of the tendency of all physical systems to become more and more disordered as time goes on.

HEAT ENGINES

Thermodynamics has as its basic concern the transformation of heat into mechanical energy. A device or system that converts heat into mechanical energy is called a *heat engine*, and the principles that govern its operation are the same whether it is an automobile engine whose heat source is the burning of gasoline, a steam turbine whose heat source is a nuclear reactor, the earth's atmosphere whose heat source is the sun, or a human being whose heat source is the food he eats.

Heat is the easiest and cheapest form of energy to obtain, since all we need do to liberate it is to burn a fuel such as wood, coal, or oil. The real problem is to turn heat into mechanical energy so it can power cars, ships, airplanes, electric generators, and machines of all kinds. To appreciate the problem, we recall that heat consists of the kinetic energies of moving atoms and molecules. In order to change heat into a more usable form, we must extract some of the energy of the random motions of atoms and molecules and convert it into regular motions of a piston or a wheel. Such conversions cannot take place efficiently, for the same reason that it is easier to shatter a wineglass than to reassemble the fragments: the natural tendency of all physical systems is toward increasing disorder. The second law of thermodynamics is an expression of this tendency, whose role in the evolution of the universe is quite as central as are those of the various conservation principles.

FIRST LAW OF THERMODYNAMICS

Three characteristic processes take place in all heat engines:

1. Heat is absorbed from a source at a high temperature;
2. Mechanical work is done;
3. Heat is given off at a lower temperature.

Different heat engines carry out these processes in different ways, but the general pattern of operation is always the same. (1)

There are two broad principles that have been found to apply to all heat engines. The *first law of thermodynamics* expresses the conservation of energy:

Energy cannot be created or destroyed, but may be converted from one form to another.

In terms of a heat engine, this law states that

Net heat input = work output + change in internal energy of engine.

If the engine operates in a cycle, energy may be alternately stored and released from storage, but the engine does not experience a net change in its internal energy. In this case,

Net heat input = work output.

The net heat input equals the amount of heat the engine takes in from a reservoir at high temperature minus the amount of heat it exhausts to a reservoir at low temperature. In a steam engine the high-temperature reservoir is the boiler, and the low-tem-

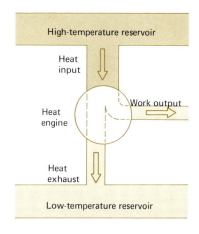

perature reservoir is the escaping steam; in a gasoline engine the high-temperature reservoir is the exploding mixture of air and gasoline vapor in each cylinder, and the low-temperature reservoir is the exhaust gas; in the earth's atmosphere the ultimate high-temperature reservoir is the sun, and the ultimate low-temperature reservoir is the rest of the universe.

SECOND LAW OF THERMODYNAMICS

The first law of thermodynamics prohibits an engine from operating without a source of energy, but it does not tell us anything about the character of possible sources of energy. For instance, there is an immense amount of internal energy in the atmosphere and the

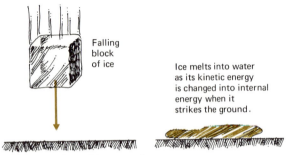

Falling block of ice

Ice melts into water as its kinetic energy is changed into internal energy when it strikes the ground.

A possible process that conserves energy

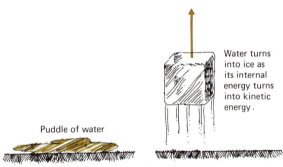

Puddle of water

Water turns into ice as its internal energy turns into kinetic energy.

An impossible process that conserves energy

oceans, yet we know that more work must be done to extract this energy than can be performed with its help. Or, to give an extreme case, it is energetically possible for a puddle of water to rise spontaneously into the air, cooling and freezing into ice as internal energy changes into potential energy. After all, a block of ice dropped from a sufficient height melts when it strikes the ground, its initial potential energy first being converted to kinetic energy and then into heat. Needless to say, water does not rise upward of its own accord, and we must find an appropriate way of expressing this conclusion. (2)

The *second law of thermodynamics* is the physical principle, independent of the first law and not derivable from it, that supplements the first law in limiting our choice of heat sources for our engines. It can be stated in a number of equivalent ways, a common one being as follows:

It is impossible to construct an engine, operating in a cycle (that is, continuously), which does nothing other than take heat from a source and perform an equivalent amount of work.

According to the second law of thermodynamics, then, no engine can be completely efficient—some of its heat input *must* be ejected. The greatest efficiency any heat engine is capable of depends upon the temperatures of its heat source and of the reservoir to which it exhausts heat: the greater the difference between these temperatures, the more efficient the engine. The second law is a consequence of the empirical fact we have already noted:

The natural direction of heat flow is from a reservoir of heat at a high temperature to a reservoir of heat at a low temperature, regardless of the total internal energy of each reservoir.

The latter statement, in fact, may be regarded as an alternative expression of the second law. (3)

If we are to utilize the internal energy of the atmosphere or the oceans, we must first provide a reservoir at a lower temperature than theirs in order to extract heat from them. There is no reservoir in nature suitable for this purpose, for if there were heat would flow into it until its temperature reached that of its surroundings. To establish a low-temperature reservoir, we must employ a refrigerator (which is a heat engine running in reverse by using up energy to extract heat),

Heat flows naturally
from a hot reservoir
to a cold one.

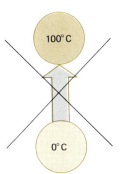

Heat cannot flow
by itself from a
cold reservoir to
a hot one.

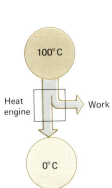

Some of the heat
flow can be con-
verted to work by
an engine.

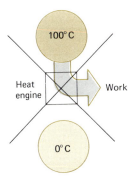

All of the heat that
leaves a reservoir
cannot be converted
into work; some of
the heat must flow
into a cold reservoir.

3

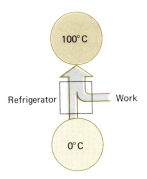

4

and in so doing we will perform more work than we can successfully obtain from the internal energy of the atmosphere or oceans. (4)

The laws of thermodynamics can be summarized by saying that the first law prohibits us from getting something for nothing, while the second law prohibits us from doing as well as breaking even.

ENGINE EFFICIENCY

All heat engines behave in the same way: they absorb heat at a high temperature, convert some of it into work, and exhaust the rest at a low temperature. We may therefore simplify the task of analyzing their principles of operation by imagining an ideal heat engine free of the complications involved in actual engines, for instance friction, but which naturally obeys all physical laws.

The efficiency of any engine is

$$\text{Eff} = \frac{\text{work output}}{\text{energy input}}.$$

A detailed calculation shows that the efficiency of an ideal heat engine depends *only* on the temperatures at which it absorbs and ex-

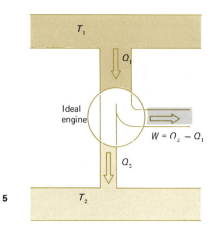

5

hausts heat. If the engine takes in heat at the absolute temperature T_1 and ejects heat at the absolute temperature T_2, then its efficiency is

$$\text{Eff} = 1 - \frac{T_2}{T_1}. \quad (5)$$

The smaller the ratio between the absolute temperatures T_2 and T_1, the more efficient the engine. No engine can be 100% efficient, because no reservoir can have an absolute temperature of $0°K$. (Even if such a reservoir could somehow be created, the exhaust of heat to it by the engine would raise its temperature above $0°K$ at once.) (6)

Problem. Steam enters the most modern steam turbines at a temperature of $565°C$ and emerges into a partial vacuum at a temperature of $93°C$. What is the upper limit to the efficiency of such an engine?

Solution. The equivalents of $565°C$ and $93°C$ in the absolute temperature scale are $838°K$ and $366°K$ respectively. The efficiency of an

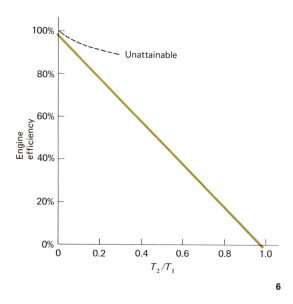

6

ideal engine operating between these two absolute temperatures is

$$\text{Eff} = 1 - \frac{T_2}{T_1}$$

$$= 1 - \frac{366°K}{838°K}$$

$$= 1 - 0.44 = 0.56$$

which is 56%. The efficiency of an ideal engine is the maximum possible for an engine operating between a given pair of temperatures. An actual steam turbine operating between 565°C and 93°C would have an efficiency of less than 40% because of the inevitable presence of friction and heat losses to the atmosphere.

THE REFRIGERATOR

A *refrigerator* is a heat engine operating in reverse, as mentioned earlier. A heat engine absorbs heat from a high-temperature reservoir and exhausts it to a low-temperature one, with an output of mechanical work being provided in the process. In a refrigerator, on the other hand, heat is transferred from a low-temperature reservoir (typically a storage chamber) to a high-temperature one (the outside world), and mechanical energy must be supplied in order to do this. In effect, heat must be "pushed uphill" if it is to go from a cold region to a warm one. It is important to keep in mind that a refrigerator does not "produce cold," since cold is a relative deficiency of internal energy and not something in its own right, as heat is. What a refrigerator does is to remove internal energy from a specific region and transport it elsewhere. (7)

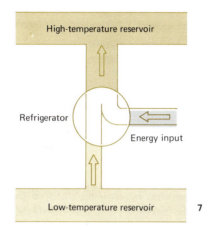

High-temperature reservoir

Refrigerator

Energy input

Low-temperature reservoir   7

STATISTICAL MECHANICS

According to the second law of thermodynamics, it is impossible to convert heat into any other form of energy efficiently. Some of the heat input to an engine *must* be lost. Why? The reason lies in the nature of heat, which is molecular kinetic energy; the temperature of a body is a measure of the average kinetic energy of each of its constituent molecules.

Let us see how the microscopic picture of matter as consisting of molecules in motion leads to the second law of thermodynamics. Although the argument will be based on molecules in a gas, the essential ideas hold for matter in any state.

The molecules of a gas are in constant random motion and undergo frequent collisions with

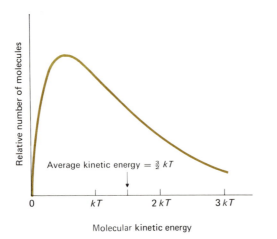

Average kinetic energy $= \frac{3}{2}kT$

Relative number of molecules

0   kT $2\,kT$ $3\,kT$

Molecular kinetic energy

8

9

1 cm³

one another. While we cannot hope to follow an individual gas molecule in its vicissitudes, it is possible to predict on the basis of statistical arguments what fraction of the time it will have any specified amount of kinetic energy. Hence we can calculate the distribution of molecular energies in a gas sample at a particular temperature. This distribution, which has been confirmed by experiment, has the form shown and holds for all equilibrium conditions in which each molecule has the same average energy over a period of time. A molecule that moves more swiftly than usual at one instant will move less swiftly at a later instant after a number of collisions have taken place. **(8)**

An equilibrium condition is the most probable one according to *statistical mechanics*, a branch of physics which mathematically deduces the behavior of assemblies of so many particles that deviations from statistically probable behavior are not significant. If we toss a coin a dozen times, it is unlikely that heads and tails will come up equally often, but if we toss it a million times, the percentage

deviation from an equal number of heads and tails will be minute. To appreciate why departures from the equilibrium distribution of molecular energies for more than the briefest instant are so unlikely, let us look at a cubic centimeter—a thimbleful—of air at atmospheric pressure and room temperature. There are 2.7×10^{19} molecules in the cubic centimeter, and each molecule undergoes an average of 4×10^9 collisions per second (equivalent to every person in the world colliding with every other person, one at a time, in each second). **(9)**

Let us consider a heat reservoir at a high temperature and a heat reservoir at a low temperature. The molecules of each are in equilibrium and have the molecular energy distributions shown in (a) and (b). If we consider the two reservoirs as a single system, the molecular energy distribution in the system is like that of (c). This distribution is, in a statistical sense, very improbable; if they were mixed together, the molecules of the two reservoirs would soon blend their energies in collisions to attain the equilibrium distri-

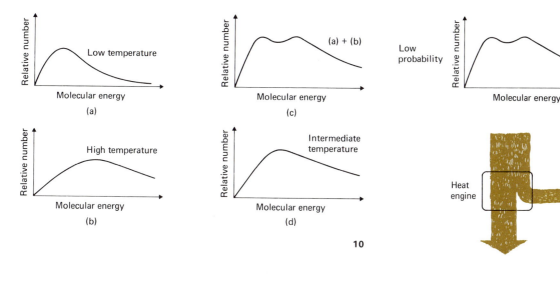

(a) Low temperature — Relative number vs Molecular energy

(b) High temperature — Relative number vs Molecular energy

(c) (a) + (b) — Relative number vs Molecular energy

(d) Intermediate temperature — Relative number vs Molecular energy

10

Low probability — Relative number vs Molecular energy

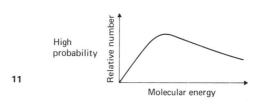

Heat engine → Work

High probability — Relative number vs Molecular energy

11

bution of (d), which corresponds to a temperature intermediate between the initial ones of the two reservoirs. (10)

We note the important fact that the total energy contents of the distributions of (c) and (d) are identical; the only distinction between them is the manner in which the energy is allotted to the molecules on the average. However, it is just this distinction with which the second law of thermodynamics is concerned, because this law states that a system of two heat reservoirs at different temperatures can be made to yield a net work output, while a single heat reservoir, no matter how much energy it contains, cannot be made to perform any net work. A system of molecules whose energies are distributed in the most probable way is "dead" thermodynamically, while a system having a different distribution of molecular energies is capable of doing mechanical work as it progresses to an equilibrium state. (11)

The universe may be thought of as a single system of molecules, and its evolution is powered by the flow of energy from high-temperature reservoirs (the stars) to low-temperature reservoirs (everything else). Ultimately the entire universe will be at the same temperature and all its constituent particles will have the same average energy, a condition sometimes called the "heat death" of the universe.

TIME AND THE SECOND LAW

The second law of thermodynamics is evidently an unusual kind of physical principle. It does not apply to the interactions of individual particles, only to trends in the evolution of assemblies of many particles. In fact, the second law is hardly a basic principle in the usual sense, since it is the result of combining the laws of mechanics with the theory of probability, and cannot be used to predict anything specific.

However, the second law has the unique property of being correlated with the direction of time. Events that involve individual particles are always reversible—the same laws of motion apply to the billiard balls seen in the film of a game whether the film is run forward or backward. But events that involve systems of large numbers of particles are not always reversible—the film of an egg being cracked makes no sense at all when run backward. The transformation of a broken egg into a whole egg is not totally impossible, but it *is* exceedingly unlikely.

The second law of thermodynamics is thus a statement of probability: as time goes on, order becomes disorder. The sequence can be reversed in parts of the universe now and then—after all, heat engines do turn heat into work—but in the universe as a whole, increasing disorder is inevitable.

EXERCISES

1. Is the first law of thermodynamics a special case of the second, just as the first law of motion is a special case of the second?

2. An attempt is made to cool a kitchen during the summer by leaving the refrigerator door open and closing the door and windows. What is the result?

3. Does a refrigerator exhaust more, less, or the same amount of heat as it absorbs from its contents?

4. The sun's corona is a very dilute gas at a temperature of about 10^6 °K that is believed to extend into interplanetary space well past the earth's orbit. Why can the corona not be used as the high-temperature reservoir of a heat engine in an earth satellite?

5. An engine operating between 300°C and 60°C is 15% efficient. What would its efficiency be if it were an ideal engine?

6. An ideal engine takes in 10^3 kcal of heat from a reservoir at 327°C and exhausts heat to a reservoir at 127°C. How much work does it do?

7. The total drop of the Wollomombi Falls in Australia is 482 m. What would be the efficiency of an ideal engine operating between the top and bottom of the falls if the water temperature at the top were 10°C and all the potential energy of the water at the top were converted to heat at the bottom?

8. Three designs for a heat engine to operate between 450°K and 300°K are proposed. Design A is claimed to require a heat input of 0.2 kcal for each 1000 J of work output, design B a heat input of 0.6 kcal, and design C a heat input of 0.8 kcal. Which design would you choose and why?

PART 4

ELECTROMAGNETISM

21

ELECTRIC CHARGE

"There are two distinct electricities, very different from each other: one of these I call vitreous electricity; the other resinous electricity . . . The characteristic of these two electricities is that a body of, say, the vitreous electricity repels all such as are of the same electricity; and on the contrary, attracts all those of the resinous electricity." Charles Du Fay (1698–1739)

"Electricity is of two kinds, positive and negative. The difference is, I presume, that one comes a little more expensive, but is more durable; the other is a cheaper thing, but the moths get into it." Stephen Leacock (1869–1944)

Electricity is familiar to all of us as the name for that which causes our light bulbs to glow, many of our motors to turn, our telephones and radios to communicate sounds, our television screens to communicate pictures. But there is more to electricity than its technological uses. Electrical forces bind electrons to nuclei to form atoms, and they hold atoms together to form molecules, solids, and liquids. All of the chief properties of matter in bulk—with the notable exception of mass—can be traced to the electrical nature of its constituent particles.

ELECTRICITY

The success of the laws of motion, of the law of gravitation, and of the kinetic-molecular theory of matter might tempt us into thinking that we now have, at least in outline, a complete picture of the workings of the physical universe. To dispel this notion all we need do is perform a simple experiment: on a dry day, we run a hard rubber comb through our hair, and find that the comb is now able to pick up small bits of paper and lint. The attraction is surely not due to gravity, because the gravitational force between comb and paper is far too small and should not, in any event, depend upon whether the comb is run through our hair or not. What has been revealed by this experiment is an *electrical* phenomenon, so called after *elektron*, the Greek word for amber, a substance used in the earliest studies of electricity.

Because electricity and magnetism are intimately related, they are considered as a single type of interaction, one of the four interactions (no more) that govern the behavior of the entire physical world from atoms to galaxies of stars. The list of these interactions is worth repeating:

1. *Gravitational*—acts between all masses, determines structures of planets, stars, galaxies;
2. *Electromagnetic*—acts between all charged particles, determines structures of atoms, molecules, and solids, and is an important factor in the astronomical universe;
3. *Strong nuclear*—acts between protons and neutrons, determines structures of atomic nuclei;
4. *Weak nuclear*—acts between elementary particles, helps determine compositions of atomic nuclei.

POSITIVE AND NEGATIVE CHARGE

Let us begin our study of electricity by examining a few basic experiments.

Here is the first experiment: **(1)**

A pith ball suspended by a fine string is touched by a hard rubber rod. Nothing happens.

The rubber rod is stroked with a piece of fur.

The pith ball is again touched by the rubber rod.

Now the ball flies away from the rod.

We conclude that being stroked by fur has somehow caused something—called by convention *negative electric charge*—to be added to the rubber rod. Part of the negative charge on the rod flowed to the pith ball when it was touched, and the fact that the ball then flew away from the rod suggests that *negative electric charges repel each other.*

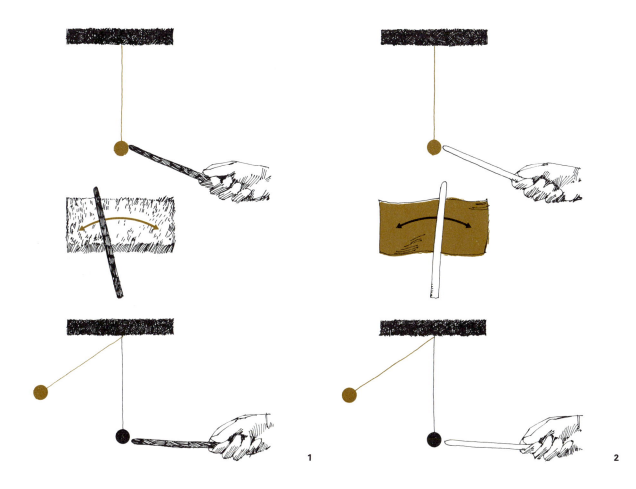

The next experiment is very similar: (2)

A second pith ball is touched by a glass rod. Nothing happens.

The glass rod is stroked with a silk cloth.

The pith ball is again touched by the glass rod.

Now the ball flies away from the rod.

Being stroked by a silk cloth has somehow caused something—called by convention *positive electric charge*—to be added to the glass rod. Part of the positive charge on the rod flowed to the pith ball when it was touched, and the fact that the ball then flew away from the rod suggests that *positive electric charges repel each other*.

Why is it assumed that the electric charge on the glass rod is different from that on the rubber rod? The reason lies in the result of another experiment: **(3)**

A pith ball is touched by a charged rubber rod and another one is touched by a charged glass rod.

The pith balls then swing toward each other.

The attraction of the two pith balls means (1) that the charges they carry are different, since like charges repel, and (2) that *unlike charges attract each other.*

The preceding observations can be summarized very simply:

Like charges repel; unlike charges attract.

HOW A CHARGED OBJECT ATTRACTS AN UNCHARGED ONE

A hard rubber comb that has been charged by being passed through our hair on a dry day is able to attract small bits of paper. Since the paper bits were originally uncharged, how could the comb exert a force on them?

The explanation is shown in the drawing. When the negatively charged comb is brought near the paper, some of the negative charges in the paper which are not tightly bound in place move as far away as they can from the comb, while some of the positive charges which are not tightly bound move toward the comb. Because electrical forces vary inversely with distance, the attraction between the comb and the closer positive charges is greater than the repulsion between the comb and the farther negative charges, and so the paper moves toward the comb. Only a small amount of charge separation actually occurs, and so, with little force available, only very light objects can be attracted in this way. **(4)**

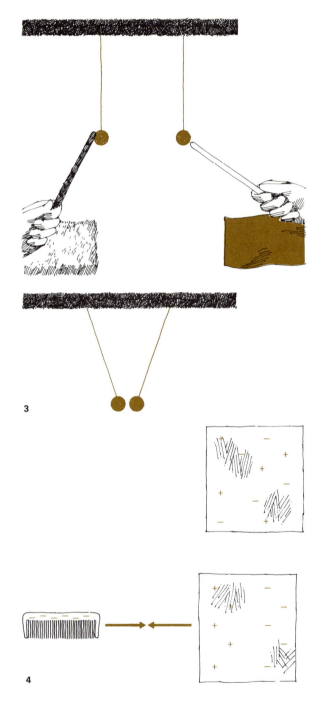

3

4

186   Electric charge

CONSERVATION OF CHARGE

Where do the charges come from when one substance is stroked with another?

When we charge one pith ball with a rubber rod and another with the fur the rod was stroked with, we find that the two balls *attract*. Since the rubber rod is negatively charged, this experiment indicates that the fur is positively charged. A similar experiment with a glass rod and a silk cloth indicates that the cloth acquires a negative charge during the stroking. Evidently the process of stroking serves to separate charges. We might infer that rubber has an affinity of some kind for negative charges and fur an affinity of some kind for positive charges, so that, when rubbed together, each tends to acquire a different kind of charge. (5)

A great many experiments with a variety of substances have shown that there are only the two kinds of electric charge, positive and negative, that we have spoken of. All electrical phenomena involve either or both kinds of charge. An "uncharged" body of matter actually possesses equal amounts of positive and negative charge, so that appropriate treatment—mere rubbing is sufficient for some substances—can leave an excess of either kind on the body and thereby cause it to exhibit electrical effects.

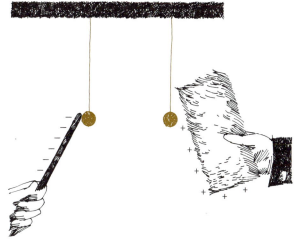

What is electric charge? All that can be said is that charge, like rest mass, is a fundamental property of certain of the elementary particles of which all matter is composed. Three types of particle are found in atoms, the positively charged *proton*, the negatively charged *electron*, and the neutral (that is, uncharged) *neutron*. The proton and electron have exactly equal amounts of charge, though of opposite

5

sign. An atom normally contains equal numbers of protons and electrons, so it is electrically neutral unless disrupted in some way.

The *principle of conservation of charge* states:

The net electric charge in an isolated system remains constant.

By "net charge" is meant the algebraic sum of the charges present—the total positive charge minus the total negative charge. Net charge can be positive, negative, or zero.

Every known physical process in the universe conserves electric charge, just as energy is conserved in every known process. Separating or bringing together charges does not affect their magnitudes, so such rearrangements leave the net charge unaffected. Under certain circumstances matter can be created from energy, but whenever this happens, the numbers of positively charged and negatively charged particles are always exactly the same. Under other circumstances matter can be completely converted into energy, and in such events the numbers of positively and negatively charged particles that disappear are again always exactly the same. Unlike rest mass, charge is invariably conserved.

Electric charge is the fourth quantity we have studied which is conserved in every physical process. The others are energy, linear momentum, and angular momentum, and their conservation can be traced to fundamental symmetry properties of nature. These properties are, respectively, the independence of physical laws to shifts in time, in space, and in orientation. Conservation of charge is also associated with a symmetry property of nature, although this property is too abstract to be described here.

COULOMB'S LAW

In order to arrive at the law of gravitation,

$$F = G \, \frac{m_A m_B}{r^2},$$

Newton had to make use of astronomical data and an indirect argument because gravitational forces are appreciable only when the masses involved are very large. However, the law that electrical forces obey can be readily determined in the laboratory because these forces are so much greater in magnitude than gravitational ones.

The law of force between charges was first published by the eighteenth century French scientist Charles Coulomb, and is called Coulomb's law in his honor. (Actually, other investigators, notably Henry Cavendish, had come to the same conclusion some years earlier than Coulomb, but Coulomb is given credit because of the care with which he conducted his experiments and his publication of the results. Cavendish's equally meticulous work remained hidden in his notebooks for a century.)

If we use the symbol q for electric charge, Coulomb's law for the magnitude F of the force **F** between two charges q_A and q_B the distance r apart states that

$$F = k \, \frac{q_A q_B}{r^2}. \quad \text{Coulomb's law}$$

The force between two charges is proportional to the product of the charges and is inversely proportional to the square of the distance between them. The quantity k is a constant whose value depends upon the units employed and upon the medium (air, vacuum, oil, and so forth) in which the charges are located.

Electric force is, of course, a vector quantity, and the above formula only gives its magnitude. The direction of F is always along the line joining q_A and q_B, and the force is attractive if q_A and q_B have opposite signs and repulsive if they have the same signs. **(6)**

THE COULOMB

The unit of electric charge is the *coulomb* (abbreviated C).

The formal definition of the coulomb is expressed in terms of magnetic forces and is given in Section 25. A more realistic way to think of the coulomb is in terms of the number of individual elementary charges that add up to this amount of charge. All charges, both positive and negative, occur only in multiples of 1.60×10^{-19} C; no elementary particle with a charge of other than $\pm 1.60 \times 10^{-19}$ C—or 0—has ever been found. The electron has a charge of -1.60×10^{-19} C and the proton has a charge of $+1.60 \times 10^{-19}$ C. Hence if we could assemble 6.25×10^{18} electrons, we would have a total charge of -1 C, and if we could assemble 6.25×10^{18} protons, we would have a total charge of $+1$ C. **(7)**

Because electric charge always occurs in multiples of $\pm 1.60 \times 10^{-19}$ C, this amount of charge has been given a special name, the electron (or electronic) charge, and a special symbol, e:

$$e = 1.60 \times 10^{-19} \text{ C.} \qquad \text{Electron charge}$$

Thus a charge of $+1.60 \times 10^{-19}$ C is abbreviated $+e$, and one of -3.20×10^{-19} C is abbreviated $-2e$.

In most processes that lead to a net charge on some object, electrons are either added to it or removed from it. Hence we can think of an

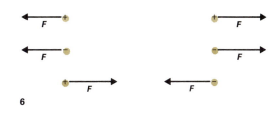

6

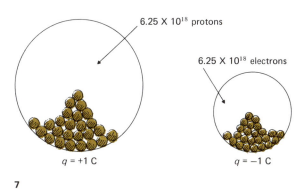

6.25 X 10^{18} protons

6.25 X 10^{18} electrons

$q = +1$ C

$q = -1$ C

7

object whose charge is -1 C as having 6.25×10^{18} electrons more than its normal number, and of an object whose charge is $+1$ C as having 6.25×10^{18} electrons less than its normal number. (By "normal number" is meant a number of electrons equal to the number of protons present, so that the object has no net charge.)

When q_A and q_B in Coulomb's law are expressed in coulombs and r in meters, the constant k has the value in vacuum of

$$k = 9.0 \times 10^9 \ \frac{\text{N-m}^2}{\text{C}^2}.$$

The value of k in air is very slightly greater.

The force between two charges of 1 C each that are 1 m apart is

$$F = k \frac{q_A q_B}{r^2}$$

$$= 9.0 \times 10^9 \ \frac{\text{N-m}^2}{\text{C}^2} \times \frac{1 \ \text{C} \times 1 \ \text{C}}{1 \ \text{m}^2}$$

$$= 9.0 \times 10^9 \ \text{N},$$

which is about 2 billion lb! Evidently even the most highly charged objects that can be produced seldom contain more than a minute fraction of a coulomb of net charge of either sign. (8)

electron is always $-e$ regardless of whether the electron is at rest or is moving. This invariance is an experimental observation, like the similar invariance of the velocity of light.

A significant consequence of charge invariance is the electrical neutrality of atoms and molecules, which contain equal numbers of protons and electrons that are all in motion but at different velocities. If charge varied with velocity, as mass does, atoms and molecules would not, in general, be electrically neutral. The most accurate measurements to date indicate that the proton and electron charges in the hydrogen molecule are equal in magnitude to within at least one part in 10^{10}, very strong evidence in favor of charge invariance.

EXERCISES

1. Electricity was once regarded as a weightless fluid, an excess of which was "positive" and a deficiency of which was "negative." What phenomena can this hypothesis still explain? What phenomena can it not explain?

2. An insulating rod has a positive charge at one end and a negative charge of the same magnitude at the other. How will the rod behave when it is placed near a fixed positive charge that is initially equidistant from the ends of the rod?

3. When two bodies attract each other electrically, must both of them be charged? When two bodies repel each other electrically, must both of them be charged?

4. Why do people sometimes get small electric shocks when they leave automobiles?

5. An inflated rubber balloon is charged by stroking it with a piece of fur. Will it stick better to a plaster wall or to a metal wall? Why?

CHARGE INVARIANCE

As we know, measurements of length, time, and mass depend upon the frame of reference from which the measurements are made. If the observer is in relative motion with respect to the object whose properties he is examining, he will find shorter lengths, longer time intervals, and larger masses than an observer at rest relative to the object.

Electric charge, on the other hand, is relativistically invariant: a charge whose magnitude is found to be q in one frame of reference will be found to be q in all other frames of reference. For example, the charge on an

6. How do we know that the inverse square force holding the earth in its orbit around the sun is not an electrical force?

7. Two electric charges originally 8 cm apart are brought closer together until the force between them is greater by a factor of 16. How far apart are they now?

8. Two charges repel each other with a force of 10^{-6} N when they are 10 cm apart. What is the force between them when the charges are 2 cm apart?

9. If equal amounts of positive charge were to be placed on the earth and moon until their repulsion exactly balanced the gravitational attraction between earth and moon, how much net charge would each body have?

10. Ten thousand electrons are removed from a neutral pith ball. What is the resulting charge on the ball?

11. A point charge of $+10^{-9}$ C is placed 5 cm from another point charge of $+3 \times 10^{-9}$ C. Find the force between them.

12. Two charges attract each other with a force of 4×10^{-6} N when they are 0.4 m apart. Find the force between them when their separation is increased to 0.8 m.

13. Two charges of unknown magnitude and sign are observed to repel one another with a force of 0.1 N when they are 5 cm apart. What will the force be when they are

a) 10 cm apart?
b) 50 cm apart?
c) 1 cm apart?

14. A test charge of -5×10^{-5} C is placed between two other charges so that it is 5 cm from a charge of -3×10^{-5} C and 10 cm from a charge of -6×10^{-5} C. The three charges lie along a straight line. What is the magnitude and direction of the force on the test charge?

15. A particle carrying a charge of $+6 \times 10^{-5}$ C is located halfway between two other charges, one of $+1 \times 10^{-4}$ C and the other of -1×10^{-4} C, that are 40 cm apart. All three charges lie on the same straight line. What is the magnitude and direction of the force on the $+6 \times 10^{-5}$ C charge?

16. Two metal spheres, one with a charge of $+2 \times 10^{-5}$ C and the other with a charge of -1×10^{-5} C, are 10 cm apart.
a) What is the force between them?
b) The two spheres are brought into contact, and then separated again by 10 cm. What is the force between them now?

17. How far apart should two electrons be if the force each exerts on the other is to equal the weight of an electron?

18. According to one model of the hydrogen atom, it consists of a proton circled by an electron whose orbit has a radius of 5.3×10^{-11} m. How fast must the electron be moving if the orbit is to be a stable one?

22

THE ELECTRIC FIELD

Electric forces, like gravitational forces, act between objects that may be widely separated. An appropriate way to regard such forces involves the concept of a force field that was discussed in Section 12. When a charge is present somewhere, the properties of space in its vicinity can be considered to be so altered that another charge brought to this region will experience a force there. The "alteration in space" caused by a charge at rest is called its electric field, and any other charge is thought of as interacting with the field and not directly with the charge responsible for it.

ELECTRIC FIELD

An electric field exists wherever an electric force acts on a charge. The field may be due to the presence of a single charge or to the presence of many charges. We would like to specify electric field in such a way that it will be possible to determine the force acting on any arbitrary charge at any point in the field. Accordingly the electric field E at a point in space is defined as the ratio between the force F on a charge q at that point and the magnitude of q:

$$E = \frac{F}{q}. \qquad \text{Electric field}$$

The units of E are newtons per coulomb. Electric field is a vector quantity that possesses both magnitude and direction.

Once we know what the electric field E is somewhere, from its definition we see that the force the field exerts on a charge q there is

$$\text{Force} = \text{charge} \times \text{electric field},$$

$$F = qE.$$

Problem. The electric field between the electrodes of the gas discharge tube used in a

"To the top of the upright stick of the kite is to be fixed a very sharp pointed wire . . . This kite is to be raised when a thunder-gust appears to be coming on . . . and from the electric fire thus obtained spirits may be kindled, and all the other electric experiments be performed, which are usually done by the help of a rubbed globe or tube, and thereby the sameness of the electric matter with that of lightning completely demonstrated." Benjamin Franklin (1706–1790)

certain neon sign has the magnitude 5.0×10^4 N/C. Find the acceleration of a neon ion of mass 3.3×10^{-26} kg in the tube if it carries a charge of $+ e$.

Solution. Since $e = 1.6 \times 10^{-19}$ C, the force on the ion has the magnitude

$$F = qE$$

$$= 1.6 \times 10^{-19} \text{ C} \times 5.0 \times 10^4 \frac{\text{N}}{\text{C}}$$

$$= 8.0 \times 10^{-15} \text{ N}.$$

Hence the acceleration of the ion is

$$a = \frac{F}{m}$$

$$= \frac{8.0 \times 10^{-15} \text{ N}}{3.3 \times 10^{-26} \text{ kg}}$$

$$= 2.4 \times 10^9 \frac{\text{m}}{\text{s}^2}$$

which is 250 million times greater than the acceleration of gravity. (**1**)

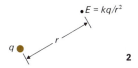

1

We can use Coulomb's law to determine the magnitude of the electric field around a single charge q. First we find the force F that q exerts upon a test charge q_0 at the distance r away, which is

$$F = k \frac{qq_0}{r^2}.$$

Since $E = F/q_0$ by definition, we have

$$E = \frac{F}{q_0}$$

$$= k \frac{q}{r^2}.$$

This formula tells us that the electric field magnitude the distance r from a point charge is directly proportional to the magnitude q of the charge and is inversely proportional to the square of r. (**2**)

•$E = kq/r^2$

q •⟋ r

2

ELECTRIC LINES OF FORCE

In Section 12 the use of *lines of force* to picture a force field was discussed in connection with gravitational fields. Lines of force are also convenient in visualizing electric fields. In drawing the lines of force that correspond to a particular electric field we follow two rules:

1. The direction of a line of force at any point is that in which a positive charge would move if placed at that point. Thus lines of force are to be thought of as leaving positive charges and entering negative ones;

2. The spacing of lines of force is such that they are close together where the field is strong and far apart where the field is weak.

Here is how the electric fields around a positive and a negative charge look in terms of lines of force. (**3**)

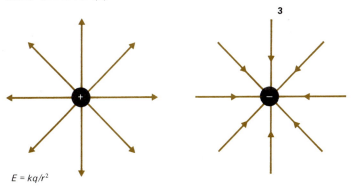

3

$E = kq/r^2$

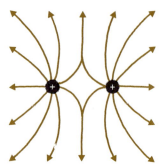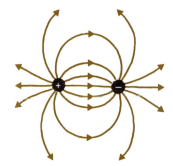

4

When the electric field in a region is produced by more than one charge, the resultant intensity E at any point is the vector sum of the intensities E_1, E_2, E_3, . . . produced by each of the charges separately. That is,

$$E = E_1 + E_2 + E_3 + \ldots.$$

The electric fields around a pair of like charges and around a pair of unlike charges have these forms. **(4)**

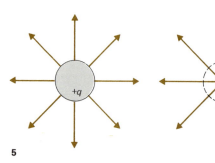

5

Electric lines of force around a charged sphere

Electric lines of force around a point charge

DISTRIBUTIONS OF CHARGE

Here is an example of how the notion of lines of force can guide our thinking in a specific situation. Suppose that, instead of a point charge q, we have a spherical distribution of charge whose total amount is also q. Such a distribution is quite easy to arrange—all we need do is to add the charge q to a metal sphere. Since metals are good conductors of electricity, the mutual repulsion of the individual charges that make up q causes them to spread out uniformly over the surface of the sphere; in this way the charges are as far apart as possible. Evidently it doesn't matter whether the sphere is hollow or solid.

Whatever the number of lines of force we choose to represent the electric field around a point charge q, the same number must emerge from a body that carries the same net charge q. In the case of a charged sphere, the lines of force are symmetrically arranged, which means that the pattern of lines of force outside the sphere is identical with that around a point charge of the same magnitude. **(5)**

Inside a charged conducting object of any shape the electric field is 0 everywhere; the electric fields due to the individual charges on the surface all cancel out in the interior, although they don't outside the object. We can see why this must be true even without a formal calculation. Suppose there *were* an electric field E in the interior of the object. Then the charges inside the object that are free to move (electrons in the case of a metal) would do so under the influence of the field E. But no currents are observed in a charged conducting object except for a moment after the charge is placed on it; indeed, since energy is needed to maintain an electric current, an infinite supply of energy would be needed for currents to persist in such an object. The only

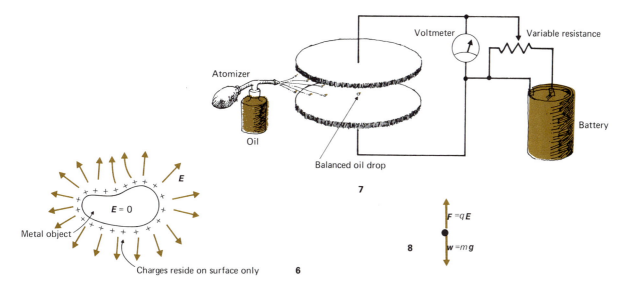

Atomizer

Oil

Voltmeter

Variable resistance

Battery

Balanced oil drop

7

$$E$$

$$E = 0$$

Metal object

Charges reside on surface only

6

$$F = qE$$

$$w = mg$$

8

conclusion is that the interior of a conducting object is always free of electric field. (**6**)

QUANTIZATION OF CHARGE

Is electric charge continuous, so that it can be divided into smaller and smaller portions indefinitely? Or do the charges found in nature and in the laboratory actually consist of units of some minimum, indivisible charge? Beginning in 1906 the American physicist Robert A. Millikan performed a series of careful experiments which showed that electric charge is indeed granular, or *quantized*: all charges, both positive and negative, occur only in multiples of 1.60×10^{-19} C, and no charge smaller than $\pm 1.60 \times 10^{-19}$ C has ever been found. Today the concept that electric charge is quantized fits in nicely with the picture of all matter as being composed of definite, individual elementary particles, but at the beginning of this century it was an unproven idea yet to be firmly established.

Millikan's apparatus is shown schematically in the figure above. (**7**) An atomizer is used to spray tiny oil drops between two metal plates. In passing through the atomizer nozzle, the drops become electrically charged. A uniform electric field is established by charging the upper and lower plates oppositely, in this case with the help of a battery. If this field has the proper magnitude and direction, it can exert an upward electrical force on a charged oil drop that exactly counterbalances the downward gravitational force on the drop.

The electric force on an oil drop whose charge is q is

$$F = qE$$

in an electric field E, while the gravitational force on the drop is just its weight

$$w = mg.$$

When $F = w$, the drop is acted upon by no net force and remains stationary. If this is true,

$$qE = mg,$$

and the charge on the drop is

$$q = \frac{mg}{E}. \quad (8)$$

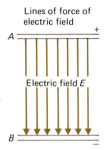

Lines of force of
electric field

Electric field E

Lines of force of
gravitational field

Acceleration of
gravity g

The earth

9

difference, turns out to be especially convenient in electrical problems.

Let us examine the potential energy of a charge in a uniform electric field in parallel with a gravitational analogy. At the left is a uniform electric field E between two parallel, uniformly charged plates A and B, and at the right is a region near the earth's surface in which the gravitational field is also uniform. (**9**)

Millikan could measure directly or indirectly all the quantities on the right-hand side of the last equation, which permitted him to determine the precise magnitudes of a great many charges. The advantage of this method is that extremely small values of q can be determined, whereas experiments using objects such as pith balls suspended from strings can only be used for relatively large values of q.

Millikan found many different charges on the various oil drops he studied, but they shared a singular property: all the charges were multiples of a single value, 1.60×10^{-19} C. The measurements he made of q ranged from 1.60×10^{-19} C up to 27.2×10^{-19} C, in every case equal to 1.60×10^{-19} C multiplied by a whole number (27.2×10^{-19}, for instance, is $17 \times 1.60 \times 10^{-19}$). Hence he felt justified in concluding that the smallest electric charge in nature is 1.60×10^{-19} C.

ELECTRIC POTENTIAL ENERGY

In our study of mechanics we found the related concepts of work and potential energy to be useful in analyzing a wide variety of situations. These concepts are equally useful in the study of electrical phenomena, in particular electric current. Instead of potential energy itself, a related quantity, *potential*

Now we place a particle of charge q in the electric field and a particle of mass m in the gravitational field. The charge is acted upon by the electric force

$$F_{\text{elec}} = qE,$$

and the mass is acted upon by the force

$$F_{\text{grav}} = mg. \ (\mathbf{10})$$

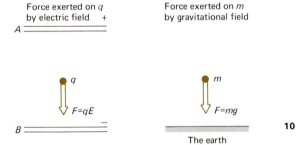

Force exerted on q
by electric field

Force exerted on m
by gravitational field

q

$F = qE$

m

$F = mg$

The earth

10

If the charge is on plate B and we want to move it to plate A, we must apply a force of magnitude qE to it because we have to push against a force of this magnitude exerted by the electric field. When the charge is at A, we will have performed the amount of work

$$\text{Work} = \text{force} \times \text{distance},$$

$$W = qEs,$$

on it, where s is the distance the charge has

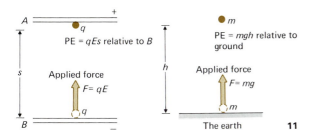

PE = qEs relative to B

Applied force

$F = qE$

PE = mgh relative to ground

Applied force

$F = mg$

The earth **11**

moved. Similarly, to raise the mass from the ground to a height h, we must apply a force of magnitude mg to it, and the work we do is

$$W = mgh. \text{ (11)}$$

At plate A the charge has the potential energy

$$\text{PE} = qEs$$

with respect to B. If we let it go, the potential energy will become kinetic energy as the electric field E accelerates the charge, and when the charge is back at B it will have a kinetic energy equal to qEs. In the same way, the mass has potential energy *with respect to the ground* in its new location. This potential energy is equal to the work done in raising it through the height h, and is

$$\text{PE} = mgh.$$

If we let the mass go, it will fall to the ground with a final kinetic energy of mgh. **(12)**

KE = qEs

KE = mgh

The earth **12**

To summarize, the amount of work qEs must be performed to move a charge q from B to A the distance s apart in an electric field E. At A the charge accordingly has the potential energy qEs. If the charge is released at A, the force qE acting on it produces an acceleration such that the charge has the kinetic energy qEs when it is back at B. Thus the work done in moving the charge in the field becomes potential energy which in turn becomes kinetic energy when it is released.

There is no change in the energy of a charge moved perpendicular to an electric field, just as there is no change in the energy of a mass moved perpendicular to a gravitational field (for instance, along the earth's surface).

POTENTIAL DIFFERENCE

The quantity *potential difference* is introduced to describe the situation of a charge in an electric field in a quantitative way. The potential difference V_{AB} between two points A and B is defined as the ratio between the work that must be done to take a charge q from A to B and the value of q:

Potential difference = work per unit charge,

$$V_{AB} = \frac{W_{AB}}{q}.$$

The unit of potential difference is the joule per coulomb. Because this quantity is so frequently used, its unit has been given the name *volt* (V). Thus

$$1 \text{ V} = 1 \, \frac{\text{J}}{\text{C}}.$$

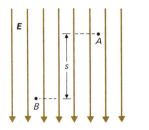

13

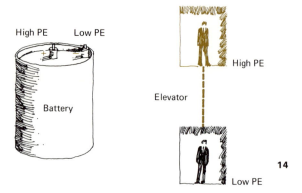

High PE   Low PE

Battery

Elevator

High PE

Low PE

14

In a uniform electric field $W_{AB} = qEs$, with the result that the potential difference between A and B is qEs/q or

$$V_{AB} = Es. \qquad \text{Potential difference in uniform electric field}$$

In a uniform electric field, the potential difference between two points is the product of the field magnitude E and the separation s of the two points in a direction parallel to that of E. (13)

A positive potential difference means that the energy of the charge is *greater* at B than at A; a negative potential difference means that its energy is *less* at B than at A. If V_{AB} is positive, then a charge at B tends to return to A, whereas if V_{AB} is negative, the charge tends to move further away from A.

One advantage of specifying the potential difference between two points in an electric field, rather than the magnitude of the field between them, is that an electric field is normally created by imposing a difference of potential between two points in space. A *battery* is a device that uses chemical means to produce a potential difference between two terminals. A "six-volt" battery is one that has a potential difference of 6 V between its

terminals. A *generator* is another device for producing a potential difference. Batteries and generators are to electric charges what elevators are to masses: all of them increase the potential energy of what they act upon. (14)

When a charge q goes from one terminal of a battery whose potential difference is V to the other, the work

$$W = qV$$

is done on it regardless of the path taken by the charge and regardless of whether the actual electric field that caused the motion of the charge is strong or weak. Given V we can find W at once, no matter what the details of the process are. Hence the notion of potential difference permits us to simplify our analyses of electrical phenomena, just as the notion of potential energy permitted us to simplify our analysis of mechanical phenomena.

Problem. The diagram shows a tube that has a source of electrons at one end and a metal plate at the other. A 100-V battery is connected between the electron source and the plate, so that there is a potential difference of 100 V between them. The negative terminal of the battery is connected to the electron

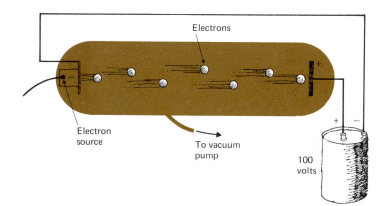

Electrons

Electron source

To vacuum pump

100 volts

+ −

15

source. What is the velocity of the electrons when they arrive at the metal plate? (The tube is evacuated to prevent collisions between the electrons and air molecules.) **(15)**

Solution. To find the kinetic energy, we note that the work done by the electric field within the tube on an electron is

$$W = qV$$
$$= 1.6 \times 10^{-19} \text{ C} \times 100 \text{ V}$$
$$= 1.6 \times 10^{-17} \text{ J}.$$

Since the KE of the electron is equal to the work done on it, its KE after passing through the entire field is 1.6×10^{-17} J. Hence

$$\text{KE} = W = \tfrac{1}{2}mv^2,$$

$$v = \sqrt{\frac{2W}{m}}$$

$$= \sqrt{\frac{2 \times 1.6 \times 10^{-17} \text{ J}}{9.1 \times 10^{-31} \text{ kg}}}$$

$$= 5.9 \times 10^6 \, \frac{\text{m}}{\text{s}}.$$

Here a nonrelativistic calculation is satisfactory.

THE ELECTRON VOLT

The electron volt (abbreviated eV) is a widely used energy unit in atomic physics. By definition, 1 eV is the energy acquired by an electron that has been accelerated through a potential difference of 1 V. Hence

$$W = qV,$$

$$1 \text{ eV} = 1.60 \times 10^{-19} \text{ C} \times 1.00 \text{ V},$$

and so

$$1 \text{ eV} = 1.60 \times 10^{-19} \text{ J}. \qquad \text{Electron volt}$$

Typical quantities expressed in eV are the ionization energy of an atom (which is the work needed to remove one of its electrons) and the binding energy of a molecule (which is the work needed to break it apart into separate atoms). Thus the ionization energy of nitrogen is usually stated to be 14.5 eV and the binding energy of the hydrogen molecule, which consists of two hydrogen atoms, is usually stated to be 4.5 eV.

The eV is too small a unit for nuclear physics, where its multiples the MeV (10^6 eV) and the GeV (10^9 eV) are commonly used. The M and G respectively signify *mega* ($= 10^6$) and *giga*

$(= 10^9)$ and are used in connection with other units as well, for instance the megabuck (10^6) and the gigawatt (10^9 W). The GeV was formerly called the BeV, where the B stood for "billion," but this proved confusing since in Europe a billion is 10^{12} whereas in the United States a billion is 10^9.

A typical quantity expressed in MeV is the energy liberated when the nucleus of a uranium atom splits into two parts. Such *fission* of a uranium nucleus releases an average of 200 MeV; this is the process that powers nuclear reactors and atomic bombs.

The rest energies m_0c^2 of the electron, proton, and neutron expressed in electron volts are as follows:

Particle	Rest energy
Electron	0.51 MeV $\approx \frac{1}{2}$ MeV
Proton	938 MeV $\approx$ 1 GeV
Neutron	939 MeV $\approx$ 1 GeV

In atomic and nuclear physics the energy in electron volts of a moving particle is usually specified rather than its velocity. If we want to find the velocity of the particle, we can tell whether to make a relativistic or a non-relativistic calculation by comparing its KE with its rest energy. When $m_0c^2 \gg$ KE, then $v \ll c$ and the particle behaves according to classical physics; otherwise the relativistic formulas must be used.

Problem. What is the velocity of a neutron ($m_0 = 1.7 \times 10^{-27}$ kg) whose kinetic energy is 50 eV?

Solution. The rest energy of a neutron is about 1 GeV, which is 10^9 eV, so we can safely ignore relativistic considerations. The first step is to convert 50 eV to its equivalent in joules. Since 1 eV $= 1.6 \times 10^{-19}$ J,

$$KE = 50 \text{ eV} \times 1.6 \times 10^{-19} \frac{J}{eV}$$

$$= 8.0 \times 10^{-18} \text{ J}.$$

Now we proceed in the usual way:

$$KE = \tfrac{1}{2}m_0v^2,$$

$$v = \sqrt{\frac{2KE}{m_0}}$$

$$= \sqrt{\frac{2 \times 8.0 \times 10^{-18} \text{ J}}{1.7 \times 10^{-27} \text{ kg}}}$$

$$= 9.7 \times 10^4 \frac{m}{s}.$$

EXERCISES

1. An insulating rod has a positive charge at one end and a negative charge of the same magnitude at the other. How will the rod behave when it is placed in a uniform electric field whose direction is
   a) parallel to the rod?
   b) perpendicular to the rod?

2. Four charges of $+1 \times 10^{-8}$ C are at the corners of a square 1 m on each side. Find the electric field intensity at the center of the square.

3. Two charges, one of 1.5×10^{-6} C and the other of 3×10^{-6} C, are 0.2 m apart. Where is the electric field in their vicinity equal to zero?

4. What is the electric field intensity 0.4 m from a charge of $+7 \times 10^{-5}$ C?

5. Two charges of $+4 \times 10^{-6}$ C and $+8 \times 10^{-6}$ C are 2 m apart. What is the electric field intensity halfway between them?

6. A body whose mass is 10^{-6} kg carries a charge of $+10^{-6}$ C. What is the magnitude of an electric field that can hold the body suspended in equilibrium?

7. A particle carrying a charge of 10^{-5} C starts moving from rest in a uniform electric field whose intensity is 50 N/C.
   a) What is the force on the particle?
   b) How much kinetic energy will the particle have after it has moved 1 m?

8. A cloud is at a potential of 8×10^6 V relative to the ground. A charge of 40 C is transferred in a lightning stroke between the cloud and the ground. Find the energy dissipated.

9. In charging a certain 20-kg storage battery, a total of 2×10^5 C is transferred from one set of electrodes to another. The potential difference between the electrodes is 12 V.
   a) How much energy is stored in the battery?
   b) If this energy were used to raise the battery above the ground, how high would it go?

c) If this energy were used to provide the battery with kinetic energy, what would its speed be?

10. The potential difference between two parallel metal plates 0.5 cm apart is 10^4 V. Find the force on an electron between the plates.

11. It is necessary to do 13.6 eV of work to remove the electron from a hydrogen atom.
    a) How much energy in joules would be required to remove the electrons from the 6×10^{26} hydrogen atoms in a kg of hydrogen?
    b) How many kcal is this?

12. Typical chemical reactions absorb or release energy at the rate of several eV per molecular change. What change in mass is associated with the absorption or release of 1 eV?

13. What is the energy in electron volts of an electron whose speed is 10^6 m/s?

14. What is the energy in electron volts of a potassium atom of mass 6.5×10^{-26} kg whose speed is 10^6 m/s?

15. What is the speed of a 50-eV electron?

16. What is the speed of a 26-eV electron?

23

ELECTRICITY AND MATTER

"I remember . . . Geiger coming to me in great excitement and saying, 'We have been able to get some of the alpha-particles coming backward.' It was quite the most incredible event that has ever happened to me in my life. It was almost as incredible as if you fired a 15-inch shell at a piece of paper and it came back and hit you . . . It was then that I had the idea of an atom with a minute massive center carrying a charge." Ernest Rutherford (1871–1937)

An atom consists of a tiny, positively charged nucleus surrounded at some distance by electrons. The nucleus consists of protons and neutrons held tightly together by nuclear forces, and the number of electrons equals the number of protons so the atom as a whole is electrically neutral.

ELECTRICITY AND MATTER

Coulomb's law for the electric force between charges is very similar to Newton's law for the gravitational force between masses. The most striking difference is that electric forces may be either attractive or repulsive, whereas gravitational forces are always attractive. The latter fact means that matter in the universe tends to come together to form large bodies, such as stars and planets, and these bodies are always found in groups, such as galaxies of stars and families of planets.

There is no comparable tendency for electric charges of either sign to come together; quite the contrary. Unlike charges attract strongly, which makes it hard to separate neutral matter into portions of opposite sign. Furthermore, like charges repel, so it becomes harder and harder to add further charge to an already charged object. Hence the large-scale structure of the universe is largely governed by gravitational forces.

On an atomic scale, though, the relative importance of gravity and electricity is reversed. Elementary particles are so tiny that the gravitational forces between them are insignificant, whereas their electric charges are sufficiently great for electric forces to govern the structures of atoms, molecules, liquids, and solids.

Problem. The hydrogen atom has the simplest structure of all atoms. It consists of a

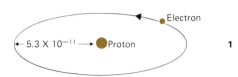

A crude model of the hydrogen atom

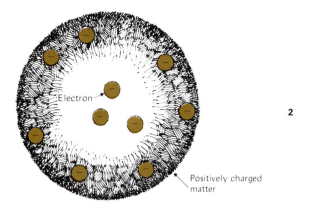

Electron

Positively charged matter

proton (mass 1.7×10^{-27} kg, charge $+ 1.6 \times 10^{-19}$ C) and an electron (mass 9.1×10^{-31} kg, charge $- 1.6 \times 10^{-19}$ C) whose average separation is 5.3×10^{-11} m. (For the time being we can think of the electron as circling the proton much as the moon circles the earth. A more realistic model of the hydrogen atom—but one that is harder to visualize—will be given later.) Compare the electrical and gravitational forces between the proton and the electron in a hydrogen atom. (1)

Solution. The electrical force between the electron and proton is

$$F_e = k\frac{q_e q_p}{r^2}$$

$$= \frac{9.0 \times 10^9 \text{ N-m}^2/\text{C}^2 \times (1.6 \times 10^{-19} \text{ C})^2}{(5.3 \times 10^{-11} \text{ m})^2}$$

$$= 8.2 \times 10^{-8} \text{ N},$$

while the gravitational force between them is

$$F_g = G\frac{m_e m_p}{r^2}$$

$$= \frac{6.7 \times 10^{-11} \text{ N-m}^2/\text{kg}^2 \times 9.1 \times 10^{-31} \text{ kg} \times 1.7 \times 10^{-27} \text{ kg}}{(5.3 \times 10^{-11} \text{ m})^2}$$

$$= 3.7 \times 10^{-47} \text{ N}.$$

The electrical force is over 10^{39} times greater than the gravitational force! Clearly the electrical forces which subatomic particles exert upon one another are so much stronger than their mutual gravitational ones that the latter can be neglected completely.

THE THOMSON MODEL

By the beginning of this century a substantial body of evidence had been accumulated in support of the idea that the chemical elements consist of atoms. The nature of the atoms themselves, however, was still a mystery, although a significant clue had been discovered. This clue was the fact that electrons are constituents of atoms, which suggests that electrical forces are involved in atomic phenomena. J. J. Thomson, whose work had led to the identification of the electron, proposed in 1898 that atoms are spheres of positively charged matter that contain embedded electrons, much as a fruitcake is studded with raisins. (2)

The most direct way to find out what is inside a fruitcake is simply to plunge a finger into it. In essence this is the classic experiment performed in 1911 by Geiger and Marsden at the suggestion of Ernest Rutherford. The probes

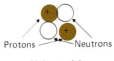

Protons — Neutrons

Alpha particle

3

they used were fast *alpha particles* spontaneously emitted by certain radioactive elements. For the present all we need to know about alpha particles is that they consist of two neutrons and two protons held tightly together, so that each one has a charge of $+2e$. **(3)**

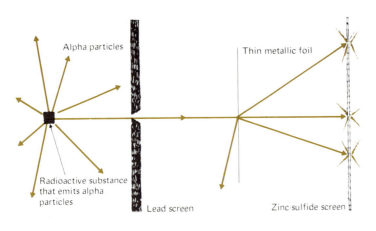

Alpha particles

Thin metallic foil

Radioactive substance that emits alpha particles

Lead screen

Zinc-sulfide screen

4

Geiger and Marsden placed a sample of an alpha-emitting substance behind a lead screen with a small hole in it, so that a narrow beam of alpha particles was produced. On the other side of a thin metal foil in the path of the beam they placed a movable zinc sulfide screen which gave off a flash of light when struck by an alpha particle, thus indicating the extent to which the alpha particles were scattered from their original direction of motion. **(4)**

Geiger and Marsden expected to find that most of the alpha particles go through the foil without being affected by it, with the remainder receiving only slight deflections. This anticipated behavior follows from the Thomson atomic model, in which the positive and negative electric charges within an atom are assumed to be spread more or less evenly throughout its volume. If the Thomson model is correct, only weak electric forces would be exerted on alpha particles passing through a thin foil, and their momenta would be enough to carry them through with only minor departures—at most 1° or so—from their original paths. **(5)**

THE RUTHERFORD MODEL

What Geiger and Marsden actually found was that, while most of the alpha particles indeed emerged unaffected from the foil, the others underwent deflections through very large angles, in some cases even being scattered in the backward direction. Since alpha particles are relatively heavy (almost 8000 times more massive than electrons) and have fairly high initial velocities (typically 2×10^7 m/s), it was clear that strong forces had to be exerted upon them to cause such marked deflections. To explain the results, Rutherford adopted the hypothesis that an atom is composed of a tiny *nucleus* in which its positive charge and nearly all its mass are concentrated, with the electrons some distance away. **(6)**

With the atom largely empty space, it is easy to see why most alpha particles proceed right through a thin foil. On the other hand, an alpha particle that happens to come near a nucleus experiences a strong electric force, and is likely to be scattered through a large angle. (The atomic electrons, being very light,

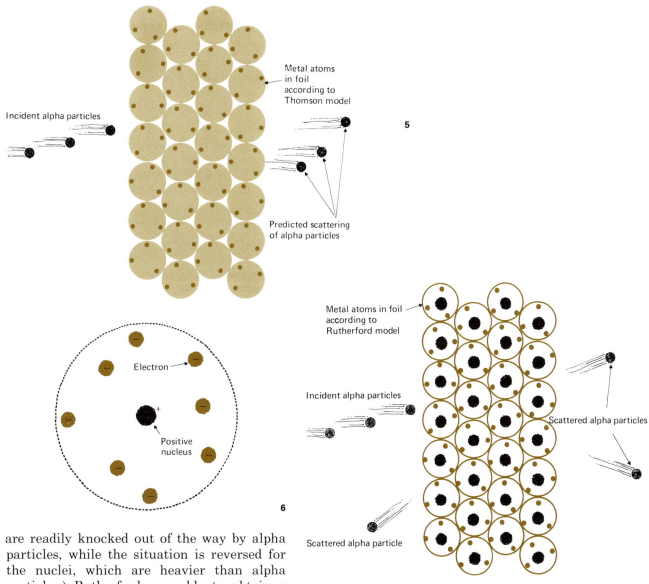

Incident alpha particles

Metal atoms
in foil
according to
Thomson model

5

Predicted scattering
of alpha particles

Electron

Positive
nucleus

6

Metal atoms in foil
according to
Rutherford model

Incident alpha particles

Scattered alpha particles

Scattered alpha particle

7

are readily knocked out of the way by alpha particles, while the situation is reversed for the nuclei, which are heavier than alpha particles.) Rutherford was able to obtain a formula for the scattering of alpha particles by thin foils on the basis of his hypothesis that agreed with the experimental results. He is therefore credited with the discovery of the nucleus. (7)

The obvious question to ask at this point is, what keeps the electrons away from the nucleus? The almost equally obvious answer would seem to be that the electrons revolve around the nucleus in stable orbits, just as the planets do around the sun. The trouble is that an electron moving in a circle is accelerated, and, as we shall find in Section 29, all accelerated charged particles radiate electromagnetic waves at the expense of their kinetic energies. So it would seem that there can be no such thing as a stable orbit for an electron in an atom, only a spiral path leading to the nucleus. Yet the Rutherford model agrees with experiment. The resolution of this paradox is one of the central problems of modern physics, and we shall see how it is done in later chapters.

CONDUCTION IN SOLIDS

An electric current is a flow of charge from one place to another. Nearly all substances fall into two categories: *conductors*, through which charge can flow easily; and *insulators*, through which charge can flow only with great difficulty. Metals, many liquids, and plasmas (gases whose molecules are charged) are conductors, whereas nonmetallic solids, certain liquids, and gases whose molecules are electrically neutral are insulators. Several substances, called *semiconductors*, are intermediate in their ability to conduct charge.

In a solid metal, each atom gives up one or more electrons to a common "gas" of freely-moving electrons that pervades the entire metal. These electrons can migrate quite readily through the crystal structure of the metal, so if one end of a metal wire is given a positive charge and the other end a negative charge, electrons will flow through the wire from the negative to the positive end. This

flow, of course, constitutes an electric current. By supplying new electrons to the negative end of the wire and removing electrons from the positive end as they arrive there—which can be done by connecting the wire to a battery or to a generator—a constant current can be maintained in the wire.

In nonmetallic solids, such as salt, glass, rubber, minerals, wood, and plastics, all the atomic electrons are bound to particular atoms or groups of atoms and cannot move from place to place. Such solids are accordingly classed as insulators. Actually, nonmetallic solids do conduct very small amounts of current, but their abilities to do this are vastly inferior to those of metals. For instance, when identical bars of copper and sulfur are connected to the same battery, about 10^{23} times more current flows in the copper bar!

As mentioned earlier, there are a few substances called semiconductors through which current flows more readily than through insulators but still with distinctly more difficulty than through conductors. Thus about 10^7 times more current flows in a germanium bar connected to a battery than in a sulfur bar of the same size, but this is still about 10^{16} times less current than in a copper bar.

At temperatures near absolute zero ($0°K$, which is $-273°C$) certain metals, alloys, and chemical compounds lose all of their resistance to the flow of electric current. This phenomenon, called *superconductivity*, was discovered by Kamerlingh Onnes in Holland in 1911. For example, aluminum is superconducting at temperatures under $1.20°K$, lead at temperatures under $7.22°K$, and CuS (copper sulfide) at temperatures under $1.6°K$. If a current is set up in a closed wire loop at room temperature, it will die out in less than a second even if the wire is made of a good conductor such as copper or silver, whereas if the wire is made of a superconducting material and is kept cold enough, the current will continue indefinitely. Currents have persisted in superconducting loops with no apparent diminution for several years.

IONIZATION AND RECOMBINATION

The mechanism of electrical conduction in liquids and gases is different from that in metals. The current in a metal consists of a flow of electrons past the stationary atoms in its structure. The current in a fluid medium other than a liquid metal, however, consists of a flow of entire atoms or molecules that are electrically charged. An atom or molecule that carries a charge is called an *ion*, and both positive and negative ions participate in the conduction process in liquids and gases.

An atom or molecule becomes a positive ion when it loses one or more of its electrons; if it gains one or more electrons in addition to its usual complement, it becomes a negative ion. The fundamental positive charges in matter are protons, which are so tightly bound in the nucleus of every atom that they can be dislodged only under exceptional circumstances. Atomic electrons, however, are held more loosely, and one or two of them can be detached from an atom with relative ease. Thus the oxygen and nitrogen gases in ordinary air become ionized when a spark occurs, in the presence of a flame, and by the passage of x-rays or even ultraviolet light. These processes so disturb the air molecules that some electrons are dislodged, leaving behind positive ions. The liberated electrons almost at once become attached to other nearby molecules to create negative ions. (8)

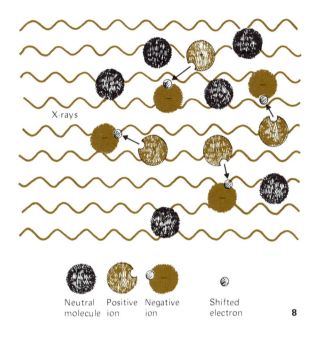

X-rays

Neutral molecule Positive ion Negative ion Shifted electron **8**

The electrical attraction between positive and negative charges in time brings the ions together, and the extra electrons on the negative ions become reattached to the positive ions. The gas molecules are then neutral, as they were originally. This *recombination* is rapid at normal atmospheric pressure and temperature.

Polar molecule

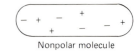

Nonpolar molecule

10

sufficient for us to note that certain molecules have asymmetrical (nonsymmetrical) distributions of charge and behave as though negatively charged at one end and positively charged at the other. A molecule of this kind is called a *polar molecule. A nonpolar molecule,* on the other hand, has a uniform distribution of charge. All molecules are normally electrically neutral, and the distinction between the polar and nonpolar varieties lies solely in the way their electrons are arranged. (**10**)

The fact that polar molecules exist helps to explain a number of familiar phenomena. The behavior of compounds in solution is a good example. Water readily dissolves such compounds as salt and sugar, but cannot dissolve fats or oils. Gasoline readily dissolves fats and oils, but cannot dissolve salt or sugar. The key to these differences lies in the strongly polar nature of water molecules and the nonpolar nature of gasoline molecules. Water molecules tend to form aggregates under the influence of the electric forces between the ends of adjacent molecules. (**11**)

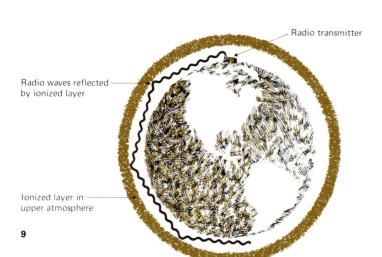

Radio transmitter

Radio waves reflected by ionized layer

Ionized layer in upper atmosphere

9

In the upper atmosphere, where air molecules are so far apart that the recombination of ions is a slow process, the continual bombardment of x-rays and ultraviolet light from the sun maintains a perceptible proportion of ions at all times. The layer of ions in the upper atmosphere is called the *ionosphere,* and it makes possible long-range radio communication by its ability to reflect radio waves. (**9**)

11

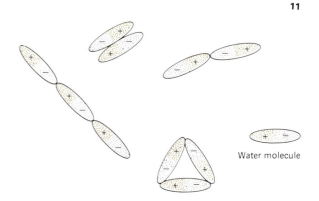

Water molecule

POLAR MOLECULES

Many liquids contain positive and negative ions at all times and hence are able to conduct electricity. Let us look into how the ions in a liquid come into being and how they are able to resist the recombination that occurs so readily in a gas.

When atoms join together to form a molecule, their electrons are shifted in such a manner that electrostatic forces hold the atoms together. We shall consider the details of the binding process later, but for the moment it is

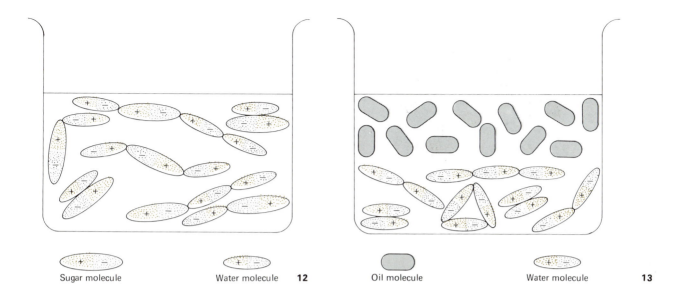

Sugar molecule   Water molecule   **12**

Oil molecule   Water molecule   **13**

Polar molecules of other substances, such as sugar, can join in the aggregates of water molecules. Such substances are therefore easily dissolved by water. (**12**)

The nonpolar molecules of fats and oil, however, do not interact with water molecules. If samples of oil and water are mixed together, the attraction of water molecules for one another acts to squeeze out the oil molecules, and the mixture soon separates into layers of each substance. (**13**)

Fat and oil molecules dissolve only in liquids whose molecules are similar to theirs, which is why gasoline, whose molecules are nonpolar, is a solvent for these compounds. In general, then, "like dissolves like." (**14**)

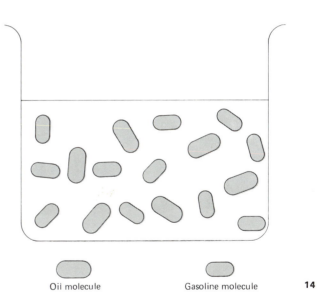

Oil molecule   Gasoline molecule   **14**

ELECTROLYTES

Many solid compounds have structures that consist of ions rather than of neutral atoms. Thus the sodium chloride (NaCl) of ordinary

Na⁺

Cl⁻

15

salt consists of Na$^+$ and Cl$^-$ ions in a regular geometrical array. (The symbol Na$^+$ refers to a sodium atom that has lost an electron to leave it with a net charge of $+e$, and the symbol Cl$^-$ refers to a chlorine atom that has gained an electron to give it a net charge of $-e$.) (**15**)

When a crystal of an ionic compound such as NaCl is placed in water, the water molecules cluster around the crystal's ions with their positive ends toward negative ions and their negative ends toward positive ions. The attraction of several water molecules is usually sufficient to pull an ion from the rest of the crystal, and it moves away surrounded by water molecules. The resulting solution contains ions rather than molecules of the dissolved compound. (**16**)

Substances that separate into free ions when dissolved in water are called *electrolytes* since they are able to conduct electric current by the migration of positive and negative ions. Here is how a solution of NaCl conducts electricity. (**17**)

All ionic compounds soluble in water and some covalent compounds (see Section 37), such as hydrochloric acid (HCl), are electrolytes. Other covalent compounds, such as sugar, are nonelectrolytes even though they are soluble in water.

Since the outer electron structure of an ion may be very different from that of a neutral atom of the same species, it is not surprising that the ions of an element may behave very differently from its atoms or molecules. Thus gaseous chlorine is greenish in color, has a strong, irritating taste, and is very active chemically, whereas a solution of chlorine ions is colorless, has a mild, pleasant taste, and is only feebly active chemically.

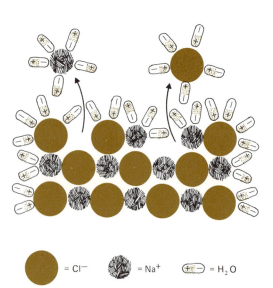

= Cl⁻   = Na⁺   = H₂O

16

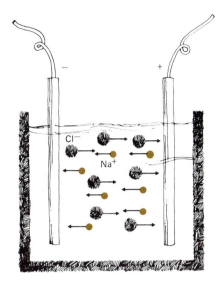

EXERCISES

1. Nearly all the mass of an atom is concentrated in its nucleus. Where is its charge located?

2. In what ways do the Thomson and Rutherford models of the atom agree? In what ways do they disagree?

3. At what distance apart (if any) are the electric and gravitational forces between two electrons equal in magnitude? between two protons? between an electron and a proton?

4. Distinguish between a molecular ion and a polar molecule.

5. Does water dissolve only ionic compounds?

6. Which is a better solvent for an ionic crystal such as NaCl, water or gasoline? Why?

7. How does the flow of electric current through air differ from its flow through a copper wire?

24

ELECTRIC CURRENT

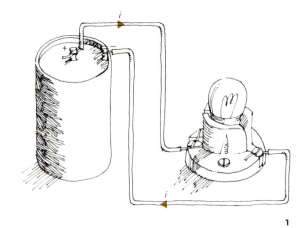

1

"Electromotive action is manifested by two sorts of effects . . .
I shall call the first electric tension, the second electric
current." André Marie Ampère (1775–1836)

A flow of electric charge from one place to another is called an electric current. Electric currents and not static charges are involved in nearly all the practical applications of electricity because such currents transport energy. Electric current provides the most convenient means of carrying energy from a source to a point of application, since only a pair of wires is needed, and it is widely used to supply energy to motors, lamps, heaters, and so forth.

THE AMPERE

The magnitude of an electric current, denoted i, is the rate at which charge passes a given point. If the net charge q goes past in the time interval Δt, then the current is

$$\text{Current} = \frac{\text{charge}}{\text{time interval}},$$

$$i = \frac{q}{\Delta t}.$$

The unit of electric current is the *ampere* (abbreviated A), where

$$1 \text{ A} = 1 \frac{\text{C}}{\text{s}}.$$

The direction of a current is, by convention, taken as that in which *positive* charges would have to move in order to produce the same effects as the observed current. Thus a current is always assumed to proceed from a point of high electric potential to one of low potential, from the positive terminal of a battery or generator to its negative terminal. (**1**)

Despite the above convention, nearly all actual electric currents consist of flows of electrons, which carry negative charges. However, a current that consists of negative particles moving in one direction is electrically the same as a current that consists of positive particles moving the other way. Since there is no overwhelming reason to prefer one way of designating current to the other, we shall follow the usual practice of considering current as a flow of positive electric charge.

Problem. How many electrons per second pass any point in a wire that carries a current of 1 A?

Solution. Since the charge on an electron is 1.6×10^{-19} C, a 1-A current corresponds to a flow of

$$\frac{1 \text{ C/s}}{1.6 \times 10^{-19} \text{ C/electron}} = 6.3 \times 10^{18} \frac{\text{electrons}}{\text{s}}.$$

This figure does not mean, if we have a copper wire in which there is a current of 1 A, that 6.3×10^{18} electrons flow from one end of the wire to the other each second. What it does signify is that this many electrons enter one end of the wire and the same number leave the other end each second, but they do not have

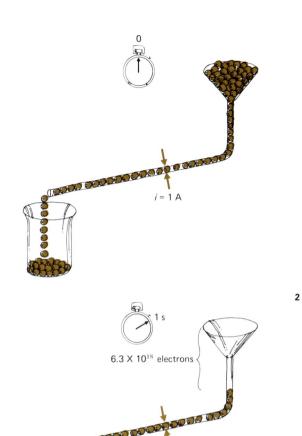

to be the same electrons. A legitimate analogy is with the flow of water in a pipe. (**2**)

RESISTANCE

There are two conditions that must be met in order for an electric current to exist between two points.

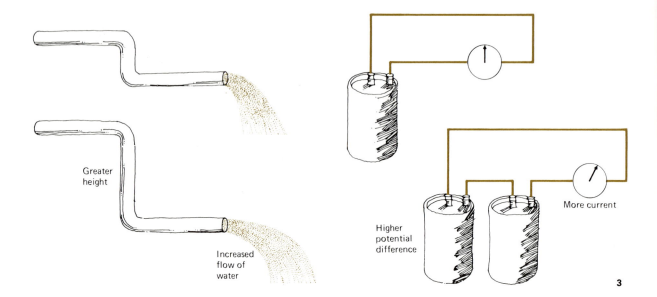

Greater height

Increased flow of water

Higher potential difference

More current

3

1. There must be a path between the two points along which charge can flow. As was discussed earlier, metals, many liquids, and plasmas (gases whose molecules are charged) allow charge to pass through them readily and are classed as conductors. Nonmetallic solids, certain liquids, and gases whose molecules are electrically neutral allow charge to pass through them only with great difficulty and are classed as insulators. A few substances have an intermediate ability to permit the flow of charge and are classed as semiconductors.

2. There must be a difference of potential between the two points. Just as the rate of flow of water between the ends of a pipe depends upon the difference of pressure be-

tween them, so the rate of flow of charge between two points depends upon the difference of potential between them. A large potential difference means a large "push" given to each charge.

The analogy between electric current and water flow is a close one. The rate of flow of water in a pipe may be increased by having the water fall through a greater height, which increases the pressure in the pipe and thereby leads to a greater force on each parcel of water. Similarly the current in a wire may be increased by increasing the potential difference across it, which means a greater electric field intensity in the wire and thus a greater force on the moving charges that constitute the current. (3)

A particular conducting path—for instance, a copper wire, a light bulb, an electric heater, a transistor—is usually called a conductor, even though this is also the name of the class of substances through which current flows readily. The *resistance* of a conductor is the ratio between the potential difference V across it and the resulting current i that flows:

$$R = \frac{V}{i},$$

$$\text{Resistance} = \frac{\text{potential difference}}{\text{current}}.$$

The unit of resistance is the *ohm* (Ω) where

$$1\,\Omega = 1\,\frac{V}{A}.$$

A conductor in which there is a current of 1 A when a potential difference of 1 V exists across it has a resistance of 1 Ω.

The resistance of a conductor depends in general both upon its properties—its nature and its dimensions—and upon the potential difference applied across it. In some conductors R increases when V increases, in others R decreases when V increases, and in still others R depends upon which way the current flows.

OHM'S LAW

Metallic conductors usually have constant resistances, so that i is directly proportional to V in them. This relationship is called *Ohm's law*, since it was first verified experimentally by the German physicist Georg Ohm (1787–1854). Ohm's law states that

$$i = \frac{V}{R}. \qquad \text{Ohm's law}$$

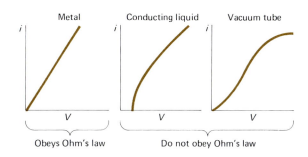

Metal | Conducting liquid | Vacuum tube

Obeys Ohm's law | Do not obey Ohm's law

4

Despite its name, Ohm's law is not a fundamental physical principle, but rather an empirical relationship obeyed by most metals under a wide range of circumstances.

Ohm's law must be distinguished from the definition of resistance,

$$R = \frac{V}{i}.$$

Ohm's law only holds for conductors in which the ratio V/i is constant regardless of the values of V and i, whereas $R = V/i$ can be evaluated regardless of whether it is constant or not as the voltage across a certain conductor is changed. **(4)**

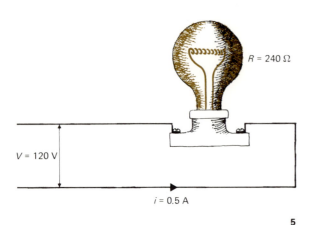

R = 240 Ω

V = 120 V

i = 0.5 A

5

Problem. A light bulb has a resistance of 240 Ω. Find the current that flows through it when it is placed in a 120-V circuit.

Solution. From Ohm's law

$$i = \frac{V}{R}$$

$$= \frac{120 \text{ V}}{240 \text{ Ω}}$$

$$= 0.5 \text{ A. } \textbf{(5)}$$

The resistance of a conductor that obeys Ohm's law depends upon three factors:

1. The material of which it is composed. The ability to carry an electric current varies more than almost any other physical property of matter.

2. Its length L. The longer the conductor, the greater its resistance.

3. Its cross-sectional area A. The thicker the conductor, the less its resistance.

Once again we note the correspondence to

water flowing through a pipe: the longer the pipe, the more chance friction against the pipe wall has to slow down the water, and the wider the pipe, the larger the volume of water that can pass through per second when everything else is the same. **(6)**

ELECTRIC POWER

Electrical energy in the form of electric current is converted into heat in an electric stove, into radiant energy in a light bulb, into chemical energy when a storage battery is charged, and into mechanical energy in an electric motor. The widespread use of electrical energy is due as much to the ease with which it can be transformed into other kinds of energy as to the ease with which it can be transported through wires.

The work that must be done to take a charge q through the potential difference V is, by definition,

$$W = qV.$$

Since a current i carries the amount of charge $q = i\Delta t$ in the time Δt, the work done is

$$W = iV\Delta t.$$

The energy input to a device of any kind through which the current i flows when the potential difference V is placed across it is equal to the product of the current, the potential difference, and the time span.

We recall from Section 13 that *power* is the term given to the rate at which work is being done, so that

$$P = \frac{W}{\Delta t},$$

$$\text{Power} = \frac{\text{work done}}{\text{time interval}}.$$

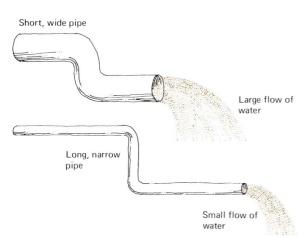

Short, wide pipe

Large flow of water

Long, narrow pipe

Small flow of water

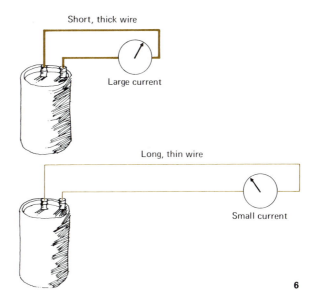

Short, thick wire

Large current

Long, thin wire

Small current

6

When the work is done by an electric current, $W = iV\Delta t$, and

$$P = iV,$$

Electric power = current × potential difference.

The unit of power is the watt, and when i and V are in amperes and volts respectively, P will be in watts.

Problem. How much current is drawn by a $\frac{1}{2}$ hp electric motor operated from a 120-V source of electricity? Assume that all of the electrical energy absorbed by the motor is turned into mechanical work.

Solution. Since 1 hp = 746 W, we find from $P = iV$ that

$$i = \frac{P}{V}$$

$$= \frac{\frac{1}{2}\,\text{hp} \times 746\,\text{W/hp}}{120\,\text{V}}$$

$$= 3.1\,\text{A.} \qquad (7)$$

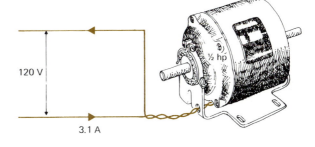

120 V

½ hp

3.1 A

7

If the device through which current passes does obey Ohm's law—that is, if its resistance is a constant—then the power it consumes may be expressed in the alternative forms

$$P = iV,$$

$$P = i^2R,$$

$$P = \frac{V^2}{R}.$$

Depending upon which quantities are known in a specific case, any of the above expressions for P may be used.

Problem. Find the power consumed by a 240-Ω light bulb when the current through it is 0.5 A.

Solution. The formula $P = i^2R$ is easiest to use here. We have

$$P = i^2R$$

$$= (0.5 \text{ A})^2 \times 240 \ \Omega$$

$$= 60 \text{ W}.$$

Owing to the resistance that all conductors offer to the flow of current through them, electrical power is dissipated whenever a current is maintained, regardless of whether the current also supplies energy that is converted to some other form. Electrical resistance is analogous to friction, and so the power consumed in causing a current to flow is dissipated as heat. If too much current flows in a particular wire, it becomes so hot that it may start a fire or even melt. To prevent this from happening, nearly all electrical circuits are protected by fuses or circuit breakers which interrupt the flow of current when i exceeds a certain value. For example, a 15-A fuse in a 120-V power line means that the maximum power that can be carried is

$$P = iV$$

$$= 15 \text{ A} \times 120 \text{ V}$$

$$= 1800 \text{ W}.$$

EXERCISES

1. Would you expect bends in a wire to affect its electrical resistance?

2. It is sometimes said that an electrical appliance "uses up" electricity. What does such an appliance actually use in its operation?

3. Why are two wires used to carry electric current instead of a single one?

4. Currents of 3 A flow through two wires, one which has a potential difference of 60 V across its ends and another which has a potential difference of 120 V across its ends. Compare the rates at which charge passes through each wire.

5. A certain piece of copper is to be shaped into a conductor of minimum resistance. Should its length and cross-sectional area be, respectively, L and A, $2L$ and $A/2$, or $L/2$ and $2A$?

6. A current of 3 A flows through a wire. How much charge passes through the wire per minute?

7. About 10^{20} electrons in each cm participate in carrying a 1-A current in a certain wire. What is the average speed of these electrons?

8. A certain 12-V storage battery is rated at 80 A-hr, which means that when it is fully charged it can deliver a current of 1 A for 80 hr, 2 A for 40 hr, 80 A for 1 hr, etc.
   a) How many coulombs of charge can this battery deliver?
   b) How much energy is stored in it?

9. What is the current in a 60-W, 120-V light bulb when it is operated at 80 V?

10. What is the power rating of an electric motor that draws a current of 3 A when operated at 120 V?

11. How many electrons flow through the filament of a 120-V, 60-W light bulb per second?

12. When a certain 1.5-V battery is used to power a 3-W flashlight bulb, it is exhausted after an hour's use.
    a) How much charge has passed through the bulb in this period of time?
    b) If the battery costs $0.30, find the cost of a kilowatt-hour of electric energy obtained in this way. How does this compare with the cost of the electric energy supplied to your home?

13. The starting motor of a certain car requires a current of 100 A from a 12-V battery to turn over the engine. How many horsepower does this represent? (1 hp = 746 W)

14. An electric water heater has a resistance of $12\,\Omega$ and is operated from a 120-V power line. If no heat escapes from it, how much time is required for it to raise the temperature of 40 kg of water from $15\,°C$ to $80\,°C$?

15. In a Van de Graaff generator an insulating belt is used to carry charges to a large metal sphere. In a typical generator of this kind, the potential difference between the sphere and the source of the charges is 5×10^6 V.
    a) If the belt carries charge to the sphere at a rate of 10^{-3} A, how much power is required?
    b) How much energy in eV will an electron have if it is accelerated by such a potential difference?
    c) Express the answer to (b) in joules.

25

THE MAGNETIC FIELD

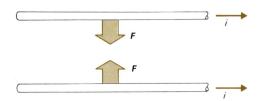

Magnetism and electricity are both manifestations of the same basic interaction between electric charges. Charges at rest relative to an observer exert only electric forces upon one another. When the charges are in motion relative to the observer, he finds that the forces acting between them are different, and these differences are customarily attributed to "magnetic" forces. In reality, magnetic forces represent relativistic corrections to electric forces due to the motion of the charges involved.

MAGNETIC FORCES

Electric charges in motion exert forces upon one another quite different from those they exert while at rest. For instance, if we place a current-carrying wire parallel to another current-carrying wire, with the currents in the same direction, we find that the wires attract each other. (1)

If the currents are in opposite directions, the forces on the wires are repulsive. (2)

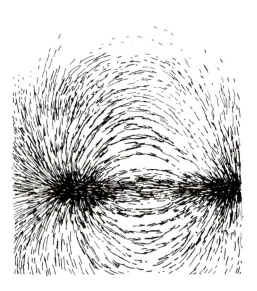

"What led me more or less directly to the special theory of relativity was the conviction that the electromotive force acting on a body in motion in a magnetic field was nothing else but an electric field." Albert Einstein (1879–1955)

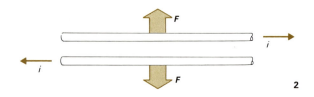

These observations cannot be accounted for unless the motion of the charges is taken into account. Gravitational forces cannot be responsible, since they are never repulsive, and electrical forces apparently cannot be, since there is no net charge on a wire when a current is present in it. The forces that come into being when electric currents interact are called *magnetic forces*. All magnetic effects can ultimately be traced to currents or, more exactly, to moving electric charges. Of course, the word "magnetic" suggests ordinary magnets and their familiar attraction for iron objects, but, as we shall see, this is but one aspect of the whole subject of magnetism.

THE ELECTROMAGNETIC INTERACTION

The gravitational force between two masses and the electrical force between two charges at rest are both *fundamental forces* in the sense that they cannot be accounted for in terms of anything else. On the other hand, the force a bat exerts on a ball is not fundamental because it can be traced to electrical forces between the atoms of the bat and the atoms of the ball.

What about magnetic forces? It is an important fact that whatever it is in nature that manifests itself as an electrical force between stationary charges *must*, according to the principles of relativity, also manifest itself as a magnetic force between moving charges. It is impossible, even in principle, to have one without the other. There is only a single interaction between charges, the *electromagnetic interaction*, and the distinction we make between electric and magnetic forces is an artificial one for the sake of convenience only.

It is hardly obvious that when we pick up a nail with a magnet, we are witnessing a consequence of relative motion. The relativistic effects thus far considered in this book— length contraction, time dilation, mass increase—are imperceptible in everyday life because the speeds of the objects around us are so small compared with the speed of light. Even though experiments show that there are moving electrons in the atoms of the nail and the magnet, their speeds are nowhere near that of light. The puzzle is underscored when we consider that the effective speeds of the electrons that carry a current in a wire are less than 1 mm/s—slower than a caterpillar— yet current-carrying wires do give rise to appreciable magnetic effects, as anyone who has seen an electric motor in operation can testify.

If we think about the matter for a moment, though, the idea that there is a connection between electricity and magnetism via relativity becomes less implausible. For one thing, electrostatic forces are extremely strong, so even a small alteration in their character due to relative motion (which is what magnetic forces represent) may have large consequences. As we saw in Section 23, the electrical force between the electron and proton in a hydrogen atom is more than 10^{39} times greater than the gravitational force between them. Second, though the individual charges involved in magnetic forces usually do move slowly, there may be such enormous numbers of them that the total effect is not negligible; for example, even a modest current in a wire involves the motion of 10^{20} electrons in each centimeter of the wire.

FORCES BETWEEN PARALLEL CURRENTS

As an illustration of how relativity accounts for the magnetic forces between moving charges, let us look into how the forces

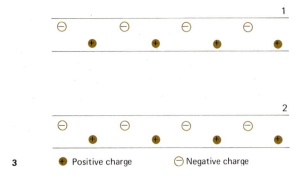

1

2

3   ⊕ Positive charge   ⊖ Negative charge

between two parallel currents come into being. In doing so, we must keep in mind that, as mentioned in Section 21, electric charge is relativistically invariant, so that a charge whose magnitude is found to be q in one frame of reference will be found to be q in all other frames of reference regardless of their relative velocities.

Here are two parallel conductors when no current is present. They contain equally-spaced positive and negative charges that are at rest. The conductors are electrically neutral. (3)

The conductors are now carrying currents in the same direction. The positive charges move to the right at the velocity u and the negative charges move to the left at the same velocity u, as seen from the laboratory frame of reference. The spacing of the charges is smaller than before owing to the Lorentz contraction, which as we know reduces lengths in the direction of motion by the factor $\sqrt{1 - v^2/c^2}$. Since the charges of both signs have the same velocity, the contractions in their spacings are the same, and the conductors are still neutral to an observer in the

laboratory frame of reference. There is an attractive force between the conductors: How does it arise? (4)

We begin by looking at conductor 2 from the frame of reference of one of the negative charges in conductor 1. To this charge, the negative charges in conductor 2 are at rest, since they are (as we see the situation from the outside) all moving at the same velocity u as it is. The spacing of the negative charges is not Lorentz contracted, as it is to an observer in the laboratory, so they are farther apart than in the previous diagram. However, in this frame the positive charges in conductor 2 are moving at the velocity $2u$, and their spacing accordingly exhibits a greater Lorentz contraction. Conductor 2 therefore appears positively charged to the negative charge in conductor 1, and there is an attractive electric force on this charge in its own frame of reference. (5)

From the frame of reference of one of the positive charges in conductor 1, the positive charges in conductor 2 are at rest and their spacing, in the absence of any Lorentz contraction, is greater than we find in the laboratory. The negative charges in conductor 2 have the velocity $2u$ and they are accordingly closer together than in the laboratory frame of reference. There is a net negative charge on conductor 2 as seen by a positive charge in conductor 1, and it is attracted electrically to conductor 2. (6)

An identical argument shows that both the negative and positive charges in conductor 2 are attracted to conductor 1. To any of the charges in either conductor, the force on it is an "ordinary" electric force that occurs because the charges of opposite sign in the other conductor are closer together than the charges of the same sign, yielding a net

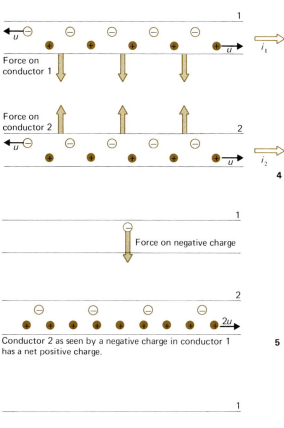

Force on conductor 1

Force on conductor 2

i_1

i_2

4

Force on negative charge

Conductor 2 as seen by a negative charge in conductor 1 has a net positive charge.

5

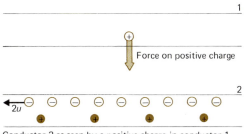

Force on positive charge

Conductor 2 as seen by a positive charge in conductor 1 has a net negative charge.

6

attractive force. To an observer in the laboratory, both conductors are electrically neutral, and he therefore finds it natural to ascribe the force to a special "magnetic" interaction between the currents in the conductors.

As in everything else where there is relative motion, the frame of reference from which a phenomenon is viewed is an essential part of the description of the phenomenon. Although for many purposes it is convenient to think of magnetic forces as something different from electric ones, it is worth keeping in mind that both are manifestations of a single electromagnetic interaction that occurs between charges.

A similar approach accounts for the repulsive force between parallel currents in opposite directions. Again the "magnetic force" turns out to be an inevitable consequence of Coulomb's law, charge invariance, and the principles of special relativity.

Actual currents in metal wires consist of flows of electrons only, with the positive ions remaining in place. The advantage of considering the idealized currents above, which are electrically equivalent to actual currents, is that they are easier to analyze; the results are exactly the same in both cases.

As we have seen, a current-carrying conductor which is electrically neutral in one frame of reference might not be neutral in another frame. But this observation does not apply to the *entire* circuit of which the conductor is a part. Every electric circuit in which a current exists more than momentarily is a closed circuit, so for every current element in one direction that a moving observer finds to have a positive charge there must be another current element in the opposite direction

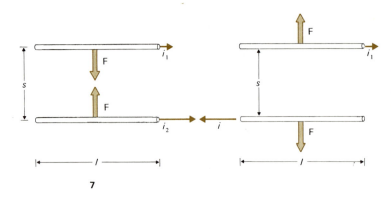

7

which the same observer finds to have a negative charge. Hence we would expect magnetic forces to occur between different parts of a circuit, which is experimentally observed, even though all observers agree on the electrical neutrality of the circuit as a whole. The latter agreement, of course, is required by charge invariance.

THE AMPERE

When the ideas pictured in the preceding section are worked out in detail, the results show that the magnetic force acting on each one of a pair of parallel current-carrying wires is proportional to the magnitudes of the currents i_1 and i_2, proportional to the length l of the wires, and inversely proportional to their separation s. In symbols,

$$F = k' \frac{i_1 i_2}{s} l. \qquad \text{Force between parallel currents}$$

The force is attractive when the currents are in the same direction and repulsive when they are in opposite directions. **(7)**

The constant of proportionality k' has the value

$$k' = 2.00 \times 10^{-7} \frac{\text{N}}{\text{A}^2}.$$

The above equation is used to define the ampere: An ampere is that current flowing in each of two parallel wires 1 m apart that produces a force on each one of exactly 2.00×10^{-7} N per meter of length. In turn, the coulomb is defined in terms of the ampere and the second as that amount of charge transferred per second by a current of 1 A.

Problem. The cables that connect the starting motor of a car with its battery are 1.0 cm apart for a distance of 40 cm. Find the force on each cable when the current in them is 300 A.

Solution. The currents are in opposite directions, and hence the forces are repulsive. The magnitude of the force on each cable is

$$F = k' \frac{i_1 i_2}{s} l$$

$$= 2.0 \times 10^{-7} \cdot \frac{\text{N}}{\text{A}^2} \times \frac{300 \text{ A} \times 300 \text{ A} \times 0.40 \text{ m}}{10^{-2} \text{ m}}$$

$$= 0.72 \text{ N}$$

which is 0.16 lb, a perceptible amount.

THE MAGNETIC FIELD

It is always possible to separate the force on a charge into an electric part, which is independent of its motion, and a magnetic part, which is proportional to its velocity relative to the observer. These forces are additive, so that, for example, it is entirely possible for the magnetic force on a charge to exactly cancel the electric force on it under the proper circumstances. Because fields are always defined in terms of the forces they exert, the additive character of electric and magnetic forces permits a similar separation of an electromagnetic field into electric and magnetic parts.

In our study of electricity, we defined the electric field E at a given location in terms of the force F the field exerts on a stationary charge q placed there. The direction of E is taken as the direction of the force F, and its magnitude is

$$E = \frac{F}{q}$$

by definition. Given E, we can then find the magnitude and direction of the electric force on *any* charge at that location.

The symbol for *magnetic field* is B; this quantity is also known as *magnetic induction* and as *magnetic flux density*. Magnetic forces only act on moving charges, so it is appropriate to use the force F_{mag} exerted on a charge q moving at the velocity v to define B.

Experiment and theory both show that *every* magnetic force on a moving charge, regardless of the details of how the force arises, is proportional to qv. By analogy with the case of the electric field E, we therefore define the magnitude B of the magnetic field acting on a moving charge by the formula

$$B = \frac{F_{mag}}{qv}. \qquad \text{Definition of magnetic field}$$

The magnetic field a distance s from a current i turns out to be

$$B = k'\frac{i}{s}. \qquad \text{Magnetic field around a current}$$

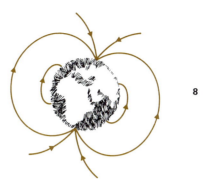

8

The stronger the current and the closer we are to it, the stronger the magnetic field.

THE TESLA

The unit of magnetic field is, from the above definition, the N/A-m, since the units of qv are C-m/s = A-m. The name *tesla* (abbreviated T) has been given to this unit:

$$1\text{ T} = 1\frac{\text{N}}{\text{A-m}}.$$

Thus a force of 1 N will be exerted on a charge of 1 C when it is moving at 1 m/s in a magnetic field whose magnitude is 1 T.

The tesla is also referred to as the weber/m^2 (weber is abbreviated Wb). A different unit of B in common use is the *gauss* (G), where

$$1\text{ G} = 10^{-4}\text{ T},$$
$$1\text{ T} = 10^{4}\text{ G}.$$

Some representative values of magnetic field may help in acquiring a feeling for the magnitude of the tesla.

The magnitude of the earth's magnetic field at sea level is about 3×10^{-5} T. (8)

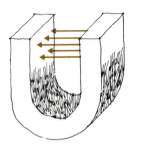

9

The magnetic field near a strong permanent magnet is about 1 T. (9)

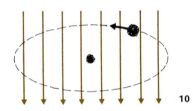

10

The magnetic field produced at the nucleus of a hydrogen atom by the electron circling around it is about 14 T. (10)

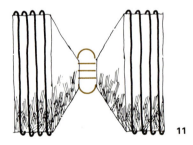

11

The most powerful magnetic fields achieved in the laboratory have magnitudes in the neighborhood of 100 T. (11)

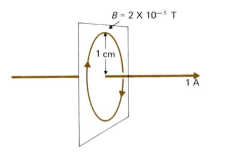

$B = 2 \times 10^{-5}$ T

1 cm

1 A

12

Problem. Find the magnetic field 1 cm from a wire that carries a current of 1 A.

Solution. Since 1 cm $= 10^{-2}$ m, we have

$$B = k' \frac{i}{s}$$

$$= 2 \times 10^{-7} \frac{\text{N}}{\text{A}^2} \times \frac{1 \text{ A}}{10^{-2} \text{ m}}$$

$$= 2 \times 10^{-5} \text{ T}.$$

This is only a little smaller than the magnitude of the earth's magnetic field. Hence great care is taken aboard ships to keep current-carrying wires away from magnetic compasses. (12)

RIGHT-HAND RULE FOR DIRECTION OF *B*

Suppose our charge q is not moving parallel to the current i: what will the magnetic force be on it then? We can investigate this question either theoretically by a relativistic analysis or experimentally with the help of a cathode-ray tube (which is like a television picture tube). Here is what we find experimentally. (13)

Evidently the magnitude and direction of F_{mag} both depend upon the orientations of i and *v*: magnetic forces are a lot more complicated than electric or gravitational ones. Experience has shown that assigning a certain direction to the magnetic field *B* around a

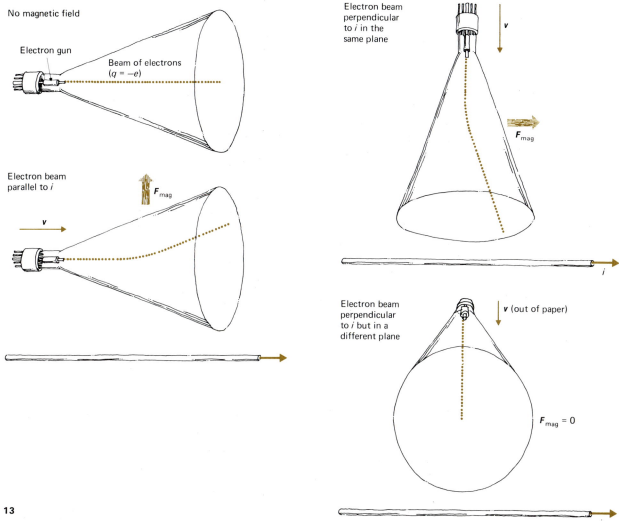

No magnetic field

Electron gun

Beam of electrons
$(q = -e)$

Electron beam
parallel to i

F_{mag}

v

Electron beam
perpendicular
to i in the
same plane

v

F_{mag}

i

Electron beam
perpendicular
to i but in a
different plane

v (out of paper)

$F_{mag} = 0$

i

13

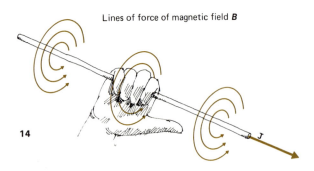

Lines of force of magnetic field **B**

14

current greatly simplifies matters. The direction chosen is such that a charge moving in this direction experiences no magnetic force. The magnetic field around a straight current then takes the form of a series of concentric circles with the current at their center. To avoid ambiguity, the sense of **B** is specified by the right-hand rule:

Grasp the wire with the right hand so the thumb points in the direction of the current; the curled fingers of that hand point in the direction of the magnetic field. (14)

Because the magnetic force on a moving charge varies in magnitude with the relative directions of **v** and **B**, it is necessary to add a proviso to the definition

$$B = \frac{F_{mag}}{qv}$$

given earlier. This proviso is that F_{mag} is the *maximum* force at a given location on the moving charge. The way in which F_{mag} depends upon **v** and **B** will be discussed in the next section.

EXERCISES

1. Why do you think the coulomb is defined in terms of the ampere instead of vice-versa?

2. A stream of protons is moving parallel to a stream of electrons. Is the force between the two streams necessarily attractive? Explain.

3. A beam of protons, initially moving slowly, is accelerated to higher and higher speeds. What happens to the diameter of the beam during this process?

4. A current is passed through a loop of flexible wire. What shape does the loop assume?

5. An observer moves past a stationary electron on the earth's surface. Do his instruments measure an electric field only, a magnetic field only, both an electric field and a magnetic field, or does the answer depend upon his speed?

6. Discuss the chief similarities and differences between electric and magnetic fields.

7. A current is flowing north along a power line. What is the direction of the magnetic field above it? below it?

8. Compare the way in which the magnetic field around a long, straight current varies with distance from the current with the way in which the electric field around a charge varies with distance from the charge.

9. A weak magnetic field exists in interplanetary space between the earth and the sun. What effect would you expect this field to have on the transit times of fast protons emitted by the sun that strike the earth's atmosphere to produce auroras?

10. What should the current be in a long, straight wire if the magnetic field 10 cm from it is to be 0.02 T?

11. At what distance from a long, straight wire carrying a current of 12 A does the magnetic field equal that of the earth?

12. The parallel wires in a lamp cord are 2.0 mm apart. What is the force per meter between them when the cord is used to supply power to a 120-V, 200-W light bulb?

13. A certain electric transmission line consists of two wires 4 m apart that carry currents of 10^4 A. If the towers supporting the wires are 200 m apart, how much force does each current exert on the other between the towers?

26

MAGNETIC FORCES

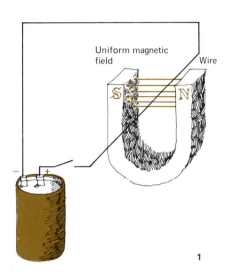

Uniform magnetic field

Wire

S N

1

"It was not until 1820 that a Danish philosopher, Mr. Oersted, succeeded in discovering the relation that had so long been sought after between magnetism and electricity; but it was not where it had been constantly thought to exist that he discovered it. Electricity acts upon a magnet; and a magnet in its turn acts upon electricity; but only when the electricity is in motion . . . We may with Ampère represent the action of a current upon a magnetized needle under a form very convenient for the memory. We have only to conceive a man lying down in the portion of the circuit under considera-tion in such a manner that the current enters by his feet and goes out consequently by his head: furthermore, we have but to conceive that this man has always his face turned towards the needle, so as to look at it; the action is always found to be such that the north pole of the needle is deviated to the left of this man." August de la Rive (nineteenth century)

At this point we shift our concern from the nature of a magnetic field and how it is described to the forces it exerts on currents and moving charges. The emphasis will be on understanding the interaction between a uniform magnetic field and a current in a straight wire and between such a field and a moving charge. This background will prepare us for the more exotic fields and current con-figurations that are the subject of the next section.

MAGNETIC FORCE ON A CURRENT

A few very simple experiments can lead us to the relationships that govern the force a uniform magnetic field exerts on a straight current. First, we note that if we hold a length of wire perpendicular to a magnetic field, nothing happens. (**1**)

When we connect the wire to a battery so that a current flows in it, a force appears that is perpendicular to both the current and the magnetic field. (2)

When we reverse the direction of the current or reverse the direction of the magnetic field, the direction of the force on the wire is also reversed. (3)

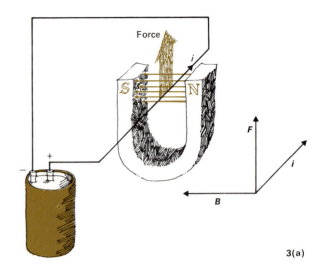

3(a)

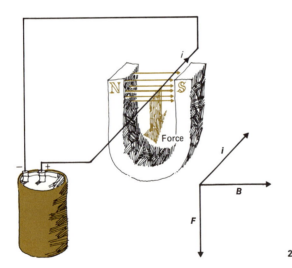

2

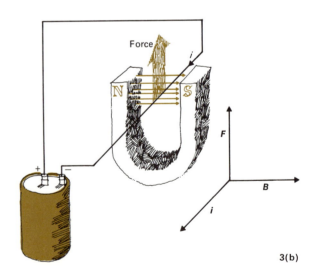

3(b)

Reversing *both* the direction of the current and the direction of the magnetic field leaves the direction of the force on the wire the same as it originally was. (4)

RIGHT-HAND RULE FOR MAGNETIC FORCE

If we look closely at the above pictures, we see that in every case the directions of i, B, and F have the same relationship. Another right-hand rule provides an easy way to remember what this relationship is:

Open your right hand so the fingers are together and the thumb sticks out. When your thumb is in the direction of the current and your fingers are in the direction of the magnetic field, your palm faces in the same direction as the force acting on the current. (5)

MAGNITUDE OF THE FORCE

Experiments show that the magnitude F of the force on a current-carrying wire in a magnetic field depends upon three factors:

1. F is proportional to i, the current in the wire. This is reasonable, since two wires that carry the current i in a magnetic field jointly experience twice the force on one of them alone, and these wires are equivalent to one wire with the current $2i$.

2. F is proportional to l, the length of the wire that is in the magnetic field. The same reasoning as above applies here as well: doubling the length of the wire to $2l$ is the same as having two wires l long in the field, and the result is twice the force.

3. F is proportional to $B_\perp$, the component of the magnetic induction B perpendicular to the wire. The force is a maximum when the wire is perpendicular to B and is zero when the wire is parallel to B. Doubling $B_\perp$ doubles the force,

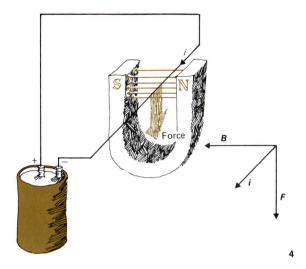

4

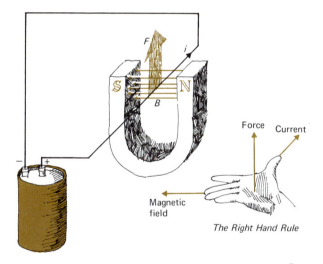

The Right Hand Rule

5

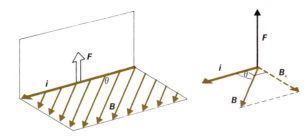

6　When **B** is at the angle θ with respect to i, the force on the current varies with $B_\perp$.

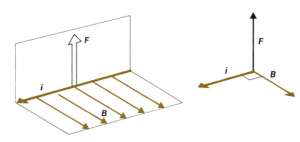

7　When **B** is perpendicular to i, $B_\perp = B$ and the force on the current is a maximum.

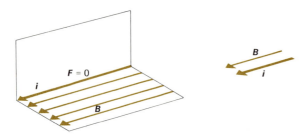

8　When **B** is parallel to i, $B_\perp = 0$ and there is no force on the current.

which makes sense since we would expect that two identical magnets produce twice as much force on a certain wire as either of them alone.

In equation form,

$$F = ilB_\perp. \qquad \text{Magnetic force on a current}$$

The accompanying diagrams illustrate how F depends upon $B_\perp$. (**6**, **7**, and **8**)

Problem. A wire carrying a current of 100 A due east is suspended between two towers 50 m apart. The earth's magnetic field there is in a northerly direction and the magnitude of the field is 3×10^{-5} T. Find the force on the wire exerted by the earth's field.

Solution. The wire is perpendicular to **B**, and so the magnitude of the force is

$$F = ilB$$
$$= 100 \text{ A} \times 50 \text{ m} \times 3 \times 10^{-5} \text{ T}$$
$$= 0.15 \text{ N}.$$

This force is equivalent to about $\frac{1}{2}$ oz. The force is directed upward. (**9**)

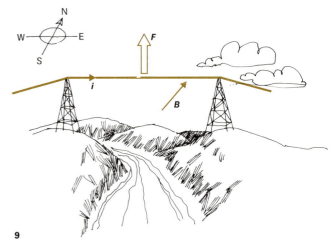

9

MAGNETIC FORCE ON A MOVING CHARGE

A moving charge is magnetically equivalent to a current, and a simple modification of the formula for the magnetic force on a current applies to the magnetic force on a moving charge. Many important devices make use of the magnetic force on a moving charged particle.

Let us transfer the preceding findings about the force on a current in a uniform magnetic field to the case of a moving charged particle. What we do is replace the current i in a wire l long by a particle of charge q and velocity v. This particle requires the time interval

$$\Delta t = \frac{l}{v}$$

to travel the distance l. During this time interval the charge is equivalent to a current of magnitude

$$i = \frac{q}{\Delta t}.$$

Hence we can substitute

$$il = \left(\frac{q}{\Delta t}\right)(v\Delta t) = qv$$

in the formula for the force on a current to obtain

$$F = qvB_\perp. \qquad \text{Force on a moving charge}$$

The direction of the force F is perpendicular to both v and B, and is given by the same right-hand rule as before except that the thumb must now point in the direction of v when q is positive and opposite to v when q is negative.

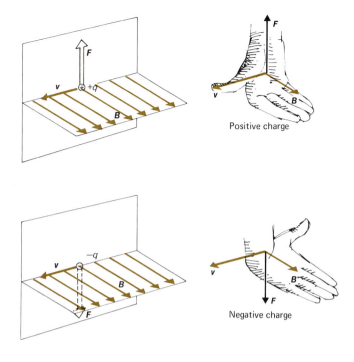

Positive charge

Negative charge

10

With the outstretched fingers in the direction of B, the palm of the right hand faces in the direction of the magnetic force F. (10)

CIRCULAR MOTION IN A MAGNETIC FIELD

A particle of charge q and velocity v that is moving in a uniform magnetic field so that v is perpendicular to B experiences a force of magnitude

$$F = qvB.$$

This force is directed perpendicular to both v

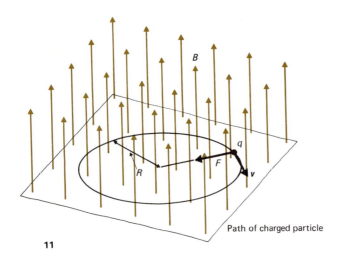

B

q

R

F

v

Path of charged particle

11

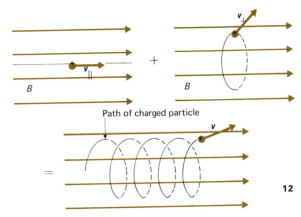

$v_\perp$

B

$v_{||}$

B

+

Path of charged particle

v

=

v

12

and **B**, so the particle therefore travels in a circular path. (11)

To find the radius R of the circular path of the charged particle, we note that the magnetic force qvB is what provides the particle with the centripetal force mv^2/R that keeps it moving in a circle. Equating the magnetic and centripetal forces yields

$$F_{\text{magnetic}} = F_{\text{centripetal}},$$

$$qvB = \frac{mv^2}{R},$$

and so, solving for R, we obtain

$$R = \frac{mv}{qB}.$$

The radius of a charged particle's orbit in a uniform magnetic field is directly proportional to its momentum mv and inversely proportional to its charge and to the magnitude of the field. The greater the momentum, the larger the circle, and the stronger the field, the smaller the circle.

The work done by a force on a body upon which it acts depends upon the component of the force in the direction the body moves. Because the force on a charged particle in a magnetic field is perpendicular to its direction of motion, the force does no work on it. Hence the particle keeps the same speed v and energy it had when it entered the field, even though it is being deflected. On the other hand, the velocity and energy of a moving charged particle in an electric field are always affected by the interaction between the field and the particle.

A charged particle moving parallel to a magnetic field experiences no force and is not deflected; the same particle moving perpendicular to the field follows a circular path. Hence a charged particle whose direction of motion is oblique with respect to **B** follows a helical (corkscrew) path. If we call $v_{||}$ the component of the particle's velocity v that is parallel to **B**, and $v_\perp$ the component of v perpendicular to **B**, then the motion of the particle is the resultant of a forward motion

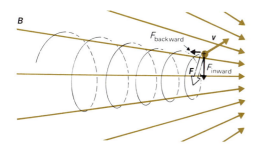

13

at the velocity $v_{\parallel}$ and a circular motion perpendicular to this whose radius is $mv_{\perp}/qB$. (12)

MAGNETIC MIRRORS

An extremely interesting phenomenon occurs when a charged particle moving in a magnetic field approaches a region where the field becomes stronger. The lines of force that describe such a field converge, since the spacing of lines of force is always proportional to the magnitude of the field they describe. The force the particle experiences now has a backward component as well as the inward component that leads to its helical path. The backward force may be strong enough and extend over a long enough distance to reverse the particle's direction of motion. A converging magnetic field can thus act as a *magnetic mirror*. (13)

Magnetic mirrors are found both in the laboratory and in nature. In the laboratory a pair of them are used as a "magnetic bottle" to contain a hot plasma (highly-ionized gas)

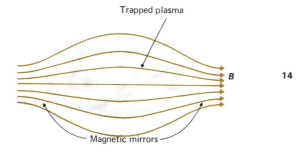

14

in research on thermonuclear fusion (see Section 39). If a solid container were used, contact with its walls would cool the plasma and the ions would then not have enough energy to interact. Magnetic bottles of this kind are somewhat leaky, because ions moving along the axis of a magnetic mirror experience no backward force and hence are able to escape. (14)

The earth's magnetic field traps electrons and protons from space in the *magnetosphere*, a giant doughnut-shaped magnetic bottle that surrounds the earth and extends from about 600 mi above the equator out to perhaps 40,000 mi. Within the magnetosphere are two

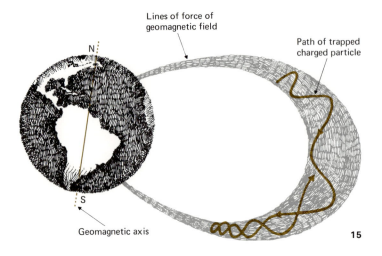

Lines of force of
geomagnetic field

N

Path of trapped
charged particle

S

Geomagnetic axis

15

regions called the *Van Allen belts* that contain at times large numbers of particles with relatively high energies—protons of several MeV in the outer belt and of as much as 100 MeV in the inner one. The diagram shows a typical particle trajectory in the magnetosphere. (**15**)

THE MASS SPECTROMETER

The mass of an atom is one of its characteristic properties and an accurate knowledge of atomic masses provides considerable insight into nuclear phenomena. A variety of instruments with the generic name of *mass spectrometers* have been devised to measure atomic masses, and we shall consider the operating principles of the particularly simple one shown here. (**16**)

The first step in the operation of this spectrometer is to produce ions of the substance under study. If the substance is a gas, ions can be formed readily by electron bombardment, while if it is a solid it often can be incorporated into an electrode that is used as one terminal

of an electric arc discharge. The ions emerge from their source through a slit with the charge $+e$ and are then accelerated by an electric field. (Ions with other charges are sometimes present but are easily taken into account.)

When the ions enter the spectrometer, as a rule they are traveling in slightly different directions with slightly different speeds. A pair of slits serves to collimate the beam, that is, to eliminate those ions not moving in the desired direction. Then the beam passes through a *velocity selector*. The velocity selector consists of uniform electric and magnetic fields that are perpendicular to each other and to the beam of ions. The electric field E exerts the force

$$F_{\text{electric}} = eE$$

on the ions to the right, whereas the magnetic field B exerts the force

$$F_{\text{magnetic}} = evB$$

on them to the left. In order for an ion to reach the slit at the far end of the velocity selector it must suffer no deflection inside the selector, which means that the condition for escape is

$$F_{\text{electric}} = F_{\text{magnetic}},$$
$$eE = evB.$$

Hence the ions that escape all have the speed

$$v = \frac{E}{B}.$$

Once past the velocity selector the ions enter a uniform magnetic field and follow circular paths of radius

$$R = \frac{mv}{eB}.$$

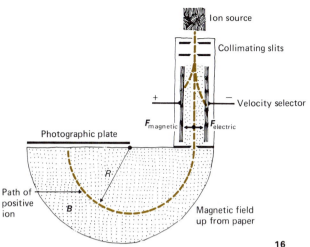

Ion source
Collimating slits
+ −
F_{magnetic} ← → F_{electric} — Velocity selector
Photographic plate
R
Path of positive ion
B
Magnetic field up from paper

16

Since v, e, and B are known, a measurement of R yields a value for m, the ion mass.

Problem. The velocity selector of a mass spectrometer consists of an electric field of $E = 40{,}000$ V/m perpendicular to a magnetic field of $B = 0.0800$ T. The same magnetic field is used to deflect the ions that have passed through the velocity selector. Ions of a certain isotope of lithium are found to have radii of curvature in the magnetic field of 39.0 cm. What is their mass?

Solution. The speed of the ions is

$$v = \frac{E}{B}$$

$$= \frac{4.00 \times 10^4 \text{ V/m}}{8.00 \times 10^{-2} \text{ T}}$$

$$= 5.00 \times 10^5 \text{ m/s}.$$

From the formula for the radius of curvature of the ion path

$$R = \frac{mv}{eB},$$

we see that

$$m = \frac{eBR}{v},$$

and so

$$m = \frac{1.60 \times 10^{-19} \text{ C} \times 8.00 \times 10^{-2} \text{ T} \times 0.390 \text{ m}}{5.00 \times 10^5 \text{ m/s}}$$

$$= 9.98 \times 10^{-27} \text{ kg.}$$

EXERCISES

1. A current-carrying wire is in a magnetic field.
   a) What angle should the wire make with B for the force on it to be zero?
   b) What should the angle be for the force to be a maximum?

2. A charged particle moving in a magnetic field follows a curved path, which means that its linear momentum is changing. How can this observation be reconciled with the principle of conservation of linear momentum?

3. A charged particle moves through a magnetic field. Which of the following quantities is never affected by the field: the particle's mass, velocity, linear momentum, or kinetic energy?

4. In a certain electric motor, wires carrying currents of 4 A are perpendicular to a magnetic field of 1.2 T. Find the force on each cm of these wires.

5. A wire 1 m long is perpendicular to a magnetic field of 0.05 T. What is the force on the wire when it carries a current of 2 A?

6. A horizontal north–south wire 5 m long is in a 0.02-T magnetic field whose direction is east. What is the magnitude and direction of the force on the wire
   a) when a 4-A current flows north in it?
   b) when a 4-A current flows south in it?

7. A vertical wire 2 m long is in a 10^{-2}-T magnetic field whose direction is northeast. What is the

magnitude and direction of the force on the wire
a) when a 5-A current flows upward in it?
b) when a 5-A current flows downward in it?

8. A copper wire whose linear density is 10 g per m is stretched horizontally perpendicular to the direction of the horizontal component of the earth's magnetic field at a place where the magnitude of that component is 2×10^{-5} T. What must the current in the wire be in order that its weight be supported by the magnetic force on it? What do you think would happen to such a wire if this current were passed through it?

9. What is the radius of the path of a 4×10^4-eV electron in a 0.02-T magnetic field?

10. Compute the radius of curvature in the earth's magnetic field ($B = 3 \times 10^{-5}$ T) of
a) a proton whose speed is 2×10^7 m/s, and
b) an electron of the same speed.

11. A velocity selector uses a magnet to produce a magnetic field of 0.05 T and a pair of parallel metal plates 1 cm apart to produce a perpendicular electric field. What potential difference should be applied to the plates to permit singly charged ions of speed 5×10^6 m/s to pass through the selector?

12. An electron moving through an electric field of intensity 500 V/m and a magnetic field of induction 0.1 T experiences no force. The two fields and the electron's direction of motion are all mutually perpendicular. What is the speed of the electron?

13. A mass spectrometer employs a velocity selector consisting of a magnetic field of 0.04000 T perpendicular to an electric field of 50,000 V/m. The same magnetic field is then used to deflect the ions. Find the radius of curvature of singly charged lithium ions of mass 1.16×10^{-26} kg in this spectrometer.

27

CURRENT LOOPS

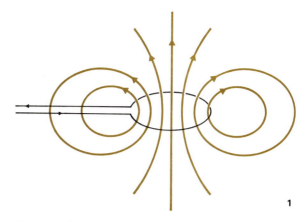

1

"From about a magnetical body the virtue magnetical is poured out on every side . . . it has no intercourse with air, water, or any non-magnetick; nor does it move a magnetick with any motion by forces rushing upon it, but being present in an instant, it invites friendly bodies." William Gilbert (1540–1603)

Current loops are common in nature and in technology: the magnetic fields of an electron and of the earth are the same as those produced by a current loop, and the operation of an electric motor depends upon the interaction between a current loop and a magnetic field. The origin of the magnetic properties of iron can be traced to current loops on a sub-atomic scale.

MAGNETIC FIELD OF A CURRENT LOOP

The magnetic field around a circular current loop has the configuration shown here. (**1**)

The direction of *B* can be remembered with the help of yet another right-hand rule:

*Grasp the loop so the curled fingers of the right hand point in the direction of the current; the thumb of that hand then points in the direction of **B** inside the loop.* (**2**)

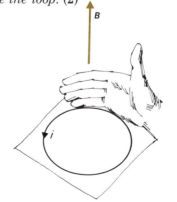

2

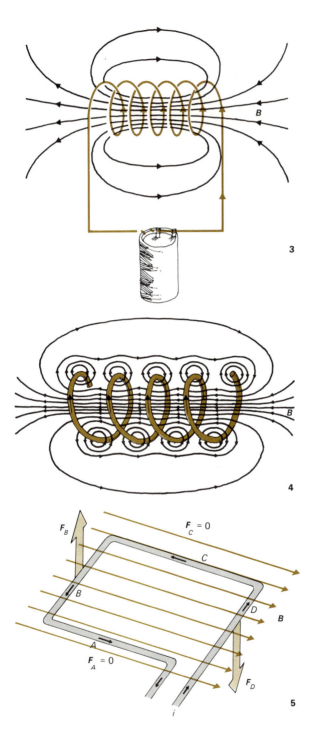

A *solenoid* is a coil of wire in the form of a helix. If the turns are close together and the solenoid is long relative to its diameter, then the magnetic field within it is uniform and parallel to its axis except near the ends. The direction of the field inside a solenoid is given by the same right-hand rule that gives the direction of B inside a current loop. (3)

Here is an expanded view of a solenoid showing how the magnetic fields of the individual turns add together to yield a uniform field inside it. In an actual solenoid the turns are as close together as possible. (4)

TORQUE ON A CURRENT LOOP

A straight current-carrying wire is acted upon by a force when it is in a magnetic field, provided that it is not parallel to the direction of B. A loop of current in a uniform magnetic field experiences no net force, but instead a torque (twisting force) occurs that tends to rotate the loop so that its plane is perpendicular to B. This is the principle that underlies the operation of all electric motors, from the tiniest one in a clock to the giant of many thousand horsepower in a locomotive.

Let us examine the forces on each side of a rectangular current-carrying wire loop whose plane is parallel to a uniform magnetic field B.

Sides A and C of the loop are parallel to B and so there is no magnetic force on them. Sides B and D are perpendicular to B, however, and each therefore experiences a force. To find the directions of the forces on B and D we can use the right-hand rule: with the fingers of the right hand in line with B and the outstretched thumb in line with i, the palm faces the same way as F. What we find is that F_B is opposite in direction to F_D. (5)

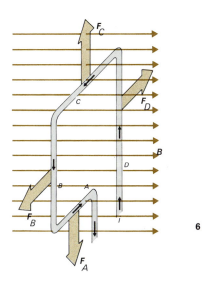

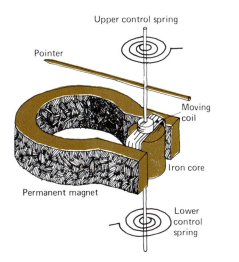

The forces F_B and F_D are the same in magnitude, so there is no net force on the current loop. But F_B and F_D do not act along the same line, and hence they exert a torque on the loop that tends to turn it. This is a perfectly general conclusion that holds for a current loop of any shape in a magnetic field.

If the loop is perpendicular to the magnetic field instead of parallel to it, there is neither a net force nor a net torque on it. This is easy to verify from a diagram, bearing in mind the right-hand rule for the direction of the force on each side of the loop. Evidently F_A and F_C cancel each other out, and F_B and F_D also cancel each other out. There is no tendency to turn now because F_A and F_C have the same line of action and F_B and F_D have the same line of action. (6)

The above results can be summarized by saying that *a current loop in a magnetic field always tends to turn so that it is perpendicular to the field.*

THE GALVANOMETER

We have seen that a current-carrying wire loop tends to rotate in a magnetic field. The *galvanometer* capitalizes upon this behavior to furnish a means for measuring current. The diagram shows the basic construction of a galvanometer. A U-shaped permanent magnet is used to provide a magnetic field, and between its poles is a small coil wound on an iron core to enhance the torque developed when the unknown current is passed through it. The coil assembly is held in place by two bearings that permit it to rotate, and a pair of hairsprings keeps the pointer at 0 when there is no current in the coil. (7)

When a current flows, there is a torque on the coil because of the interaction between the current and the magnetic field, and the coil rotates as far as it can against the opposing torque of the springs. The more the current, the stronger the torque, and the farther the coil turns.

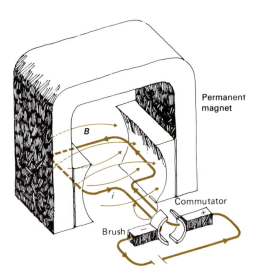

Permanent magnet

Commutator

Brush

8

Galvanometers of the above type can be constructed that are able to respond to currents of as little as 0.1 microampere (10^{-7} A), though ordinary commercial meters are less sensitive. Even greater sensitivity can be attained if the moving coil is suspended by a thin wire to which a small mirror is attached: bearing friction is avoided in this way, and the mirror deflects a light beam so that the "pointer" may be a meter or more long instead of a few centimeters. Laboratory galvanometers like this can be used to measure currents of 10^{-10} A.

THE DC ELECTRIC MOTOR

The torque which a magnetic field exerts on a current loop disappears when the loop turns so that its plane is perpendicular to the field direction. If the loop swings past this position, the torque on it will be in the opposite sense and will return the loop to the perpendicular orientation. In order to construct a motor capable of continuous rotation, then, the current in the loop must be automatically reversed each time it turns through 180°. The method by which this reversal is accomplished is shown in the diagram. The current is led to the loop by means of graphite rods called *brushes* which press against a split ring called a *commutator*. As the loop rotates, the current is reversed twice per turn as the commutator segments make contact alternately with the brushes. The torque is always in the same direction, except at the moments of switching when it is zero because the loop is perpendicular to the field. However, the angular momentum of the loop carries it past this point, and it can continue to turn indefinitely. (8)

While actual direct-current electric motors are the same in principle as the simple device described above, they employ a number of stratagems to increase the available torque. Electromagnets rather than permanent magnets provide the field, and there are six or more different coils with many turns each on a slotted iron core called an *armature*, instead

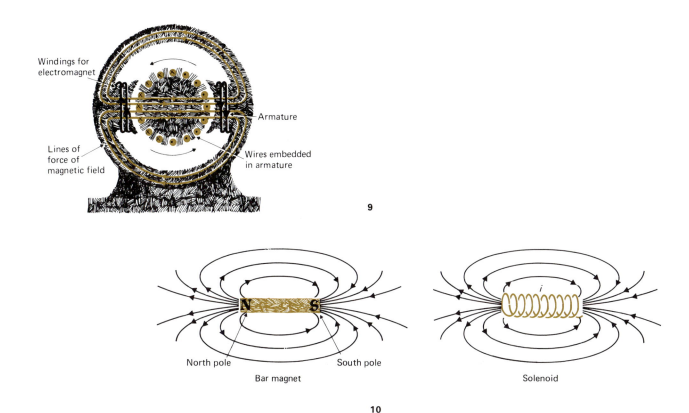

Windings for electromagnet

Armature

Lines of force of magnetic field

Wires embedded in armature

9

North pole

South pole

Bar magnet

Solenoid

10

of a single loop. A commutator with a pair of segments for each coil is provided so that only those coils parallel to the magnetic field receive current at any time, which means that the maximum torque is developed continuously. (9)

MAGNETIC POLES

It may seem strange that we have not yet referred to the "magnetic poles" that figure so prominently in elementary discussions of magnetism. The reason is that all magnetic fields, including those of permanent magnets, originate in electric currents, and all magnetic forces arise from interactions between currents (or, more precisely, electric charges in motion) and magnetic fields; to understand electromagnetic phenomena, it is necessary to start directly from these fundamental concepts.

The magnetic field of a bar magnet is identical with that of a solenoid, which is not surprising since all permanent magnets owe their character to an alignment of atomic current loops no different in principle from the alignment of the current loops in a solenoid. Hence we should expect to be able to predict the behavior of permanent magnets from the ideas of this section, and the results of these predictions do indeed agree with experiment. (10)

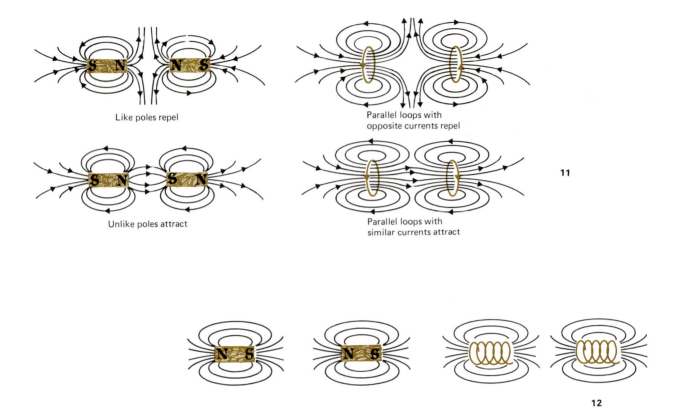

Like poles repel

Parallel loops with
opposite currents repel

Unlike poles attract

Parallel loops with
similar currents attract

11

12

Because the external magnetic field of a bar magnet seems to originate in its ends, these are by custom called its *poles*. At one time it was believed that the poles are magnetic charges analogous to electric charges, and that the field of the magnet is due to these poles. This belief was reinforced by the repulsion of like poles and the attraction of unlike ones, phenomena that have their true explanation in the forces between parallel and antiparallel currents. (**11**)

An important difference between magnetic poles and electric charges is that the former invariably come in pairs of equal strength and opposite polarity: if a magnet is sawn in half, the poles are not separated but instead two new magnets are created. Magnetic poles are therefore only superficially like electric charges, and all effects that can be attributed to them can be explained in terms of the behavior exhibited by current-carrying solenoids. (**12**)

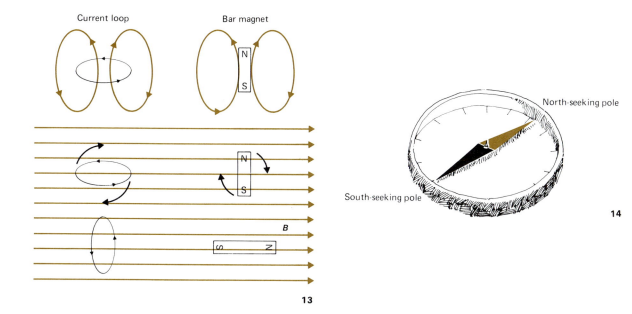

Current loop

Bar magnet

N
S

N
S

B

S   N

13

North-seeking pole

South-seeking pole

14

Curiously enough, although no known elementary particle has magnetic properties that cannot be duplicated by a suitable combination of currents, there seems to be no fundamental physical principle that prohibits the existence of a particle that would behave like a single, isolated magnetic pole. Despite much effort, no such particle has ever been found. Perhaps *magnetic monopoles* are simply very rare in nature; or perhaps there *is* a basic reason why they cannot exist. The only thing that seems certain is that no magnetic phenomenon yet discovered needs the hypothesis of a magnetic monopole for its explanation.

THE EARTH'S MAGNETIC FIELD

As we saw, a current loop tends to rotate in a magnetic field until its axis is parallel to the field. A bar magnet, too, tends to rotate in a magnetic field until it is aligned with the field direction. **(13)**

Because of the earth's magnetic field, a magnet suspended by a string turns so as to line up in an approximately north–south direction. A compass consists of a pivoted magnetized iron needle together with a card that permits directions relative to magnetic north to be determined. The end of a freely-swinging magnet that points toward the north is called its north-seeking pole, usually shortened to just *north pole*, and the other end is its south-seeking pole, or *south pole*. Magnetic lines of force leave the north pole of a magnet and enter its south pole. (The north geomagnetic pole is thus in reality a south pole, and the south geomagnetic pole is a north pole; this has been a source of confusion for several hundred years.) **(14)**

Measurements of the earth's magnetic field show that it is very much like the field that would be produced by a powerful current loop whose center is a few hundred miles from the

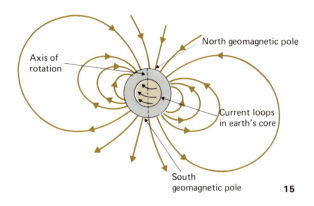

Axis of rotation

North geomagnetic pole

Current loops in earth's core

South geomagnetic pole

15

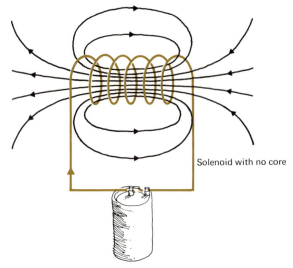

Solenoid with no core

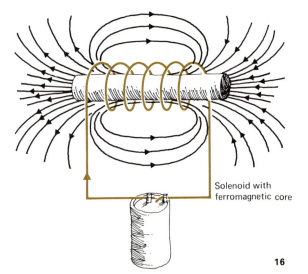

Solenoid with ferromagnetic core

16

earth's center and whose plane is tilted by 11° from the plane of the earth's equator. On the basis of geological evidence the earth is thought to have a core of molten iron 3470 km in radius (a little over half the earth's radius), and there is little doubt today that electric currents in this core are responsible for the observed geomagnetic field. The details of how these currents came into being and how they are maintained are still uncertain, but there seems little difficulty in accounting for them in a general way. The "magnetic poles" of the earth are those points on the earth's surface where the lines of force are vertical; they are near but not at the geographical north and south poles. (**15**)

FERROMAGNETISM

The magnetic field produced by a current-carrying solenoid is changed in strength when a rod of almost any material is inserted in it. Some materials increase B (for example, aluminum), others decrease B (for example, bismuth), but in almost all cases the difference

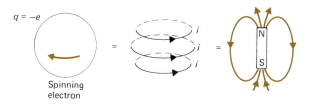

Spinning
electron

$q = -e$

17

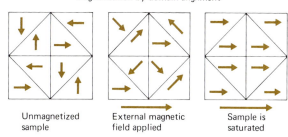

Magnetization by domain alignment

| Unmagnetized sample | External magnetic field applied | Sample is saturated |

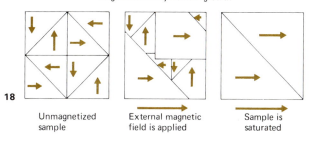

Magnetization by domain growth

18

| Unmagnetized sample | External magnetic field is applied | Sample is saturated |

is very small. However, a few substances yield a dramatic increase in B when placed in a solenoid—the new field may be hundreds or thousands of times greater in magnitude than before. Such substances are called *ferromagnetic*; iron is the most familiar example. **(16)**

Ferromagnetism can be traced to the magnetic behavior of electrons. An electron in certain respects resembles a spinning charged sphere, which we may imagine as a series of ultraminute current loops. Hence every electron has the magnetic field of a tiny bar magnet due to its spin. **(17)**

The spin magnetic fields of most or all of the electrons in an atom (depending on the element involved) cancel each other out in pairs. (If we shake a number of small bar magnets together in a box, they will also end up paired off with opposite poles together.) In a ferromagnetic material, the unpaired electrons in each atom interact strongly with their counterparts in adjacent atoms, which causes the unpaired spin magnetic fields in all the atoms to be locked together. Atoms in a ferromagnetic material are accordingly grouped together in assemblies called *domains*, each about 5×10^{-5} m across and just visible in a microscope.

In an unmagnetized sample, the directions of magnetization of the domains are randomly oriented, though within each domain the unpaired electron spins are parallel. When such a sample is placed in an external magnetic field, either the spins within the domains turn to line up with the field or, in pure and homogeneous materials, the domain walls change so that those domains lined up with the field grow at the expense of the others. The former process requires stronger fields to take place than the latter. Hence good "permanent" magnets are irregular in structure, for instance steel rather than pure iron, and, once magnetized, cannot change their magnetization by the easy process of domain wall motion. When all the unpaired spins in a ferromagnetic sample are lined up, no further increase in B is possible, and the sample is said to be *saturated*. **(18)**

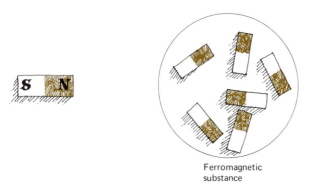

Ferromagnetic
substance

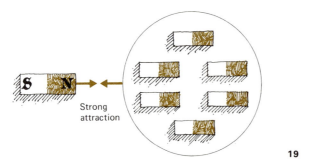

Strong
attraction

19

Above a certain temperature (760°C in the case of iron), the atoms in a domain acquire enough kinetic energy to overcome the interatomic forces that hold their spins in alignment. The spins and their magnetic fields then become randomly oriented, and the ferromagnetic material loses its special magnetic properties. Thus heating a "permanent magnet" sufficiently will cause it to become demagnetized. Hammering a permanent magnet also tends to disturb the alignment of spins, though some ferromagnetic materials are able to retain their magnetization despite almost any mechanical disturbance.

HOW A MAGNET ATTRACTS IRON

The mechanism by which a magnet or a solenoid attracts an iron object (or an object of any other ferromagnetic material such as cobalt or nickel) is very similar to the way in which an electric charge attracts an uncharged object. First the presence of the magnet causes the atomic magnets in the iron object to line up with its field, and then the attraction of opposite poles leads to a force on the object that draws it toward the magnet. (**19**)

Iron, unlike steel (which is an alloy, or mixture, of iron with carbon and other elements), tends to lose its magnetization when an external magnetic field is removed. Hence if an iron rod is placed inside a solenoid, we have a very strong magnet that can be turned on and off just by switching the current in the solenoid on and off. Such an *electromagnet* is much stronger than the solenoid itself and can be stronger than a permanent magnet as well; also, unlike a permanent magnet, its field can be controlled at will by adjusting the current in the solenoid. Electromagnets are among the most widely used electrical devices. They range in size from the tiny one in a telephone receiver that causes a steel plate to vibrate and thus produce the sounds we hear to the giant electromagnets used to pick up automobiles in scrap yards.

When iron filings are sprinkled on a card held near a current-carrying wire or a permanent magnet, they tend to line up in the direction of B by the mechanism shown earlier. Friction between the filings and the card prevents them from moving but not from rotating, which is why this method reveals in such a striking way magnetic field configurations. We must keep in mind that, despite the appearance of these patterns, lines of force do not exist as such in nature: a magnetic field is a continuous property of the region of space it occupies. Lines of force are simply aids to the imagination. Here are the patterns of iron filings formed over three bar magnets. (**20**)

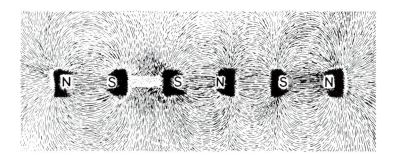

20

EXERCISES

1. A current-carrying wire loop is in a uniform magnetic field. Under what circumstances, if any, will there be no torque on the loop? no net force? neither torque nor net force?

2. A magnetic field does *not* exert a force or torque on which of the following: a stationary charge, a moving charge, a stationary permanent magnet, or a moving permanent magnet?

3. Would you expect a compass to be more accurate near the equator or in the polar regions? Why?

4. Why is a piece of iron attracted by *either* pole of a magnet?

28

ELECTROMAGNETIC INDUCTION

"I am busy just now again on electro-magnetism, and think I have got hold of a good thing, but can't say. It may be a weed instead of a fish that, after all my labour, I may at last pull up." Michael Faraday (1791–1867)

We have seen that, by causing charges to move, an electric field is able to produce a magnetic field. Is there any way in which a magnetic field can produce an electric field? This problem was unsuccessfully tackled by many of the early workers in electricity and magnetism. Finally, in 1831, Michael Faraday in England and Joseph Henry in the United States independently discovered the phenomenon of electromagnetic induction.

ELECTROMAGNETIC INDUCTION

When a stationary wire is in a magnetic field, no current flows in it. What Faraday and Henry found is that *moving* the wire causes a current to flow. It does not matter whether the wire or the source of the magnetic field is being moved, provided that a component of the motion is perpendicular to the field; the origin of the current lies in the *relative motion* between a conductor and a magnetic field. This effect is known as *electromagnetic induction*. (**1** and **2**)

INDUCED POTENTIAL DIFFERENCE

Faraday generalized his observations with the help of the notion of lines of force:

A potential difference is produced in a conductor whenever it cuts across magnetic lines of force.

It is this potential difference that leads to the current that flows whenever there is relative motion between a conductor and a magnetic field. In fact, it is not even necessary for there to be actual motion of either a wire or of a source of magnetic field, because a magnetic field that changes in strength has moving lines of force associated with it. Thus we can regard a changing magnetic field as producing an electric field, a notion we will find important in the next section when electromagnetic waves are considered.

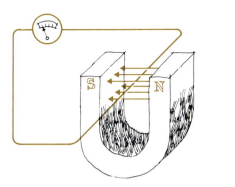

1

Stationary wire in magnetic field: no current flows.

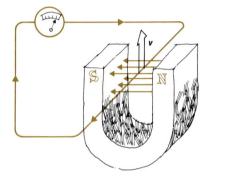

2

Wire is moved across magnetic field: a current flows. Reversing the direction of motion also reverses the direction of the current.

Moving a wire *parallel* to a magnetic field does not give rise to a current: electromagnetic induction only occurs when there is a component of the wire's velocity perpendicular to the lines of force.

Electromagnetic induction has a straightforward explanation in terms of the force exerted by a magnetic field on a moving charge. As we saw earlier, a wire carrying a

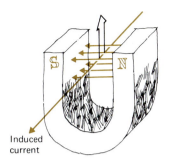

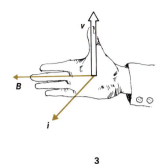

Induced current

3

current is pushed sideways in a magnetic field because of the forces exerted on the moving electrons. In Faraday's and Henry's experiments electrons are also moved through a magnetic field, but now by shifting the entire wire. As before, the electrons are pushed to the side, and in consequence move along the wire. The motion of these electrons along the wire is the electric current we measure.

The right-hand rule for the force on a moving positive charge in a magnetic field can also be used to give the direction of the induced current in a wire moving across a magnetic field. Hold your right hand so that the fingers point in the direction of B and the outstretched thumb is in the direction of v. The palm then faces in the same direction as the force on the positive charges in the wire, and hence in the direction of the conventional current. (**3**)

MOVING WIRE IN A MAGNETIC FIELD

When a wire moves through a magnetic field B with a velocity v that is not parallel to B, a potential difference V_i comes into being between the ends of the wire. It is not hard on the basis of what we already know to determine the magnitude of V_i in terms of B, v, and the length l of the wire. We shall assume that the wire, its direction of motion, and the magnetic field are all perpendicular to one another as in the previous diagrams.

According to the discussion on page 233, the force on a charge q in a wire, when the wire moves with the velocity v perpendicular to a magnetic field B, has the magnitude

$$F = qvB.$$

The direction of this force is along the wire. (4)

Let us consider a charge q that moves from one end of the wire to the other under the influence of the magnetic force qvB. The work done by this force on the charge is

$$\text{Work} = \text{force} \times \text{distance}$$

$$W = Fl$$

$$= qvBl,$$

since the wire is l long. By definition the potential difference between two points is the work done in moving a unit charge between these points, so the potential difference between the ends of the wire is W/q or

$$V_i = vBl. \qquad \begin{array}{l}\text{Induced}\\ \text{potential}\\ \text{difference}\end{array}$$

This potential difference is induced in the wire by its motion through the magnetic field.

How large will the current that flows in the wire be? The answer is given by Ohm's law, since an induced potential difference is like any other potential difference in its ability to cause a current to flow. If the total resistance in the circuit is R, the resulting current will be equal to

$$i = \frac{V_i}{R}.$$

Problem. Compute the potential difference between the wing tips of a jet airplane induced by its motion through the earth's magnetic

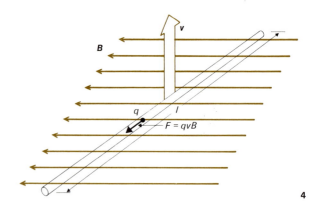

4

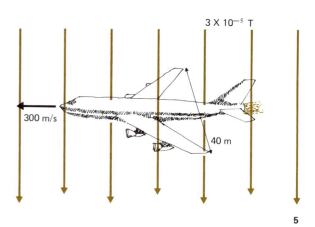

5

field. The total wing span of the airplane is 40 m and its speed is 300 m/s in a region where the vertical component of the earth's field has the magnitude 3×10^{-5} T.

Solution. Substituting the given values of v, l, and B in the formula for the induced emf yields

$$V_i = vBl$$

$$= 300 \, \frac{\text{m}}{\text{s}} \times 3 \times 10^{-5} \, \text{T} \times 40 \, \text{m}$$

$$= 0.36 \, \text{V}. \, (5)$$

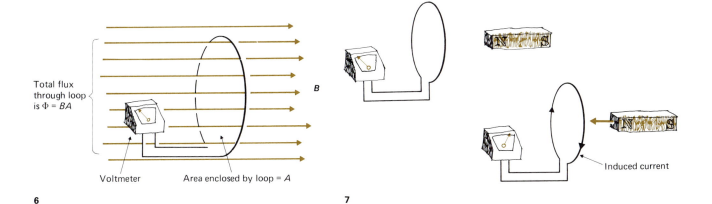

Total flux
through loop
is Φ = BA

B

Voltmeter   Area enclosed by loop = A

6

7

Induced current

FARADAY'S LAW

A convenient approach to electromagnetic induction makes use of the notion of *magnetic flux*. If the area of a wire loop is A and it is perpendicular to a magnetic field B, the total magnetic flux Φ (Greek letter *phi*) through the loop is defined as

$$\Phi = BA. \qquad \text{Magnetic flux}$$

The loop can have any shape, but it must lie in a plane. The unit of flux is the *weber*. (6)

Faraday found that the potential difference V_i in such a wire loop is *equal to the rate of change of the flux through it*. That is,

$$V_i = -\frac{\Delta\Phi}{\Delta t}, \qquad \begin{array}{l}\text{Induced}\\ \text{potential}\\ \text{difference}\end{array}$$

where $\Delta\Phi$ is the change in the flux Φ that takes place during a period of time Δt. The values of the magnetic field B and the loop area A are, in themselves, irrelevant; only the speed with which either or both of them changes is important. This conclusion is called Faraday's law of electromagnetic induction.

LENZ'S LAW

The minus sign in Faraday's formula for induced emf is a consequence of the law of conservation of energy. If the sign of V_i were the same as that of $\Delta\Phi/\Delta t$, the induced electric current would be in such a direction that its own magnetic field would *add* to that of the external field B; this additional changing field would then augment the existing rate of change of the flux Φ, and more and more current would flow even if the external contribution to Φ were to stay constant! Since energy must be supplied at all times to maintain a current, owing to the continual energy loss in the form of heat in every conductor in which there is a current, removing the source of energy (here the changing flux of external origin) must cause the current to cease. Therefore *the direction of the induced current must be such that its own magnetic field opposes the changes in flux that are inducing it*, a conclusion known as Lenz's law.

Here is a simple example of Lenz's law. The current induced in the wire loop is such that its magnetic field is opposite to the field of the permanent magnet. (7)

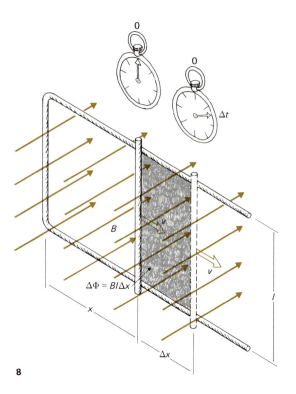

8

DERIVATION OF FARADAY'S LAW

Let us obtain Faraday's law of electromagnetic induction from the formula for the induced potential difference in a wire moved through a magnetic field. The diagram shows a moving wire that slides across the legs of a U-shaped metal frame, so that the frame completes the loop. At the moment (say $t = 0$) when the moving wire is the distance x from the closed end of the frame, the flux enclosed by the loop is

$$\Phi = BA = Blx. \quad (8)$$

At the later time $t = \Delta t$ the moving wire is $x + \Delta x$ from the closed end of the frame, so the flux now enclosed by the loop is

$$\Phi + \Delta\Phi = Blx + Bl\Delta x.$$

Hence the increase in the enclosed flux during the time Δt is

$$\Delta\Phi = Bl\Delta x.$$

The speed of the moving wire is v, and so

$$\Delta x = v\Delta t$$

and

$$\Delta\Phi = Blv\Delta t,$$

$$\frac{\Delta\Phi}{\Delta t} = Blv.$$

But Blv is the potential difference induced in the moving wire. Hence we have, inserting the minus sign required by Lenz's law,

$$V_i = -\frac{\Delta\Phi}{\Delta t},$$

which is Faraday's law.

THE BETATRON

An interesting application of electromagnetic induction is the betatron. In a betatron electrons are accelerated to high speeds through the action of an increasing magnetic field. As the magnetic field B increases, the associated electric field contributes to the energy of an electron moving in a circular path of area A by an amount per revolution equal to its increase in energy when it moves through a potential difference of

$$V_i = -\frac{\Delta\Phi}{\Delta t} = -A\frac{\Delta B}{\Delta t}.$$

That is, we can regard a circular path of this kind exactly as though it is a loop of wire insofar as electromagnetic induction is concerned. As the electron goes faster and faster, gaining the energy qV_i in each revolution, it will require a stronger magnetic field B if it is to stay in the same orbit; it is not difficult to arrange

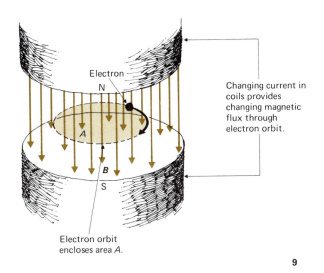

Electron
N
A
B
S

Changing current in coils provides changing magnetic flux through electron orbit.

Electron orbit encloses area *A*.

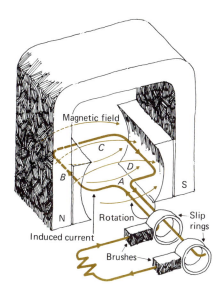

Magnetic field
C
D
B
A
N
Rotation
S
Slip rings
Induced current
Brushes

9   10

matters so that the very increase in **B** that accelerates the electron in the first place exactly keeps pace with its increasing speed, thereby maintaining a constant orbit radius. (**9**)

THE GENERATOR

Electromagnetic induction is of immense practical importance since it is the means whereby nearly all the world's electric power is produced. In a generator a coil of wire is rotated in a magnetic field so that the flux through the coil changes constantly. The resulting potential difference across the ends of the coil causes a current to flow in an external circuit, and this current can be transmitted by a suitable system of wires for long distances from its origin.

Despite its name, a generator does not *create* electrical energy; what it does is *convert* mechanical energy into electrical energy, just as a battery converts chemical energy into electrical energy.

The construction of a simple generator is shown here. A wire loop is rotated in a magnetic field, and the ends of the loop are connected to two *slip rings* on the shaft. Brushes pressing against the slip rings permit the loop to be connected to an external circuit. Only sides *B* and *D* of the loop contribute to the induced potential difference. (**10**)

Because the sides of the loop reverse their direction of motion through the magnetic field twice per rotation, the induced current is also reversed twice per rotation and for

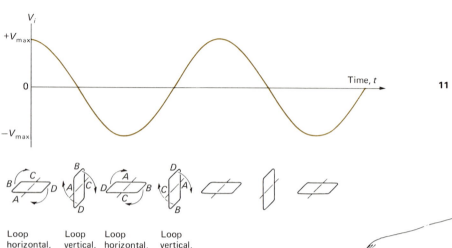

+V_{max}

0

−V_{max}

V_i

Time, t

11

| Loop horizontal, $V_i = +V_{max}$ | Loop vertical, $V_i = 0$ | Loop horizontal, $V_i = -V_{max}$ | Loop vertical, $V_i = 0$ |

this reason is called an *alternating current* (ac). The variation with time of the voltage of the above generator is shown in the accompanying graph. (**11**)

The ac output of a simple generator can be changed to dc (direct current) by substituting a commutator for the slip rings that connect the rotating coil with the external circuit. The current in the brushes is thus always in the same direction. The voltage of such a dc generator varies with time as shown in the graph; V_i is always in the same direction, but it rises to a maximum and drops to zero twice per complete rotation. (**12**)

THE TRANSFORMER

Earlier it was mentioned that a current can be induced in a wire or other conductor by a changing magnetic field as well as by relative motion between a wire and a constant magnetic field. The essential condition for an induced potential difference is that magnetic lines of force cut across the wire (or vice-

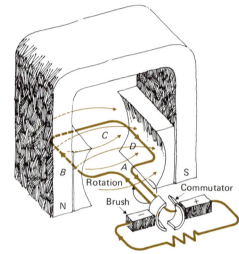

Rotation

Brush

Commutator

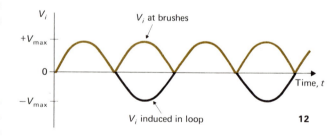

V_i

+V_{max}

0

−V_{max}

V_i at brushes

V_i induced in loop

Time, t

12

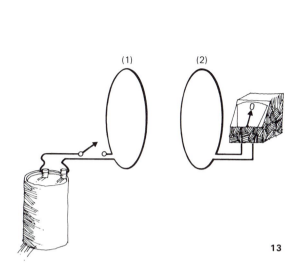

(1)　　　(2)

13

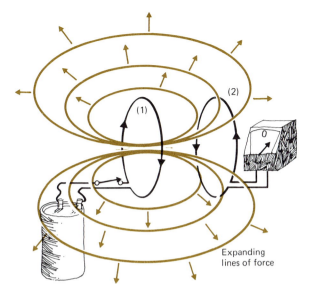

(2)

(1)

Expanding
lines of force

14

versa), and it does not matter exactly how this comes about.

A change in the current in a wire loop is accompanied by a corresponding change in its magnetic field, and if there is another wire loop nearby, a potential difference will therefore be induced in it. If an alternating current flows in the first loop, the magnetic field around it will vary periodically and an alternating potential difference will be induced in the second loop. A *transformer* is a device based upon this effect which is used to produce a voltage in a secondary ac circuit larger or smaller than the voltage in a primary circuit.

Let us see how a changing current in a wire loop induces a changing current in another wire loop not connected to it. When the switch connecting loop (1) with the battery is open, no current flows in either loop. (13)

The switch has just been closed, and the expanding lines of force resulting from the increasing current in loop (1) cut across loop (2), inducing a current in it. The current in

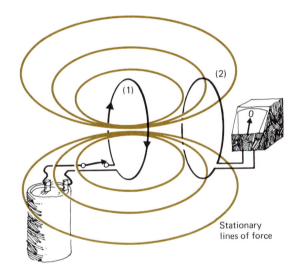

(2)

(1)

Stationary
lines of force

15

loop (2) flows in the *opposite* direction to that in loop (1) because of Lenz's law: the direction of an induced current is always such that its own magnetic field opposed the change in flux that is inducing it. (14)

Now a constant current flows in loop (1), and since the flux through it does not change, no current is induced in loop (2). (15)

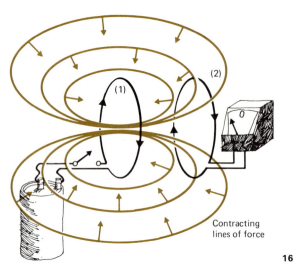

Contracting
lines of force

16

The switch has just been opened, and the contracting lines of force resulting from the decreasing current in loop (1) cut across loop (2), inducing a current in it. The current in loop (2) now flows in the *same* direction as that in loop (1), since it is the decreasing current in (1) that leads to the current in (2) and Lenz's law requires that an induced current oppose the change in flux that brings it about. **(16)**

Finally the current in loop (1) disappears, and no current flows in either loop. **(17)**

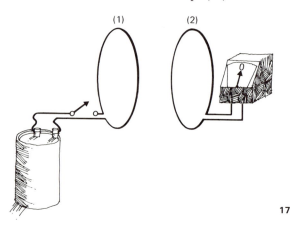

17

Instead of being switched on and off, the current input to a transformer is alternating current which reverses its direction regularly. The construction of a simple transformer is shown here. Coils consisting of many turns are used instead of single loops, and they are wound on a common iron core so that the alternating magnetic flux set up by an alternating current in one coil links with the other coil. This alternating flux induces an alternating emf in the latter coil. The *primary winding* of a transformer is the coil which is fed with an alternating current, and the *secondary winding* is the coil to which power is transferred via the changing magnetic flux. The designations of primary and secondary depend only upon how the transformer is connected; either winding may be the primary, and the other is then the secondary. **(18)**

To reduce the leakage of magnetic flux between the primary and secondary windings, transformers usually have concentric windings. Transformer cores are laminated, with many thin iron sheets arranged parallel to the flux and insulated from each other. This construction minimizes currents induced in the core itself, which are wasteful partly because of the power lost as heat and partly because the flux they establish interferes with the proper operation of the transformer. **(19)**

STEP-UP AND STEP-DOWN TRANSFORMERS

Let us consider an ideal transformer in which there is no "leakage" of flux outside the iron core and no losses within it, so that $\Delta\Phi/\Delta t$ is the same for each of the turns of both windings. The induced voltage *per turn* is therefore the same in both windings, and the total voltage in each winding is proportional to the number of turns it contains. Hence

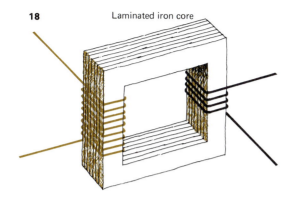

18 Laminated iron core

$$\frac{V_1}{V_2} = \frac{N_1}{N_2},$$

$$\frac{\text{Primary voltage}}{\text{Secondary voltage}} = \frac{\text{total primary turns}}{\text{total secondary turns}}.$$

When the secondary winding has more turns than the primary winding, V_2 is greater than V_1 and the result is a *step-up transformer*; when the reverse is the case and V_2 is less than V_1, the result is a *step-down transformer*.

In an ideal transformer the power output is equal to the power input. If the primary current is i_1 and the secondary current is i_2, then

$$V_1 i_1 = V_2 i_2,$$

$$\text{Power input} = \text{power output};$$

and

$$\frac{i_1}{i_2} = \frac{V_2}{V_1} = \frac{N_2}{N_1}.$$

Increasing the voltage with a step-up transformer means a proportionate decrease in the current, while dropping the voltage leads to a greater current.

Alternating current owes its wide use largely to the ability of transformers to change the voltage at which it is transmitted from one value to another. Electricity generated at perhaps 11,000 V is stepped up by transformers at the power station to as much as 635,000 V for transmission, and subsequently is stepped down by transformers at substations near the point of consumption to less than a thousand volts. High voltages are desirable because, since $P = iV$, the higher the voltage, the smaller the current for a given amount of power; and since power is dissipated as heat at the rate $i^2 R$, the smaller the current, the less the power lost due to resistance in the transmission line.

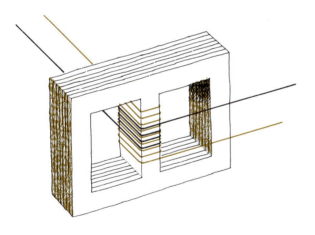

19

Problem. Find the power lost as heat when a 10-Ω cable is used to transmit 1 kW of electricity at 240 V and at 240,000 V.

Solution. The current in the cable in each case is

$$i_a = \frac{P}{V_a} = \frac{1000 \text{ W}}{240 \text{ V}} = 4.17 \text{ A},$$

$$i_b = \frac{P}{V_b} = \frac{1000 \text{ W}}{240,000 \text{ V}} = 4.17 \times 10^{-3} \text{ A}.$$

The respective rates of heat production per kilowatt are therefore

$$i_a^2 R = (4.17 \text{ A})^2 \times 10 \Omega = 174 \text{ W},$$

$$i_b^2 R = (4.17 \times 10^{-3} \text{ A})^2 \times 10 \Omega = 1.74 \times 10^{-4} \text{ W},$$

a difference of a factor of 10^6.

EXERCISES

1. A bar magnet held vertically with its north pole downward is dropped through a wire loop whose plane is horizontal. Is the current induced in the loop just before the magnet enters it clockwise or counterclockwise as seen by an observer above? What is the direction of the current when the magnet passes through the center of the loop? Just after it leaves the loop?

2. One end of a bar magnet is thrust into a coil, and the induced current is clockwise as seen from the front of the coil.
   a) Was the end of the magnet its north or south pole?
   b) What will be the direction of the induced current when the magnet is withdrawn?

3. Lenz's law is a consequence of which conservation principle?

4. A car is traveling from New York to Florida. Which of its wheels have a positive charge and which a negative charge?

5. Upon which of the following properties of a wire does the potential difference produced in it by motion through a magnetic field *not* depend: length, diameter, orientation, or composition?

6. A loop of copper wire is rotated in a magnetic field about an axis along a diameter.
   a) Why does the loop resist this rotation?
   b) If the loop were made of aluminum wire, would the resistance to rotation be different?

7. Why is it easy to turn the shaft of a generator when it is not connected to an outside circuit, but much harder when such a connection is made?

8. Why must alternating current be used in a transformer?

9. In a particular betatron the direction in which the electrons rotate is clockwise as seen from above. What is the direction of the magnetic field?

10. A rectangular wire loop 5 cm $\times$ 10 cm in size is perpendicular to a magnetic field of 10^{-3} T.
    a) What is the flux through the loop?
    b) If the magnetic field drops to zero in 3 s, what is the potential difference induced between the ends of the loop during that period?

11. A square wire loop 10 cm on a side is situated so that its plane is perpendicular to a magnetic field. The resistance of the loop is 5 Ω. How rapidly should the magnetic field change if a current of 2 A is to flow in the loop?

12. A potential difference of 1.8 V is found between the ends of a 2-m wire moving in a direction perpendicular to a magnetic field at a speed of 12 m/s. What is the magnitude of the field?

13. An automobile has a speed of 30 m/s on a road where the vertical component of the earth's magnetic field is 8×10^{-5} T. What is the potential difference between the ends of its axles, which are 2 m long?

14. A train is traveling at 80 mi/hr (35.8 m/s) in a region where the vertical component of the earth's magnetic field is 3×10^{-5} T. The rail-

way is standard gauge (that is, its rails are 4 ft $8\frac{1}{2}$ in. apart, which is 1.43 m). If the conducting path between the rails through the train has a resistance of 20 Ω, find the retarding force on the train due to its motion through the magnetic field.

15. The primary winding of a transformer has 200 turns and its secondary winding has 50 turns. If the current in the secondary winding is 40 A, what is the current in the primary winding? (Assume 100% efficiency.)

16. The primary winding of a transformer has 120 turns and the secondary winding has 1800 turns.

A current of 10 A flows in the primary when a potential difference of 550 V is applied across it. What is the potential difference across the secondary coil and what is the current in it? (Assume 100% efficiency.)

17. An electric welding machine employs a current of 400 A. The device uses a transformer whose primary winding has 400 turns and which draws 2 A from a 220-V power line.
    a) How many turns are there in the secondary winding of the transformer?
    b) What is the potential difference across the secondary? (Assume 100% efficiency.)

PART 5

WAVES AND PARTICLES

29

ELECTROMAGNETIC WAVES

"It seems we have strong reason to conclude that light itself (including radiant heat, and other radiations if any) is an electromagnetic disturbance in the form of waves propagated through the electromagnetic field according to electromagnetic laws." James Clerk Maxwell (1831–1879)

"From a long view of the history of mankind . . . there can be little doubt that the most significant event of the nineteenth century will be judged as Maxwell's discovery of the laws of electrodynamics." Richard P. Feynman (1918–)

Among the most noteworthy achievements of nineteenth-century science was the realization that light consists of electromagnetic waves. Electromagnetic waves themselves were predicted by James Clerk Maxwell in 1864 on the basis of his theory of electric and magnetic fields. The speed Maxwell calculated for these waves turned out to be the same as the speed of light. He then concluded that, since both were also transverse waves, they were the same phenomenon. This conclusion has since been verified in every detail.

MAXWELL'S HYPOTHESIS

In electromagnetic induction, a changing magnetic field induces a potential difference in a nearby wire loop or other conducting path. Thus a changing magnetic field is equivalent in its effects to an electric field.

The converse phenomenon is also true: a changing electric field is equivalent in its effects to a magnetic field. This holds even in empty space, where electric currents cannot flow. No simple experiment can directly demonstrate the latter equivalence (unlike electromagnetic induction, which is very easy to exhibit), and it was first proposed by Maxwell on the basis of an indirect argument. Electromagnetic waves occur as a consequence of these two effects—a changing magnetic field produces an electric field, and a changing electric field produces a magnetic field. The two constantly varying fields are coupled together as they travel through space.

It follows from electromagnetic induction that whenever there is a change in a magnetic field, an electric field is produced, and it follows from Maxwell's hypothesis that whenever there is a change in an electric field, a magnetic field is produced. Evidently it is impossible to have either effect occur alone. The

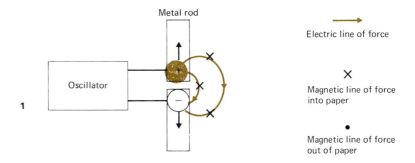

Metal rod

Oscillator

1

Electric line of force

×
Magnetic line of force
into paper

•
Magnetic line of force
out of paper

electric field that arises from a change in a magnetic field is in itself a change in the pre-existing electric field (which might have had any original value, including zero), and there-fore causes another magnetic field. The latter magnetic field, too, represents a change, and from the change an electric field is in turn produced. The process continues indefinitely, with a definite coupling between the fluc-tuating electric and magnetic fields. On the basis of his hypothesis, together with the other principles of electricity and magnetism, Max-well was able to develop a detailed picture of how these field fluctuations travel through space.

ELECTROMAGNETIC WAVES

The first idea that emerged from Maxwell's analysis was that the field fluctuations spread out in space from an initial disturbance in the same manner that waves spread out from a disturbance in a body of water; hence the name *electromagnetic waves* to describe them. If we throw a stone into a pond or otherwise alter the state of the water surface at some point, oscillations occur in which energy is continually interchanged between the kinetic energy of moving water and the potential energy of water displaced from its normal level.

These oscillations begin where the stone lands, and spread out as waves across the surface of the pond. The wave speed depends upon the properties of the pond water, varying with temperature, impurity content, and so on, but it is independent of the wave height. This is typical wave behavior. When electromagnetic waves spread out from an electric or magnetic disturbance, their energy is constantly being interchanged between the fluctuating electric fields and the fluctuating magnetic fields of the waves.

Let us connect a pair of metal rods to a source of alternating potential difference. Such a source is called an *oscillator*. For clarity we will imagine that there is only a single charge in each rod at any time. When the oscillator is switched on, a positive charge in the upper rod begins to move upward and a negative charge in the lower rod begins to move down-ward. The electric lines of force around the charges are indicated by the color lines, and the magnetic lines of force due to the motion of the charges (which are concentric circles perpendicular to the paper) are indicated by crosses when their direction is into the paper and by dots when their direction is out of the paper. (The dots represent arrowheads and the crosses, the tail feathers of arrows.) (1)

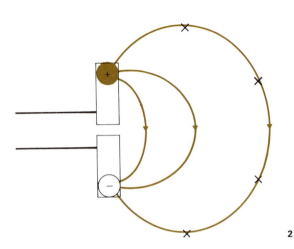

2

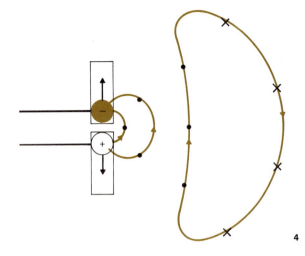

4

The charges have reached the limit of their motion and have stopped, so that they cease to produce a magnetic field. The outer magnetic lines of force do not disappear because of the finite speed at which changes in electric and magnetic fields travel. (2)

The voltage of the oscillator now begins to decrease, and the charges move toward each other. The result is a magnetic field in the opposite direction to the earlier field. The electric field is in the same direction as before. (3)

The potential difference has passed through 0 and begun to increase in the opposite sense. Consequently there is a negative charge in the upper rod and a positive one in the lower rod which begin to move apart. The electric field is therefore opposite in direction to the earlier field, but the magnetic field is in the same direction since magnetically a positive charge moving downward is equivalent to a negative charge moving upward. (4)

Owing to this sequence of changes in the fields, the outermost electric and magnetic lines of force respectively form into closed loops. These loops of force, which lie in perpendicular planes, are divorced from the oscillating charges that gave rise to them and continue moving outward, constituting an electromagnetic wave. As the charges continue oscillating back and forth, further associated loops of electric and magnetic lines of force are emitted, forming an expanding pattern of loops. The diagram shows the configuration of the electric and magnetic fields that spread outward from a pair of oscillating charges. The actual fields are present in three dimensions, so that the magnetic lines of force form

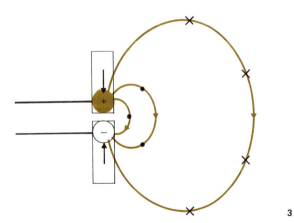

3

loops in planes perpendicular to the line joining the charges. (5)

RADIATION FROM AN ACCELERATED CHARGE

We have been considering a special kind of electromagnetic wave source. Actually, *all* accelerated charges radiate electromagnetic waves, regardless of the manner in which the acceleration occurs.

An intuitive picture of how an electromagnetic pulse is produced when a charge undergoes an acceleration may be obtained in the following way. We start with a charge $+q$ at rest at some point A at the time $t = 0$. The electric field of the charge may be represented by lines of force like those shown here. (6)

The charge then undergoes an acceleration a for the brief interval Δt in the upward direction, and its final velocity is v at the point B. The charge continues to move at the constant velocity v, and at the later time t it is at the point C. The electric field around the charge at this time is shown in the adjacent drawing. The electric field of a charge moving at constant velocity is the same as that of a stationary charge when v is much less than c, the speed of light. We shall assume that $v \ll c$ here so that we can disregard distortions of the field not relevant to the present argument. (7)

Disturbances in an electromagnetic field are propagated with the speed of light c. Hence the electric field beyond a sphere of radius ct centered at A cannot have been affected by the charge's motion: there is no way any event involving the charge that occurs after $t = 0$ can influence the outside world farther than the distance ct away. The electric field past this sphere is therefore exactly the same as in the previous figure.

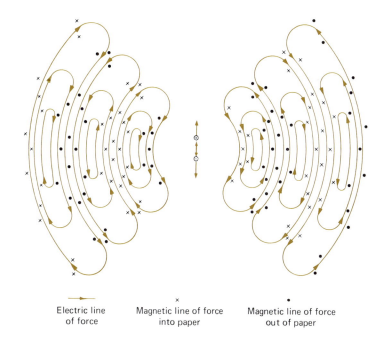

———▶	×	•
Electric line of force	Magnetic line of force into paper	Magnetic line of force out of paper

5

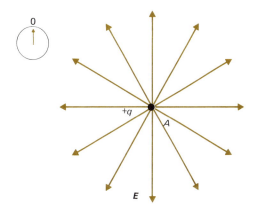

6

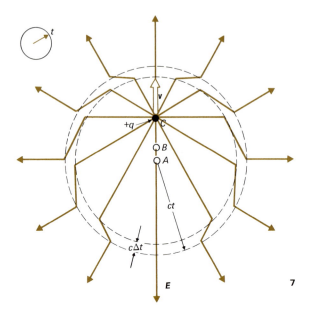

7

Inside the sphere the electric field has another radial pattern of lines of force centered at C except for a transition region $c\Delta t$ wide between the two radial fields. The effect of the charge's acceleration has thus been to put kinks in the lines of force of its electric field. More precisely, the electric field in the transition region has a transverse as well as a radial component.

The moving charge gives rise to a magnetic field as well as to an electric one, and an analysis similar to the preceding one shows that kinks appear in the magnetic lines of force during the charge's acceleration too. Accelerating the charge has given rise to an electromagnetic pulse consisting of mutually perpendicular electric and magnetic fields that

travel at the speed c. This pulse has a life of its own, so to speak, since no subsequent motion of the charge can affect it. The pulse exactly fits the specification of an electromagnetic wave.

The transverse components of E and B in the kinks—which is what the electromagnetic pulse consists of—have the important property that their magnitudes fall off with distance r as $1/r$. For this reason electromagnetic signals may be detected for quite remarkable distances from their sources—light and radio waves reach us from galaxies at the outer limit of the universe—even though the electric and magnetic fields of these sources when not accelerated decrease so rapidly with distance as to be imperceptible not very far away.

WAVES

At this point it is appropriate to review wave behavior in general.

Most waves are *periodic*, that is, they consist of a succession of identical disturbances traveling in some medium. The simplest example is a series of kinks that move down a stretched string when one end is shaken back and forth at regular intervals. The speed v of the waves depends only on the properties of the string—its tension and its density. Waves of other kinds, too, have characteristic speeds that depend only on the properties of the medium, not on the properties of the waves. Thus sound waves in air at sea level all travel at 331 m/s, regardless of the pitch of the sound or of its loudness. (8)

8

Three related quantities are useful in describing periodic waves:

1. The *wave velocity v*, which is the distance through which each wave moves per second;
2. The *wavelength λ* (Greek letter *lambda*), which is the distance between adjacent crests or troughs; and
3. The *frequency f*, which is the number of waves that pass a given point per second.

The wave velocity, wavelength, and frequency of a train of waves are not independent of one another. In every second, *f* waves (by definition) go past a particular point, with each wave occupying a distance of *λ*. Therefore a wave travels a total distance of *fλ* per second, which is the wave velocity *v*. Thus

Wave velocity = frequency × wavelength,

$$v = f\lambda,$$

which is a basic formula that applies to all periodic waves. **(9)**

Sometimes it is more useful to consider the *period T* of a wave, which is the time required for one complete wave to pass a given point. Since *f* waves pass by per second, the period of each wave is

$$T = \frac{1}{f}.$$

If there are five waves per second passing by, for example, each wave has a period of $\frac{1}{5}$s. In terms of period *T*, the formula for wave velocity is

$$v = \frac{\lambda}{T}. \ \blacksquare$$

The unit of frequency is the *cycle/s*, nowadays often called the *hertz* (Hz) after Heinrich Hertz, one of the pioneers in the study of electromagnetic waves. Multiples of the cycle/s

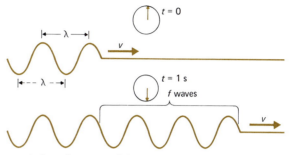

v = velocity = distance traveled per second
= (number of waves passing a point per second) × (length of each wave)
= *fλ*

9

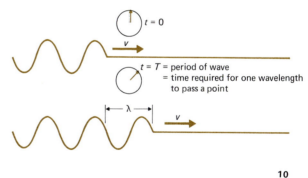

10

and of the Hz are used for high frequencies:

1 kilocycle/s (kc/s) = 1 kilohertz (kHz) = 10^3 cycles/s

1 megacycle/s (Mc/s) 1 megahertz (MHz) = 10^6 cycles/s

Thus a frequency of 50 MHz is equal to

$$50 \text{ MHz} \times 10^6 \frac{\text{Hz}}{\text{MHz}} = 5 \times 10^7 \text{ Hz} = 5 \times 10^7 \text{ cycles/s}.$$

Problem. One day at a certain place on the ocean the distance between adjacent wave crests is 160 ft and a crest passes by every 4.5 s. Find the frequency and velocity of the waves.

Solution. The frequency of the waves is

$$f = \frac{1}{T}$$
$$= \frac{1}{4.5 \text{ s}}$$
$$= 0.22 \text{ Hz},$$

and their velocity is

$$v = \frac{\lambda}{T}$$
$$= \frac{160 \text{ ft}}{4.5 \text{ s}}$$
$$= 36 \text{ ft/s}.$$

The *amplitude A* of a wave refers to the maximum displacement from their normal positions of the particles which oscillate back and forth as the wave travels by. The amplitude of a wave in a stretched string is the height of the crests above the original line of the string, or, equally well, the depth of the troughs. **(11)**

11

Longitudinal waves in a coil spring

Transverse waves in a stretched string

12

TYPES OF WAVES

Waves in a stretched string are *transverse waves* since the individual segments of the string vibrate from side to side perpendicular to the direction in which the waves travel. *Longitudinal waves* occur when the individual particles of a medium vibrate back and forth in the direction in which the waves travel. Longitudinal waves are easy to produce in a long coil spring; each portion of the spring is alternately compressed and extended as the waves pass by. Longitudinal waves, then, are essentially density fluctuations. **(12)**

Sound waves are longitudinal and consist of pressure fluctuations. The air (or other medium) in the path of a sound wave becomes alternately denser and rarer; the resulting changes in pressure cause our eardrums to vibrate with the same frequency, which produces the physiological sensation of sound. The speed of sound in air at sea level and 0°C is 331 m/s.

Most sounds are produced by vibrating objects. An example is the diaphragm of a loudspeaker. When it moves outward, it pushes the air molecules directly in front of it closer together to form a region of high pressure that spreads out in front of the loudspeaker. The diaphragm then moves backward, thereby expanding the volume available to nearby air molecules. Air molecules now flow toward the diaphragm, and consequently a region of low pressure spreads out directly behind the high-pressure region. The continued vibrations of the diaphragm thus send out successive layers of condensation and rarefaction. (13)

Waves on the surface of a body of water (or other liquid) are a combination of longitudinal and transverse waves. If we were somehow to tag individual water molecules and follow them when a train of waves pass by, we would find that their paths are like those shown in the figure. Each molecule describes a circular orbit with a period equal to the period of the wave, and does not undergo a permanent displacement. Because successive molecules reach the tops of their orbits at slightly different times, the water surface takes the form of a series of crests and troughs. At the crest of a wave the molecules move in the direction the wave is traveling, while in a trough the molecules are moving in the opposite direction.

The passage of a wave across the surface of a body of water, like the passage of a wave through any medium, involves the motion of a pattern: energy is transported by virtue of the changing pattern, but there is no transport of matter. (14)

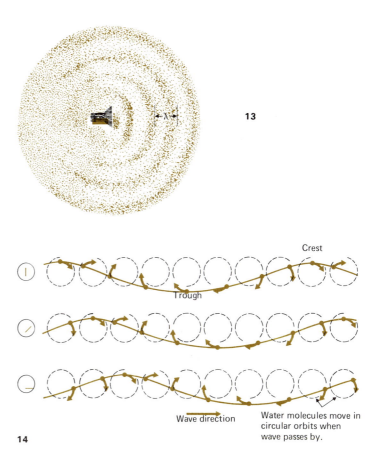

13

14

Water molecules move in circular orbits when wave passes by.

Wave direction

PROPERTIES OF ELECTROMAGNETIC WAVES

Electromagnetic waves are unique in that nothing material moves in their path. The only changes are in electric and magnetic fields. As is true for all waves, however, a flow of energy is associated with electromagnetic waves. To appreciate this, all we need do is reflect that changing electric and magnetic fields are able to exert forces on charges in their paths, and hence to perform work on them.

Nearly all the energy used by man can be traced to electromagnetic waves that come from the sun in the form of sunlight. This includes the energy in food as well as that in coal, oil, and falling water. Solar energy arrives at the earth at the rate of 1400 W/m². A single electromagnetic wave (as distinct from the complex of different individual waves that actually come from the sun) with this much power has a maximum electric field of 1030 V/m (which is appreciable) and a maximum magnetic field of 3.4×10^{-6} T (which is

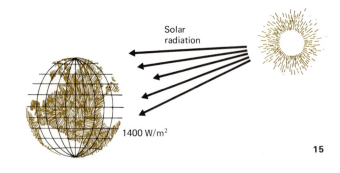

Solar
radiation

1400 W/m²

15

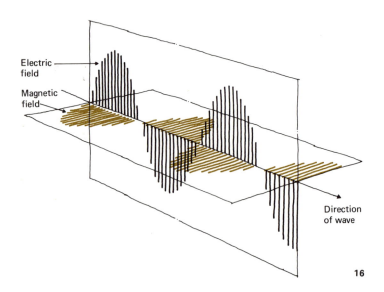

Electric
field

Magnetic
field

Direction
of wave

16

2. The directions of the electric and magnetic fields are perpendicular to each other and to the direction in which the waves are moving. Light waves are therefore transverse.

3. The speed of the waves depends only upon the electric and magnetic properties of the medium they travel in, and not upon the amplitudes of the field variations.

An attempt at portraying (1) and (2) of the above is shown here for an electromagnetic wave a long distance from its source. (**16**)

Electromagnetic waves are classified according to their frequencies as shown in the accompanying figure. (**17**)

THE SPEED OF LIGHT

Maxwell's theory of electromagnetic waves showed that their speed c depends solely upon the proportionality constants in the laws that govern the electric field between two charges and the magnetic force between two electric currents. These laws are, respectively,

$$F_{elec} = k\,\frac{q_1 q_2}{r^2}, \quad F_{mag} = k'\,\frac{i_1 i_2}{s}\,l.$$

Maxwell found that the speed c is given by

$$c = \sqrt{2\,\frac{k}{k'}}$$

$$= \sqrt{2 \times \frac{9 \times 10^9 \text{ N-m}^2/\text{C}^2}{2 \times 10^{-7} \text{ N/A}^2}}$$

$$= 3 \times 10^8 \,\frac{\text{m}}{\text{s}},$$

which is the same speed that had been experimentally measured for light waves in free space. The correspondence was too great to be

quite small). It is the electric field of an electromagnetic field that is responsible for most of the effects the wave produces. (**15**)

Three properties of electromagnetic waves are worth noting:

1. The variations occur simultaneously in both fields (except close to the oscillating charges), so that the electric and magnetic fields have maxima and minima at the same times and in the same places.

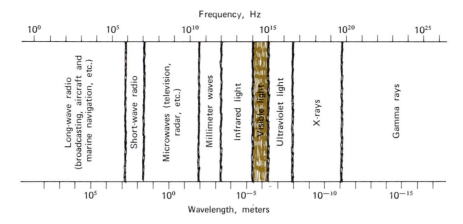

Frequency, Hz

10^0 10^5 10^{10} 10^{15} 10^{20} 10^{25}

Long-wave radio (broadcasting, aircraft and marine navigation, etc.)   Short-wave radio   Microwaves (television, radar, etc.)   Millimeter waves   Infrared light   Visible light   Ultraviolet light   X-rays   Gamma rays

10^5 10^0 10^{-5} 10^{-10} 10^{-15}

Wavelength, meters

17

accidental, and, as further evidence became known, the electromagnetic nature of light found universal acceptance.

EXERCISES

1. According to Maxwell's hypothesis, what kind of field does a changing electric field give rise to?

2. Can light waves travel through a vacuum? Can sound waves?

3. Light is said to be a transverse wave phenomenon. What is it that varies at right angles to the direction in which a light wave travels?

4. Electromagnetic radiation carries both energy and momentum. Why does the momentum of the sun not decrease with time as its energy content does?

5. Which of these quantities is transported by electromagnetic waves: wavelength, frequency, electric charge, or energy?

6. Must a charge be accelerated in order to emit electromagnetic waves?

7. An electromagnetic wave directed vertically upward has its electric field in the north-south direction. What is the direction of the magnetic field?

8. A radio transmitter has a vertical antenna. Does it matter whether the receiving antenna is vertical or horizontal?

9. Which of these properties of a wave is independent of the others: frequency, wavelength, speed, or amplitude?

10. Radio amateurs are permitted to communicate on the "10-m band." What frequency of radio waves corresponds to a wavelength of 10 m?

11. A certain radio station broadcasts at a frequency of 660 kHz. Find the wavelength of these waves.

12. A certain groove in a phonograph record moves past the needle at a speed of 0.30 m/s. If the wiggles in the groove are 0.1 mm apart, what is the frequency of the sound that is produced?

13. A violin string is set in vibration at a frequency of 440 Hz. How many vibrations does it make while its sound travels 200 m in air?

30

REFLECTION AND REFRACTION

"Light . . . spreads, as sound does, by spherical surfaces and waves: for I call them waves from their resemblance to those which are seen to be formed in water when a stone is thrown into it." Christian Huygens (1629–1695)

274

Light consists of electromagnetic waves. Remarkably enough, it is possible to understand most optical phenomena on the basis of the wave nature of light alone, without any direct reference to its electromagnetic character. Huygens' principle is a convenient approach to the behavior of light waves, and in this section it is applied to the reflection and refraction of light.

HUYGENS' PRINCIPLE

Although every aspect of the wave behavior of light can be determined from Maxwell's electromagnetic theory of light, such calculations are quite complicated. A less comprehensive but simpler and more intuitive approach to optical phenomena was devised in 1678 by Christian Huygens; it can be applied to all kinds of waves, not just light waves.

We should note in passing that in certain important respects light has the character of a stream of particles rather than that of a series of waves. This duality is the subject of Section 32.

Huygens' method of analysis concerns *wavefronts*. A wavefront is an imaginary surface that joins points where all of the waves involved are in the same phase of oscillation. Thus waves from a point source spread out in a succession of spherical wavefronts. (In the case of water waves, the wavefronts that result when a stone is dropped in a lake are circular.) (1)

At a long distance from a point source, the curvature of the wavefronts is so small that they can be considered as a succession of planes. (2)

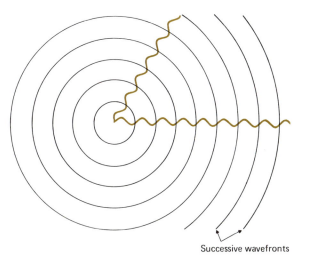

Successive wavefronts   1

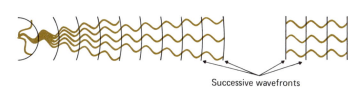

Successive wavefronts

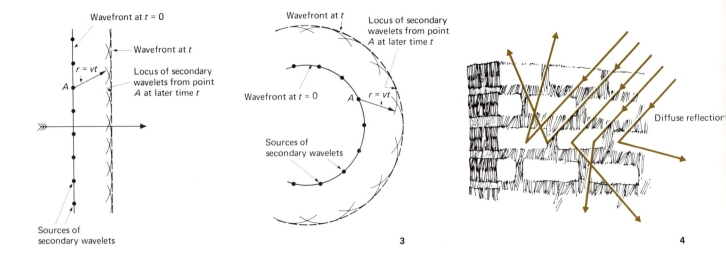

Wavefront at $t = 0$

Wavefront at t

$r = vt$

A

Locus of secondary wavelets from point A at later time t

Sources of secondary wavelets

Wavefront at t

Locus of secondary wavelets from point A at later time t

Wavefront at $t = 0$

A

$r = vt$

Sources of secondary wavelets

3

Diffuse reflection

4

Huygens' principle states:

Every point on a wavefront can be considered as a point source of secondary wavelets which spread out in all directions with the wave speed of the medium. The wavefront at any time is the envelope of these wavelets.

The diagrams show how Huygens' principle is applied to the propagation of plane and spherical wavefronts in a uniform medium. It does not matter just where on the initial wavefront we imagine the sources of the secondary wavelets to be. Despite its name, Huygens' principle is *not* in the same category as such fundamental principles as those of conservation of mass energy, momentum, and electric charge, but is rather a convenient means for studying wave motion in a geometrical manner. We shall find Huygens' principle useful in understanding a variety of optical phenomena. **(3)**

While we shall use the notion of wavefronts for the actual analysis of wave propagation, the results are commonly represented in terms of *rays*. A light ray is simply an imaginary line in the direction in which the wavefronts advance, and so it is perpendicular to the wavefronts. The picture most of us have of a light ray is a narrow pencil of light, which is perfectly legitimate. However, an approach based exclusively on rays does not reveal such characteristic wave behavior as diffraction (the bending of waves around an obstacle into the "shadow" region); hence we must remember that the motion of wavefronts is what is really significant, although it is both proper and convenient to use rays to summarize our conclusions.

SPECULAR AND DIFFUSE REFLECTION

All objects reflect a certain proportion of the light falling upon them, and it is this reflected light that enables us to see them. In most cases the surface of the object has irregularities that spread out an initially parallel beam of light in all directions to produce *diffuse reflection*. **(4)**

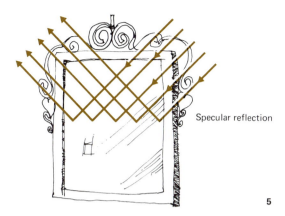

Specular reflection

5

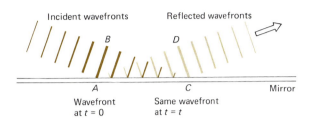

Incident wavefronts   Reflected wavefronts

B   *D*

A   *C*   Mirror

Wavefront   Same wavefront
at $t = 0$ at $t = t$

6

A surface so smooth that any irregularities in it are small relative to the wavelength of the light falling upon it behaves differently; when a parallel beam of light is directed at such a surface, it is *specularly reflected* in only one direction. **(5)**

Reflection from a brick wall is diffuse, while reflection from a mirror is specular. Often a mixture of both kinds of reflection occurs, as in the case of a surface coated with varnish or glossy enamel. Our concern here is with specular reflection only, which we shall analyze with the help of Huygens' principle.

When we see an object, what enters our eyes are light waves reflected from its surface. What we perceive as the object's color therefore depends upon two things, the kind of light falling on it and the nature of its surface. If white light is used to illuminate an object that absorbs all colors other than red, the object will appear red. If green light is used instead, the object will appear black because it absorbs green light. A white object reflects light of all wavelengths equally well, and its

apparent color depends entirely upon the color of the light reaching it. A black object, on the other hand, absorbs light of all wavelengths, and it appears black no matter what color light reaches it.

LAW OF REFLECTION

Let us examine the specular reflection of a series of plane wavefronts, which corresponds to a parallel beam of light. At $t = 0$ the wavefront AB just touches the mirror, and secondary wavelets start to spread out from A. After a time t, the end B of the wavefront reaches the mirror at C. The secondary wavelets from A are now at the point D. The wavefront which was AB at $t = 0$ is therefore CD at the later time t. **(6)**

The angle i between an approaching wavefront and the reflecting surface is called the *angle of incidence*, and the angle r between a receding wavefront and the reflecting surface is called the *angle of reflection*. The wavelet that originates at A takes the time t to reach D. Hence $AD = ct$, where c is the speed of

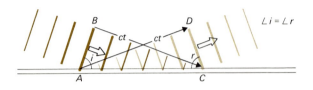

7

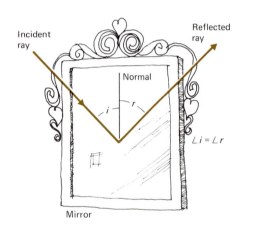

8

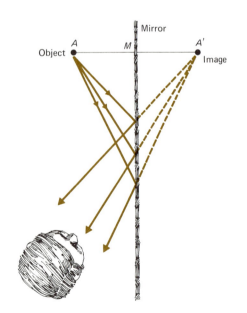

9

light. The point B of the wavefront AB also takes the time t to reach C, and so $BC = ct$ as well. Therefore $AD = BC$, which means that the triangles ABC and CDA are the same in size and shape. Hence the angles i and r are equal. Thus the basic law of reflection is:

The angle of reflection of a plane wavefront with a plane mirror is equal to the angle of incidence. (**7**)

In the ray model of light propagation, the angles of incidence and reflection are measured with respect to the *normal* to the reflecting surface at the point where the light strikes it. The normal is a line drawn perpendicular to the surface at that point. In this representation, too, the angles of incidence and reflection are equal. The incident ray, the reflected ray, and the normal all lie in the same plane. (**8**)

THE PLANE MIRROR

The image of an object we see in a plane mirror appears to be the same size as the object and as far behind the mirror as the object is in front of it. Let us try to figure out how this situation arises.

In the diagram, three typical light rays from a point object at A impinge on a mirror and are reflected, in each case of course with an angle of reflection equal to the angle of incidence. To the eye, the three diverging rays apparently come from the point A' behind the mirror. The rays that seem to come from the image do not actually pass through it, and for this reason the image is said to be *virtual*. (A *real image* is formed by light rays that pass through it; the image produced on a screen by a slide projector is a real one, for instance.) From the geometry of the figure it is clear that $AM = A'M$, so that the object and the image are the same distance on either side of the mirror. (**9**)

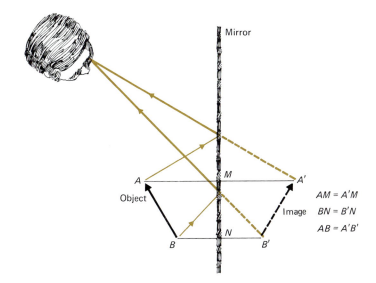

Mirror

A   M   A′

Object

Image

$AM = A'M$

$BN = B'N$

$AB = A'B'$

B   N   B′

10

In the next diagram an object of finite size is being reflected in a plane mirror. Again simple geometry shows that the object and its virtual image have the same dimensions. Every point in the image is directly behind the corresponding point on the object, and so the orientation of the image is the same as that of the object; the image is therefore *erect*. (**10**)

Although the image is erect in reflection from a plane mirror, left and right are interchanged. The "mirror image" of something is the same size and shape as the original but its transverse features are reversed. A printed page appears backward in a mirror, and what seems to be one's left hand is actually one's right hand. (**11**)

11

REFRACTION

It is a matter of experience that a beam of light passing obliquely from one medium to another, say from air to water, is deflected at the surface between the two media. The bend-

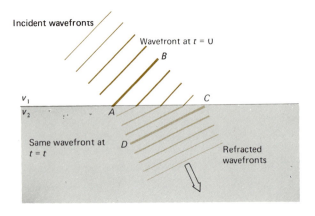

Incident wavefronts

Wavefront at $t = 0$

B

C

v_1
v_2

A

Same wavefront at
$t = t$

D

Refracted
wavefronts

12

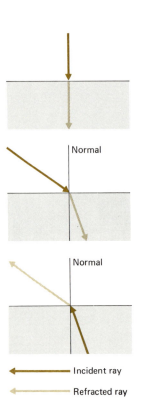

Normal

Normal

Incident ray

Refracted ray

13

ing of a light beam as it goes from one medium to another is called *refraction*, and it is responsible for such familiar phenomena as the apparent distortion of objects partially submerged in water.

Refraction occurs because light travels at different speeds in the two media. Let us see what happens to the plane wavefront AB when it passes obliquely from a medium in which its speed is v_1 to a medium in which its speed is v_2, where v_2 is less than v_1. At $t = 0$ the wavefront AB just comes in contact with the interface between the two media. After a time t, the end B of the wavefront reaches the interface at C, and the secondary wavelets from A are now at D. Since v_2 is less than v_1, the secondary wavelets generated at the interface travel a shorter distance in the same time interval than do wavelets in the first medium, and the distance AD is shorter than BC. The refracted wavefronts accordingly move in a different direction from that of the incident wavefronts. **(12)**

Several important aspects of refraction are shown here with the help of rays. Again the speed of light in the lower medium is less than its speed in the upper medium. **(13)**

When a light ray goes from one medium to another along the normal to the interface between them, it simply continues along the same straight line.

When a light ray goes obliquely from one medium to another in which its speed is smaller, it is bent *toward* the normal.

The paths of light rays are reversible; a light ray that goes obliquely from one medium to another in which its speed is greater is bent *away from* the normal.

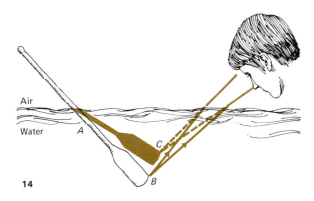

Table 30.1 Indexes of refraction

Substance	n	Substance	n
Air	1.0003	Glass, flint	1.63
Benzene	1.50	Glycerin	1.47
Carbon disulfide	1.63	Ice	1.31
Diamond	2.42	Quartz	1.46
Ethyl alcohol	1.36	Water	1.33
Glass, crown	1.52	Zircon	1.92

14

A familiar example of refraction is the apparent reduction in the depth of an object submerged in water or other transparent liquid. Thus an oar dipped in a lake seems bent upward where it enters the water because its submerged portion appears to be closer to the surface than it actually is. The diagram shows how this effect comes about. Light leaving the tip of the oar at B is bent away from the normal upon entering the air. To an observer above, who instinctively interprets what he sees in terms of the straight-line propagation of light, the tip of the oar is at C, and the submerged part of the oar seems to be AC and not AB. **(14)**

INDEX OF REFRACTION

The ratio between the speed of light c in free space and its speed v in a particular medium is called the *index of refraction* of the medium. The greater the index of refraction, the greater the extent to which a light beam is deflected upon entering or leaving the medium. The symbol for index of refraction is n, so that

$$n = \frac{c}{v}.$$

Generally the index of refraction of a medium depends to some extent upon the frequency of the light involved, with the highest frequencies having the highest values of n. In ordinary glass the index of refraction for violet light is about one percent greater than that for red light, for example. Since a different index of refraction means a different degree of deflection when a light beam enters or leaves a medium, a beam containing more than one frequency is split into a corresponding number of different beams when it is refracted. This effect, called *dispersion*, is exhibited in the familiar experiment in which a narrow pencil of white light is directed at one face of a glass prism. The initial beam separates into beams of various colors, from which we conclude that white light is actually a mixture of light of these different colors. The band of colors that emerges from the prism is known as a *spectrum*. **(15)**

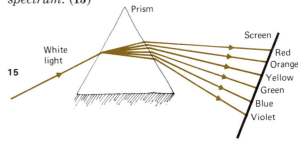

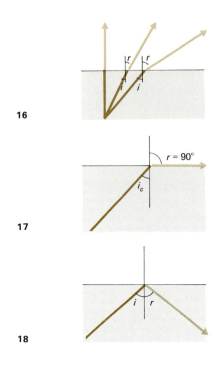

16

17

18

TOTAL INTERNAL REFLECTION

An interesting phenomenon known as *total internal reflection* can occur when light passes from one medium to another which has a *lower* index of refraction, for instance from water or glass to air. In this case the angle of refraction is greater than the angle of incidence, and a light ray is bent away from the normal. (**16**)

As the angle of incidence is increased, a certain *critical angle* i_c is reached for which the angle of refraction is 90°. The "refracted" ray now travels along the interface between the two media and cannot escape. (**17**)

A ray approaching the interface at an angle of incidence exceeding the critical angle is reflected back inside the medium it came from, with the angle of reflection being equal to the angle of incidence as in any other instance of reflection. (**18**)

The sharpness and brightness of a light beam are better preserved by total internal reflection than by reflection from an ordinary mirror, and optical instruments accordingly utilize the former in preference to the latter whenever light is to be changed in direction. Here are three typical applications for totally reflecting prisms. (**19**)

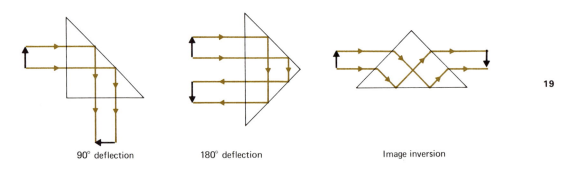

90° deflection   180° deflection   Image inversion

19

Total internal reflection makes it possible to "pipe" light from one place to another with a rod of glass or transparent plastic. As shown here, successive internal reflections occur at the surface of the rod, and nearly all the light entering at one end emerges at the other. If a cluster of narrow glass fibers is used instead of a single thick rod, an image can be transferred from one end to the other since each fiber carries intact a part of the image. (20)

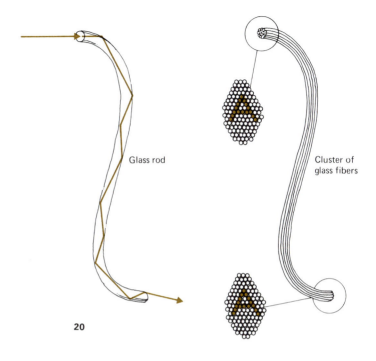

Glass rod

Cluster of glass fibers

20

EXERCISES

1. Under what circumstances is it appropriate to consider light as a wave phenomenon and under what circumstances as a ray phenomenon?

2. A man stands in front of a plane mirror. What are the properties of the image he sees?

3. What is the height of the smallest mirror in which a man 6 ft tall can see himself at full length?

4. A plane mirror is mounted on the back of a truck traveling at 20 mi/hr. How fast does the image of a man standing in the road behind the truck seem to be moving away from him?

5. Fermat's principle states that light travels between two points in such a manner that the transit time is a minimum. Is refraction in accord with this principle?

6. Why does a cut diamond show flashes of color when held in white light? What would happen if it were held in red light?

7. Why is a beam of white light not dispersed into its component colors when it passes perpendicularly through a pane of glass?

8. Flint glass and carbon disulfide have almost the same index of refraction. How does this explain the fact that a flint-glass rod immersed in carbon disulfide is nearly invisible?

9. The index of refraction of benzene is 1.5. What is the speed of light in benzene?

10. What is the index of refraction of a substance in which the speed of light is 2.3×10^8 m/s?

11. A wave of frequency f and wavelength λ passes from a medium in which its speed is v to another medium in which its speed is $2v$. What are the frequency and wavelength of the wave in the second medium? [*Hint*: Consider what happens at the interface between the two media when the wave passes through to decide whether the frequency or the wavelength or both change.]

31

LIGHT AS A WAVE

"Wherever two portions of the same light arrive at the eye by different routes, either exactly or very nearly in the same direction, the light becomes most intense when the difference of the routes is any multiple of a certain length, and the least intense in the intermediate state of the interfering portions; and this length is different for light of different colors."
Thomas Young (1773–1829)

Although they are consequences of its wave nature, the reflection and refraction of light could conceivably be explained in terms of a quite different model as well. Other optical effects make sense only on the basis of a wave model, however. Thus interference is so uniquely a wave phenomenon that the demonstration of the interference of light in 1801 provided conclusive evidence of its true character, even though over half a century had to pass before Maxwell realized that the waves involved are electromagnetic.

INTERFERENCE

Light was recognized as a wave phenomenon well before its electromagnetic character became known a century ago. The problem of the nature of light was an old one: Newton felt sure light consists of a stream of tiny particles, whereas his contemporary, Christian Huygens (1629–1695), thought it to be a succession of waves. Neither man offered any hypothesis as to what kind of particle or wave is involved. Eventually interference, diffraction, and polarization were discovered in light, and these phenomena could only be explained on the basis that light consists of transverse waves. But Newton was not completely wrong. As we shall learn in the next section light has certain distinctly particle properties in addition to its wave properties, and one of the central problems of contemporary physics has been the resolution of this apparent paradox.

When light waves from one source are mixed with those from another source, the two wave trains are said to *interfere*. As a help in understanding how interference occurs, let us consider what happens when two pulses moving in opposite directions in a stretched string meet. If the pulses are both upward, as shown here, the result is a larger pulse at the moment the two come together. Then the separate pulses reappear and continue unchanged in their original directions of motion. Each pulse proceeds as though the other does not exist. (1)

What if one of the pulses is inverted relative to the other? We might expect that, if the pulses have the same sizes and shapes, their displacements ought to cancel out when they meet, only to reappear later on after they have passed the crossing point. Such behavior is indeed observed in practice. At the instant of complete cancellation, the total energy of both pulses resides in the kinetic energy of the string segment where the cancellation occurs. (2)

The *principle of superposition* is a statement of the above behavior. This principle can be phrased as follows:

When two or more waves of the same nature travel past a point at the same time, the amplitude at that point is the sum of the instantaneous amplitudes of the individual waves.

Constructive interference refers to the reinforcement of waves in phase ("in step") with one another, and *destructive interference* refers to the partial or complete cancellation of waves out of phase with one another. (3)

Anyone with a pan of water can see for himself how interference between water waves can lead to a water surface disturbed in a variety of characteristic patterns. Two people who hum fairly pure tones slightly different in frequency will hear beats as the result of interference in the sound waves. But if we shine light from two flashlights at the same place on a screen, there is no evidence of interference: the region of overlap is merely uniformly bright.

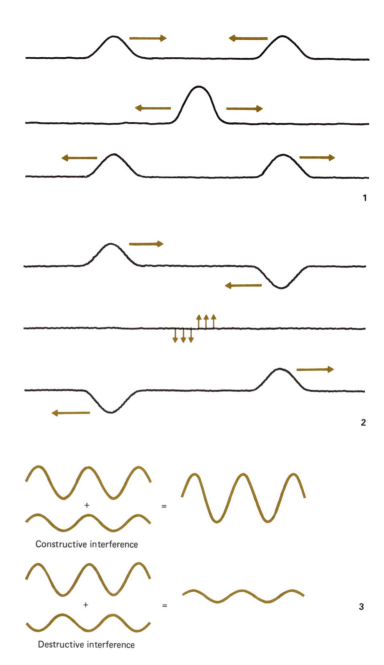

Constructive interference

Destructive interference

There are two reasons for the difficulty of observing interference in light. First, light waves have extremely short wavelengths—the visible part of the spectrum extends only from 3.8×10^{-7} m for violet light to 7.8×10^{-7} m for red light. Second, every natural source of light emits light waves only as short trains at random intervals, so that any interference that occurs is usually averaged out during even the briefest period of observation by the eye or photographic film, unless special procedures are used. Interference in light is nevertheless just as real a phenomenon as interference in water or sound waves, and there is one example of it familiar to everybody—the bright colors of a thin film of oil spread out on a water surface.

COHERENCE

Two sources of waves are said to be *coherent* if there is a fixed phase relationship between the waves they emit during the time the waves are being observed. It does not matter whether the waves are exactly in step when they leave the sources, or exactly out of step, or anything in between; the important thing is that the phase relationship stay the same. If the sources shift back and forth in relative phase while the observation is made, the phase differences average out, and there will be no interference pattern. The latter sources are *incoherent*.

The question of coherence is especially significant for light waves because an excited atom radiates for 10^{-8} s or less, depending upon its environment. Therefore a monochromatic light source such as a gas discharge tube (a neon sign is an example) does not emit a continuous wave train as a radio antenna does but instead a series of individual wave trains whose phases are random. The light from such a tube actually comes from a great many individual, uncoordinated sources, namely the individual atoms, and these individual sources are in effect being switched on and off rapidly and irregularly. (4)

Suppose we have two point sources of monochromatic light, for instance a discharge tube with a shield that has two pinholes close together. Different atoms are behind each pinhole, so they are effectively independent sources. Therefore we have at most 10^{-8} s to observe the interference of waves from the two sources. If our detecting instruments are fast enough, as some modern electronic devices are, interference can be demonstrated and the sources can be considered coherent. If we are limited to the eye and to photographic film, which average arriving light signals over times far greater than 10^{-8} s, no interference can be observed in the light from the two sources, and they must then be considered incoherent. Like beauty, coherence lies in the eye of the beholder.

Does the brief lifetime of an excited atom mean that interference patterns can never be literally seen but can only be recorded by instruments? Hardly. There are three ways to construct separate sources of light coherent for long enough periods of time to produce visible interference patterns. These are:

1. Illuminate two (or more) slits with light from one slit behind them. Then the light waves from the secondary slits are automatically coordinated.

2. Obtain coherent virtual sources from a single source by reflection or refraction. This is how interference is produced by thin films of oil.

3. Coordinate the radiating atoms in each separate source so that they always have the

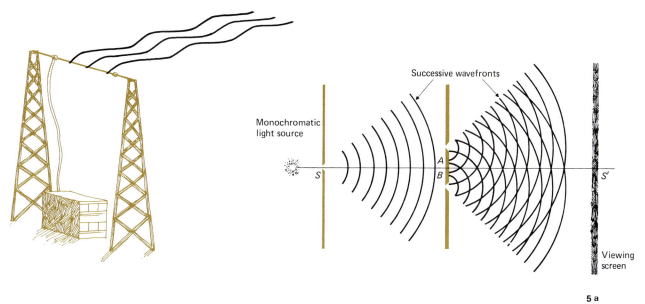

Monochromatic
light source

Successive wavefronts

S

A
B

S'

Viewing
screen

5 a

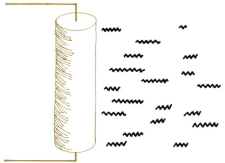

4

same phase even though different atoms are radiating at successive instants. This is done in the *laser*.

THE DOUBLE SLIT

The interference of light waves was demonstrated in 1801 by Thomas Young, who used an arrangement similar to that shown here. A source of monochromatic light (that is, light consisting of only a single wavelength)

is placed behind a narrow slit S in an opaque screen, and another screen with two similar slits A and B is placed on the other side. Light from S passes through both A and B and then to the viewing screen. If light were not a wave phenomenon, we would expect to find the viewing screen completely dark, since no light ray can reach it from the source along a straight path. What actually happens is that each slit acts as a source of secondary wavelets—we recall Huygens' principle from page 276—so that the entire screen is illuminated. Even the point S', separated from S by the opaque barrier between the slits A and B, turns out to be bright rather than dark. (5a)

Owing to interference the screen is not evenly illuminated but shows a pattern of alternate bright and dark lines. Light waves from slits A and B are exactly in phase, since A and B are the same distance from S. The centerline

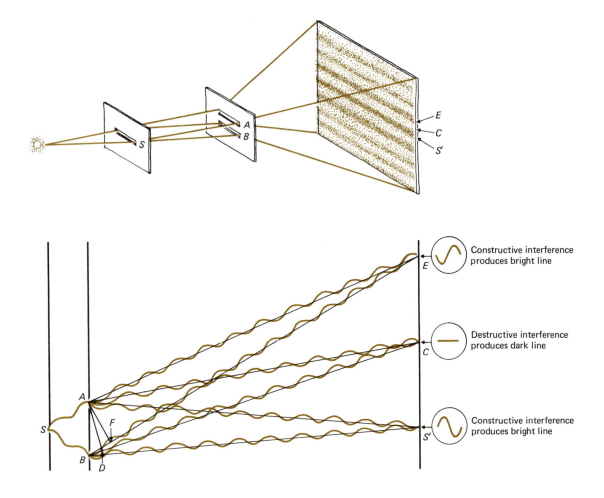

5b

Constructive interference
produces bright line

Destructive interference
produces dark line

Constructive interference
produces bright line

S' of the screen is equally distant from A and B, so light waves from these slits interfere constructively there to produce a bright line.

Let us next see what happens at the position C on the screen located to one side of S'. The distance BC is longer than the distance AC by the amount BD, which is equal to exactly half a wavelength of the light being used. That is,

$$BD = \tfrac{1}{2}\lambda.$$

When a crest from A reaches C, this difference in path length means that a trough from B arrives there at the same time, since $\tfrac{1}{2}\lambda$ separates a crest and a trough in the same wave. The two cancel each other out, the light intensity at C is zero, and dark line results on the screen there. At S' the equality of path length gives rise to constructive interference; at C the difference of $\tfrac{1}{2}\lambda$ in path length gives rise to destructive interference.

If we go past C on the screen we will come to a point E such that the distance BE is greater exactly one wavelength than the distance AE. That is, the difference BF between BE and AE is

$$BF = \lambda.$$

Consequently, when a crest from A reaches E, a crest from B also arrives there, although the latter crest left B earlier than that from A owing to the longer path it had to cover. Because $BF = \lambda$, waves arriving at E from both slits are always in the same part of their cycles, and they constructively interfere to produce a bright line at E. (**5b**)

By continuing the same analysis, we find that the alternate light and dark lines actually observed on the screen correspond respectively to locations where constructive and destructive interference occurs. Waves reaching the screen from A and B along paths that are equal or differ by a whole number of wavelengths (λ, 2λ, 3λ, and so on) reinforce, while those whose paths differ by an odd number of half wavelengths ($\frac{1}{2}\lambda$, $\frac{3}{2}\lambda$, $\frac{5}{2}\lambda$, and so on) cancel. At intermediate locations on the screen the interference is only partial, so that the light intensity on the screen varies gradually between the bright and dark lines.

THIN FILMS

We have all seen the marvelous rainbow colors that appear in soap bubbles and thin oil films. Some of us may also have observed the patterns of light and dark bands that occur when two glass plates are almost (but not quite) in perfect contact. Both phenomena owe their origins to a combination of reflection and interference.

Let us consider a beam of monochromatic light that strikes a thin film of soapy water.

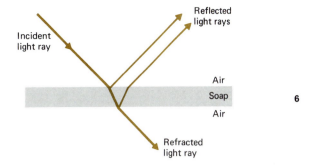

The diagram shows a ray picture of what happens. We notice that some reflection takes place at both the air–soap and soap–air interfaces. This is a general result: waves are always partially reflected when they go from one medium to another in which their speed is different. (**6**)

A light ray actually consists of a succession of wave fronts. Here is the same diagram with the wave fronts drawn in. In this particular case the two reflected wave trains are out of phase and they interfere destructively to partially or completely cancel out. Most or all of the light reaching this part of the soap bubble therefore passes right through. (**7**)

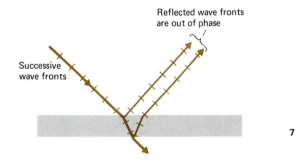

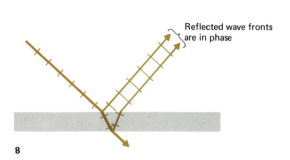

Reflected wave fronts are in phase

8

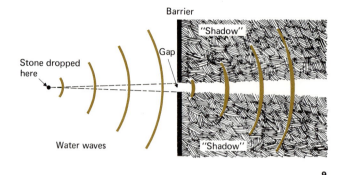

Barrier

"Shadow"

Gap

Stone dropped here

Water waves

"Shadow"

9

Another part of the soap film may have a different thickness. When the film is a little thinner than in the previous picture, the waves in the two reflected trains are exactly in phase, and they interfere constructively to reinforce one another. Light reaching this part of the soap bubble is strongly reflected. Shining monochromatic light on a soap bubble therefore yields a pattern of light and dark which results from the varying thickness of the bubble. (8)

When white light is directed at a soap bubble, light waves of each wavelength present pass through the soap film without reflection at those places where the film is exactly the right thickness for the two reflected rays to destructively interfere. Light waves of the other wavelengths are reflected to at least some extent, and give rise to the vivid colors seen. The varying thickness of the bubble means that the color of the light reflected from the bubble changes from place to place. Exactly the same effect is responsible for the coloration of thin oil films. Generally speaking, soap or oil films whose thickness is comparable with the wavelengths in visible light give rise to the most striking color effects.

DIFFRACTION

Three further important aspects of the behavior of light waves are diffraction, polarization, and scattering. Like interference, these phenomena are responsible for a number of commonly observed effects, although the connection between these effects and the wave nature of light is not always appreciated.

Waves are able to bend around the edge of an obstacle in their path, a property called *diffraction*. We all have heard sound that originated around the corner of a building from where we were standing, for example. These sound waves cannot have traveled in a straight line from their source to our ears, and refraction cannot account for their behavior. Water waves, too, diffract, as the simple experiment illustrated here shows. The waves on the far side of the gap spread out into the geometrical "shadow" of the gap's edges, though with reduced amplitude. The diffracted waves spread out as though they originated at the gap, in accord with Huygens' principle. (9)

A beam of particles behaves quite differently from a beam of waves in this respect. Bullets

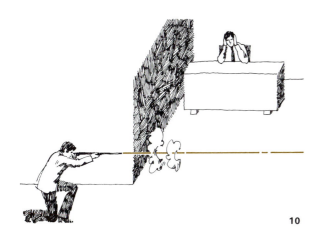

10

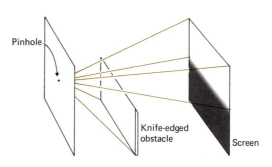

Pinhole

Knife-edged obstacle

Screen

11

fired from another side of a building cannot possibly reach us by bending around the corner, though we can hear the sound of the gun. (**10**)

If we look very carefully at the edge of the shadow cast by an obstruction in the path of the light spreading out from a pinhole or other point source, we will see that it is not sharp but smeared out. The fuzzy edges of shadows are not easy to observe because the wavelengths in visible light are so short, less than 10^{-6} m, and the extent of diffraction into the shadow zone is correspondingly small. (In contrast, a typical audible sound wave might have a wavelength of 1 m and a typical wave in a pan of water might have a wavelength of 10 cm, and it is easy to observe diffraction effects with such waves.) In fact, because he was not able to perceive any diffraction with his relatively crude apparatus, Newton felt

sure that light must be corpuscular in nature and not consist of waves at all. (**11**)

Diffraction sets a limit to the useful magnification of an optical system such as that of a telescope or microscope. Diffraction occurs whenever wavefronts of light are obstructed, and the light that enters a lens (or mirror) is affected by the finite opening which admits only part of each incident wavefront. No matter how perfect a lens is, the image of a point source of light it produces is always a tiny disc of light with bright and dark fringes around it; only if the lens has an infinite diameter can a point source give rise to a point image. The smaller the lens, the larger the image of a point source. If the lens diameter is too small, objects that are close together cannot be *resolved*—distinguished apart—no matter how high the magnification, because their images overlap.

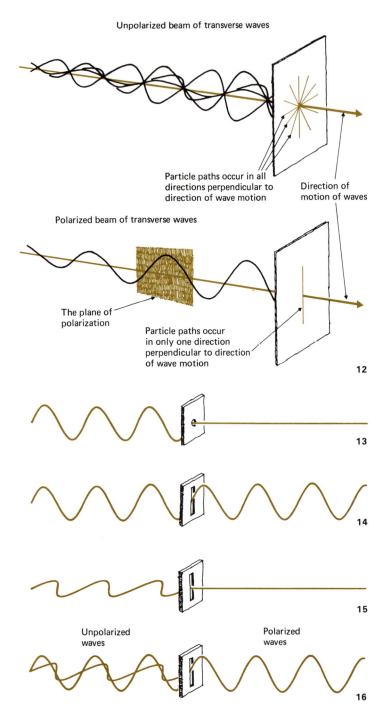

Unpolarized beam of transverse waves

Particle paths occur in all directions perpendicular to direction of wave motion

Direction of motion of waves

Polarized beam of transverse waves

The plane of polarization

Particle paths occur in only one direction perpendicular to direction of wave motion

12

13

14

15

Unpolarized waves

Polarized waves

16

POLARIZATION

A *polarized* beam of transverse waves is one whose vibrations occur in only a single direction perpendicular to the direction in which the beam travels, so that the entire wave motion is confined to a plane called the *plane of polarization*. When many different directions of polarization are present in a beam of transverse waves, vibrations occur equally often in all directions perpendicular to the direction of motion, and the beam is then said to be *unpolarized*. Since the vibrations that constitute longitudinal waves can only take place in one direction, namely that in which the waves travel, longitudinal waves cannot be polarized. (**12**)

Light waves are transverse, and it is possible to produce and detect polarized light. To clarify the ideas involved, let us first consider the behavior of transverse waves in a stretched string. If the string passes through a tiny hole in a fence, waves traveling down the string are stopped since the string cannot vibrate there at all. (**13**)

When the hole is replaced by a vertical slot, waves whose vibrations are vertical can get through the fence. (**14**)

However, waves with vibrations in other directions cannot get through. (**15**)

In a situation in which several waves vibrating in different directions move down the string, the slot stops all but vertical vibrations: an initially unpolarized series of waves has become polarized. (**16**)

The above approach can be used to determine whether a particular wave phenomenon can be polarized or not. In the case of a stretched string, what we do is erect another fence a short distance from the first. If the slot in the new fence is also vertical, those waves that can get through the first fence can also get through the second. (**17**)

If the slot in the new fence is horizontal, however, it will stop all waves that reach it from the first fence. (**18**)

Should longitudinal waves (say in a spring) go through the fence, it is possible that their amplitudes might decrease in passing through the slots, but the relative alignments of the slot would not matter. On the other hand, the alignment of the slots is the critical factor in the case of transverse waves. (**19**)

The above chain of reasoning made it possible for the polarization of light waves to be demonstrated in the last century. A number of substances, for instance quartz, calcite, and tourmaline, have different indexes of refraction for light with different planes of polarization relative to their crystal structures, and prisms can be made from them that transmit light in only a single plane of polarization. When a beam of unpolarized light is incident upon such a prism, only those of its waves whose planes of polarization are parallel to a particular plane in the prism emerge from the other side. The remainder of the waves are absorbed or deflected.

Polaroid is an artificially made polarizing material in wide use. It consists of a sheet of polyvinyl alcohol whose needle-like molecules are aligned by stretching the sheet. Iodine added to the sheet then deposits along these molecules, and in this form the iodine permits light with only one plane of polarization to pass through. To exhibit the transverse nature of light waves, we first place two Polaroid discs in line so that their axes of polarization are parallel, and note that all light passing through one disc also passes through the other. Then we turn one disc until its axis of

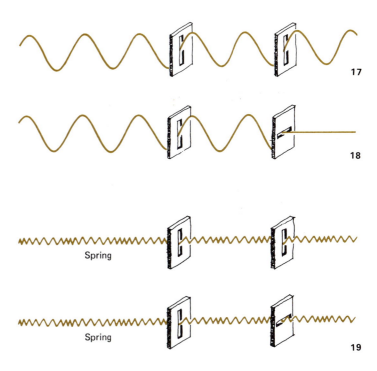

17

18

Spring

Spring

19

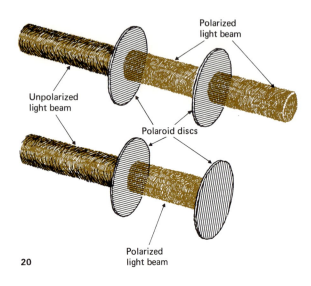

Unpolarized light beam

Polaroid discs

Polarized light beam

Polarized light beam

20

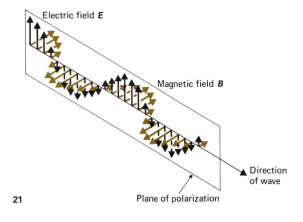

Electric field **E**

Magnetic field **B**

Direction of wave

Plane of polarization

21

polarization is perpendicular to that of the other, and note that all light passing through one disc is now *stopped* by the other. (**20**)

Just what is it whose vibrations are aligned in a beam of polarized light? As we know, light waves actually consist of oscillating electric and magnetic fields perpendicular to each other. Because it is the electric fields of light waves whose interactions with matter produce nearly all common optical effects, the plane of polarization of a light wave is considered to be that in which both the direction of its electric field and its direction of motion lie (the vertical plane in the drawing). Even though nothing material moves during the passage of a light wave, then, it is possible to establish its transverse nature and identify its plane of polarization. (**21**)

SCATTERING

When light waves encounter an obstacle, they are diffracted around its edges. If the obstacle is comparable in size with the wavelength of the light, the diffracted wavefronts are more or less spherical, and they spread out as though they originated in the obstacle. The incoming light is said to be *scattered* by the obstacle. (**22**)

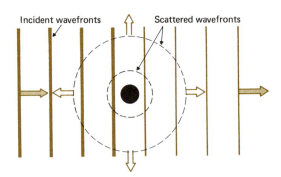

Incident wavefronts

Scattered wavefronts

22

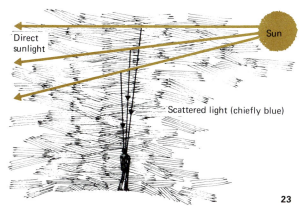

Direct sunlight

Sun

Scattered light (chiefly blue)

23

In general, the intensity of the light scattered by a small obstacle is proportional to $1/\lambda^4$, where λ is the wavelength of the light. The shorter the wavelength, the greater the proportion of the incoming light that is scattered. This is the reason why the sky is blue. When we look at the sky, what we see is light scattered by irregularities in the upper atmosphere. Blue light is scattered about ten times more readily than red light, so the scattered light is chiefly blue in color. **(23)**

At sunrise or sunset, when sunlight must make a very long passage through the atmosphere, much of its blue content is scattered out along the way, and the sun appears red in color. **(24)**

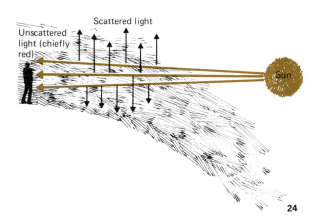

Scattered light

Unscattered light (chiefly red)

Sun

24

1. Can light from incoherent sources interfere? If so, then why is a distinction made between coherent and incoherent sources?

2. Which of the following—refraction, reflection, interference, diffraction, scattering, polarization—can occur in
   a) transverse waves, and
   b) longitudinal waves?

3. Radio waves diffract pronouncedly around buildings, whereas light waves, which are also electromagnetic waves, do not. Why?

4. A camera can be made by using a pinhole instead of a lens. What happens to the sharpness of the picture if the hole is too large? if it is too small?

5. Explain the peculiar appearance of a distant light source when seen through a piece of finely woven cloth.

6. How can a single sheet of Polaroid be used to show that skylight is partially polarized?

7. Which of these optical phenomena—refraction, interference, diffraction, polarization—are involved in Young's double-slit experiment?

8. What is the relationship between the plane of polarization of a transverse wave and its direction of propagation?

32

QUANTUM THEORY OF LIGHT

"The elementary quantum of action plays a fundamental part in atomic physics and its introduction . . . was destined to remodel basically the physical outlook and thinking of man." Max Planck (1858–1947)

Although the wave theory of light accounts very well for most aspects of its behavior, there are still other aspects whose only explanation is that light consists of a stream of tiny particles. There is no way to avoid this seeming contradiction: light—indeed, all electromagnetic waves—has both certain wave properties and certain particle properties. An appreciation of the dual nature of light is the key to understanding the physics of the atom.

THE WAVE-PARTICLE DUALITY

The formulation of the quantum theory of light early in this century led to a profound change in our ways of thinking about the physical world. Until then there seemed to be nothing ambiguous about what was meant by a particle and what was meant by a wave. Particles and waves are still entirely separate concepts in everyday life: it is impossible to confuse a stone, for instance, with the waves it produces when it is dropped into a lake. But in the microscopic realm of the atom the situation is very different. Here electromagnetic waves—which, as we know, exhibit such characteristic wave phenomena as diffraction and interference—nevertheless behave in many ways exactly as particles do, and electrons—whose particle nature is amply shown in the operation of every television picture tube—nevertheless behave in many ways exactly as waves do. On the microscopic level, then, a *wave–particle duality* replaces the distinction between waves and particles so evident on a macroscopic level.

The quantum theory of light had its genesis in the inability of classical electromagnetic theory to account for certain crucial experiments involving electromagnetic waves. This failure was ironic, because electromagnetic theory was in many ways the most successful product of classical physics and James Clerk Maxwell's masterly prediction of electromagnetic waves and his realization of the nature of light provided the theory with seemingly invincible support. Even the discovery of special relativity did not affect electromagnetic theory, although, as we have seen, the interpretation of this theory had to be changed: instead of separate electric and magnetic fields, there is only a single electromagnetic field.

But Maxwell's theory cannot explain *all* electromagnetic phenomena, and it has had to be replaced by a much more ambitious theory, called *quantum electrodynamics*, which, though not wholly satisfactory, nevertheless is able to fit a remarkable amount of data into a unified picture. One element of this picture is Maxwell's theory; another is the quantum theory of light which we are about to examine; and there are still other elements too complex to include in this book.

THE PHOTOELECTRIC EFFECT

Toward the end of the nineteenth century a number of experiments were performed that revealed the emission of electrons from a metal surface when light (particularly ultraviolet light) falls on it. This phenomenon is known as the *photoelectric effect*. It is not, at first glance, anything to surprise us, for light waves carry energy, and some of the energy absorbed by the metal may somehow concentrate on individual electrons and reappear as kinetic energy. Upon closer inspection of the data, however, we find that the photoelectric effect can hardly be explained in so straightforward a manner.

Here is how the photoelectric effect can be detected. The photoelectrons ejected from the

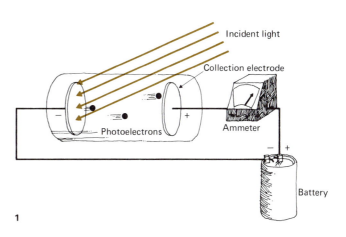

1

irradiated metal plate are attracted to the positive electrode at the other end of the tube, and the resulting current is measured with the ammeter. (**1**)

The first peculiarity of the photoelectric effect is that, even when the metal surface is only faintly illuminated, the emitted electrons (which are called *photoelectrons*) leave the surface immediately. But according to the electromagnetic theory of light, the energy content of light waves is spread out across the width of the particular light beam involved. Calculations show that a definite period of time—about *a year* in the case of a beam of very low intensity—must elapse before any individual electrons accumulate enough energy to leave the metal. Instead, the electrons are found to be emitted as soon as the light is turned on.

Another unexpected discovery is that the energy of the photoelectrons does not depend upon the intensity of the light. A bright light yields more electrons than a dim one, but their average energy remains the same. This behavior contradicts the electromagnetic

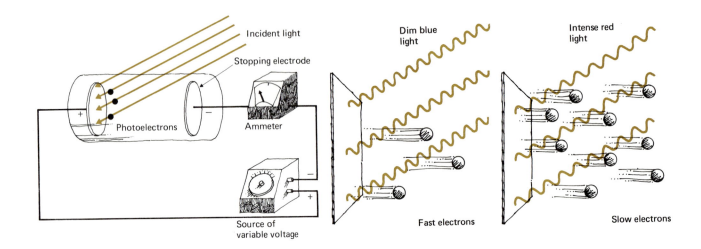

Incident light

Stopping electrode

+

Photoelectrons

Ammeter

−

Source of
variable voltage

2

Dim blue
light

Fast electrons

Intense red
light

Slow electrons

3

theory of light, which predicts that the energy of photoelectrons should depend upon the intensity of the light beam responsible for them.

An apparatus like this can be used to determine the distribution of photoelectron energies in a particular experiment. As the stopping electrode is given a higher negative potential, the slower photoelectrons are repelled before they can reach it. Finally a voltage will be reached at which no photoelectrons whatever arrive at the stopping electrode, as indicated by the current dropping to zero, and this voltage corresponds to the maximum photoelectron energy. (2)

The energies of the photoelectrons emitted from a given metal surface turn out, most surprisingly of all, to depend upon the *frequency* of the light employed. At frequencies below a certain critical one (which is characteristic of the particular metal), no electrons whatever are given off. Above this threshold frequency, the photoelectrons have a range of energies from zero to a certain maximum

value, and *this maximum energy increases with increasing frequency*. High-frequency light yields high maximum photoelectron energies; low-frequency light yields low maximum photoelectron energies. Thus dim blue light produces electrons with more energy than those produced by intense red light, although the latter results in a greater number of them. (3)

Here is a graph of maximum photoelectron energy KE_{max} plotted against the frequency f of the incident light for two different target metals. No photoelectrons are emitted at frequencies less than f_0^A in the case of metal A and less than f_0^B in the case of metal B. In both cases, however, the experimental points lie on straight lines, and these lines are parallel. It is clear that there is a simple relationship between KE_{max} and f. This relationship can be expressed in equation form as

$$KE_{max} = h(f - f_0)$$
$$= hf - hf_0,$$

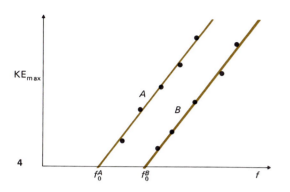

where h (the slope of the lines) has the same value in all cases but f_0, the threshold frequency below which no photoemission occurs, depends upon the nature of the target. (4)

QUANTUM THEORY OF LIGHT

Aware that the electromagnetic theory of light, despite its notable success in accounting for other optical phenomena, failed to explain the photoelectric effect, Albert Einstein in 1905 sought some other basis for interpreting it. He found what he needed in a novel assumption that Max Planck, a German physicist, had had to make a few years earlier in order to understand the origin of the radiation given off by an object so hot that it is luminous (a poker thrust in a fire, for instance). Planck found that the accepted physical laws of the time predicted the observed characteristics of this radiation *provided* that the radiation is considered as though being emitted in little bursts of energy, rather than continuously. These bursts of energy are called *quanta*.

Planck showed that the energy E of each quantum had to be related to the light frequency f by the formula

$$\text{Quantum energy} = E = hf,$$

where h is a constant, known today as Planck's constant, whose value is

$$h = 6.63 \times 10^{-34} \text{ J-s}.$$

Although the energy radiated by a heated object must be regarded as coming out intermittently, in order for theory and experiment to agree, Planck held to the conventional view that it nevertheless travels through space as continuous waves.

Einstein saw that Planck's idea could be used to interpret the photoelectric effect if light not only is emitted a quantum at a time but also propagates as separate quanta. Then the h of the photoelectric effect equation is the same as the h of the formula $E = hf$, and the significance of the former equation becomes clear when it is rewritten

$$hf = \text{KE}_{\text{max}} + hf_0. \qquad \text{Photoelectric effect}$$

What this equation states is that

Quantum energy = maximum electron energy + energy required to eject an electron.

The reason for a threshold frequency f_0 is clear: it corresponds to the energy required to dislodge an electron from the metal surface. (There must be such a minimum energy, or electrons would leave metals all the time.) And there are several plausible reasons why not all photoelectrons have the same energy even though a single frequency of light is used. For instance, not all the quantum energy hf may be transferred to a single electron, and an electron may lose some of its initial energy

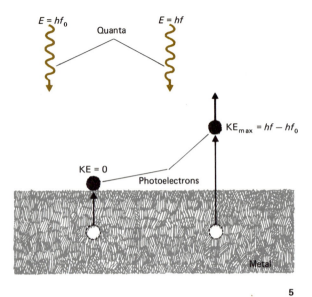

$E = hf_0$ Quanta $E = hf$

$KE_{max} = hf - hf_0$

$KE = 0$

Photoelectrons

Metal

5

in collisions with other electrons within the metal before it actually emerges from the surface. **(5)**

Problem. The threshold frequency for copper is 1.1×10^{15} Hz. When ultraviolet light of frequency 1.5×10^{15} Hz is directed on a copper surface, what is the maximum energy of the photoelectrons?

Solution. From the equation of the photoelectric effect,

$$KE_{max} = hf - hf_0 = h(f - f_0)$$

$$= 6.63 \times 10^{-34} \text{ J-s} \times (1.5 - 1.1) \times 10^{15} \text{ Hz}$$

$$= 2.7 \times 10^{-19} \text{ J}.$$

Since $1 \text{ eV} = 1.6 \times 10^{-19}$ J, the maximum photoelectron energy is

$$\frac{2.7 \times 10^{-19} \text{ J}}{1.6 \times 10^{-19} \text{ J/C}} = 1.7 \text{ eV}.$$

WHAT IS LIGHT?

Einstein's notion that light travels as a series of little packets of energy (sometimes referred to as quanta, sometimes as *photons*) is in complete contradiction with the wave theory of light. And the wave theory, as we know, has some powerful observational evidence on its side. There is no other way of explaining interference and diffraction effects, for example. According to the wave theory, light spreads out from a source in a manner analogous to the spreading out of ripples on the surface of a lake when a stone is dropped into it, with the energy of the light distributed continuously throughout the wave pattern. According to the quantum theory, light spreads out from a source as a succession of localized packets of energy, each sufficiently small to permit its being absorbed by a single electron. Yet, despite the particle picture of light that it presents, the quantum theory requires a knowledge of the light frequency f, a wave quantity, in order to determine the energy of each quantum. **(6)**

On the other hand, the quantum theory of light is able to explain the photoelectric effect. It predicts that the maximum photoelectron energy should depend upon the frequency of the incident light and not upon its intensity, precisely the opposite of what the wave theory suggests, and it is able to explain why even the feeblest light can lead to the immediate emission of photoelectrons. The wave theory can give no reason why there should be a threshold frequency below which no photoelectrons are observed, no matter how strong the light beam, something that follows naturally from the quantum theory.

Which theory is correct? The history of physics is filled with examples of physical ideas that required revision or even replace-

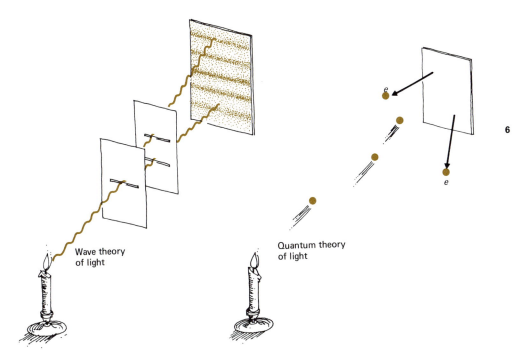

Wave theory
of light

Quantum theory
of light

6

ment when new empirical data conflicted with them, but this is the first occasion in which two completely different theories are both required to explain a single physical phenomenon. In thinking about this, it is important for us to note that, in a particular situation, light behaves *either* as though it has a wave nature *or* a particle nature. While light sometimes assumes one guise and sometimes the other, there is no physical process in which both are exhibited. The same light beam can diffract around an obstacle and then impinge on a metal surface to eject photoelectrons, but these two processes occur separately.

The electromagnetic theory of light and the quantum theory of light complement each other; by itself, each theory is "correct" in certain experiments, and there are no relevant experiments which neither can account for. Light must be thought of as a phenomenon that incorporates both particle and wave characters. Although we cannot visualize its true nature, these complementary theories of light are able to account for its behavior, and we have no choice but to accept them both.

X-RAYS

If photons of light can give up their energy to electrons, can the kinetic energy of moving electrons be converted into photons? The answer is that such a transformation is not only possible, but had in fact been discovered (though not understood) prior to the work of Planck and Einstein. In 1895 Roentgen found that a mysterious, highly penetrating radiation is emitted when high-speed electrons impinge on matter. The x-rays (so called because their nature was then unknown) caused phosphorescent substances to glow, exposed photographic plates, traveled in straight lines, and were not affected by electric or magnetic fields. The more energetic the electrons, the more penetrating the x-rays, and the greater the number of electrons, the greater the density of the resulting x-ray beam.

After over ten years of study, it was finally established that x-rays exhibit, under certain circumstances, both interference and polarization effects, leading to the conclusion that they are electromagnetic waves. From the interference experiments their frequencies were found to be very high, above those in ultraviolet light.

The basic features of an x-ray tube are shown here. Battery A sends a current through the filament, heating it until it emits electrons. The electrons are then accelerated toward a metal target by the potential difference V provided by battery B. The tube is evacuated to permit the electrons to reach the target unimpeded. The impact of the electrons causes the evolution of x-rays from the target. (7)

What is the physical process involved in the production of x-rays? It is known that charged particles emit electromagnetic waves whenever they are accelerated, and so we may reasonably identify x-rays as the radiation accompanying the slowing down of fast electrons when they strike a metal. The great majority of the incident electrons, to be sure, lose their kinetic energy too gradually for x-rays to be evolved, and merely act to heat the target. (Consequently the targets in x-ray tubes are made of metals with high melting points, and a means for cooling the target is often provided.) A few electrons, however, lose much or all of their energy in single collisions with target atoms, and this is the energy that appears as x-rays. In other words, we may regard x-ray production as an inverse photoelectric effect.

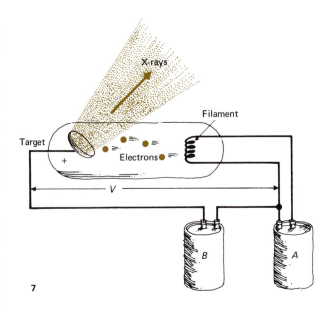

7

Since the threshold energy hf_0 needed to remove an electron from a metal is only a few eV whereas the accelerating potential V in an x-ray tube usually exceeds 10,000 V, we can neglect hf_0 here. The highest frequency f_{max} found in the x-rays emitted from a particular tube should therefore correspond to a quantum energy of hf_{max}, where hf_{max} equals the kinetic energy

$$\mathbf{KE} = eV$$

of an electron that has been accelerated through a potential difference of V. We conclude that

$$hf_{max} = eV \qquad \text{X-ray energy}$$

in the operation of an x-ray tube.

Problem. Find the highest frequency present in the radiation from an x-ray machine whose operating potential is 50,000 V.

Solution. From the formula $hf_{max} = eV$ we find that

$$f_{max} = \frac{eV}{h}$$

$$= \frac{1.6 \times 10^{-19} \text{ C} \times 5.0 \times 10^4 \text{ V}}{6.6 \times 10^{-34} \text{ J-s}}$$

$$= 1.2 \times 10^{19} \text{ s}^{-1}$$

$$= 1.2 \times 10^{19} \text{ Hz.}$$

EXERCISES

1. If light transfers energy by means of separate photons, why do we not perceive light as a series of tiny flashes?

2. If Planck's constant were smaller than it is, would quantum phenomena be more or less conspicuous than they are now in everyday life?

3. Why do you think the wave aspect of light was discovered earlier than its particle aspect?

4. What is the rest energy of a photon whose frequency is 10^{15} Hz?

5. Yellow light has a frequency of about 5×10^{14} Hz. How many photons are emitted per second by a yellow lamp radiating at a power of 10 W?

6. The eye can detect as little as 10^{-18} J of electromagnetic energy. How many photons of $\lambda = 6 \times 10^{-7}$ m does this energy represent?

7. A radio transmitter operates at a frequency of 880 kHz and a power of 10,000 W. How many photons per second does it emit?

8. What is the lowest-frequency light that will cause the emission of photoelectrons from a surface whose nature is such that 1.9 eV are required to eject an electron?

9. Photoelectrons are emitted with a maximum speed of 7×10^5 m/s from a surface when light of frequency 8×10^{14} Hz is shined on it. What is the threshold frequency for this surface?

10. Light of wavelength 5×10^{-7} m falls on a potassium surface whose nature is such that 2 eV are needed to eject an electron. What is the maximum kinetic energy in electron volts of the photoelectrons that are emitted?

11. Light from the sun reaches the earth at the rate of about 1.4×10^3 W/m² of area perpendicular to the direction of the light. Assume sunlight is monochromatic with a frequency of 5×10^{14} Hz.

    a) How many photons fall per second on each m² of the earth's surface directly facing the sun?

    b) How many photons are present in each m³ near the earth on the sunlit side?

33

MATTER WAVES

A moving object has both wave and particle properties. When the object is large, its wave properties are insignificant and it is easy to determine such particle properties as its position and momentum. But when the object is small, its wave properties become important and make the precise determination of its particle properties impossible. The uncertainty principle is an expression of the limits set by the wave nature of matter on finding the position and state of motion of a moving object.

QUANTUM MECHANICS

As we have seen, electromagnetic waves under certain circumstances have properties indistinguishable from those of particles. It requires no greater stretch of the imagination to speculate whether what we normally think of as particles might not have wave properties, too. This speculation was first made by Louis de Broglie in 1924. Soon afterward de Broglie's idea was taken up and developed by a number of other physicists (notably Heisenberg, Schrödinger, Born, Pauli, and Dirac) into the elaborate, mathematically difficult—but very beautiful—theory called *quantum mechanics*. The advent of quantum mechanics did more than provide a supremely accurate and complete description of atomic phenomena; it also altered the way in which the physicist approaches nature, so that he now thinks in terms of probabilities instead of in terms of certainties. The universe is closer in many respects to a roulette wheel than to a clock—but it is a roulette wheel that obeys certain specific rules, and the laws of physics are these rules.

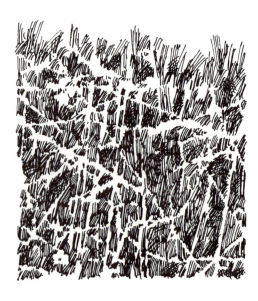

"Can we still call something with which the concepts of position and motion cannot be associated in the usual way a *thing*, a *particle*? And if not, what is the reality that our theory has been invented to describe?" Max Born (1882–1970)

"If a man will begin with certainties, he shall end in doubts; but if he will be content to begin with doubts, he shall end in certainties." Francis Bacon (1531–1626)

DE BROGLIE WAVELENGTH

The linear momentum p of a photon of light of frequency f is given by

$$p = \frac{hf}{c}.$$

Since

$$\text{Wavelength} \times \text{frequency} = \text{velocity},$$

$$\lambda f = c,$$

we can express p in terms of wavelength λ in the formula

$$p = \frac{h}{\lambda}.$$

Hence the wavelength of a photon is inversely proportional to its momentum:

$$\lambda = \frac{h}{p}.$$

De Broglie suggested that this equation for wavelength is a perfectly general one that applies to material objects as well as to photons. The linear momentum of a material object of mass m and velocity v is $p = mv$, and so its *de Broglie wavelength* is

$$\text{De Broglie wavelength} = \lambda = \frac{h}{mv}.$$

The more linear momentum an object has, the shorter its wavelength. The relativistic formula

$$m = \frac{m_0}{\sqrt{1 - v^2/c^2}}$$

must be used for m in computing de Broglie wavelengths.

How can de Broglie's hypothesis be verified? Perhaps the most striking example of wave behavior is an interference pattern, which depends upon the ability of waves both to bend around obstacles and to interfere constructively and destructively with one another. Several years after de Broglie's work, Davisson and Germer in the United States and G. P. Thomson in England independently demonstrated that streams of electrons undergo interference when they are scattered from crystals. The patterns they observed were in complete accord with the electron wavelengths predicted by de Broglie's formula.

In certain aspects of its behavior a moving object resembles a wave, and in other aspects it resembles a particle. Which type of behavior is most conspicuous depends upon how the object's de Broglie wavelength compares with its dimensions and with the dimensions of whatever it interacts with. Two examples will help us appreciate this statement.

In one of their experiments, Davisson and Germer aimed a beam of 54-eV electrons at a nickel crystal. The momentum of a 54-eV electron can be calculated nonrelativistically. Since

$$\text{KE} = \tfrac{1}{2}mv^2$$

and

$$\text{KE} = 54 \text{ eV} \times 1.6 \times 10^{-19} \frac{\text{J}}{\text{eV}}$$

$$= 8.6 \times 10^{-18} \text{ J},$$

we have

$$mv = \sqrt{2m\text{KE}}$$

$$= \sqrt{2 \times 9.1 \times 10^{-31} \text{ kg} \times 8.6 \times 10^{-18} \text{ J}}$$

$$= 4.0 \times 10^{-24} \frac{\text{kg-m}}{\text{s}}.$$

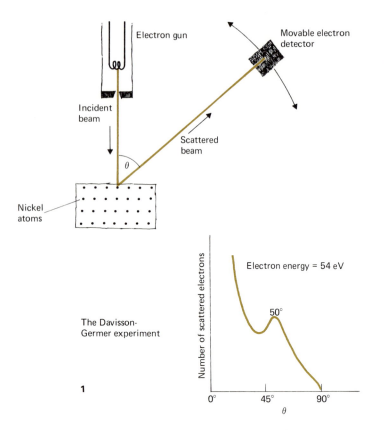

Electron gun

Movable electron detector

Incident beam

Scattered beam

θ

Nickel atoms

The Davisson-Germer experiment

Number of scattered electrons

Electron energy = 54 eV

50°

0°   45°   90°

θ

1

The electron wavelength is therefore

$$\lambda = \frac{h}{mv}$$

$$= \frac{6.63 \times 10^{-34} \text{ J-s}}{4.0 \times 10^{-24} \text{ kg-m/s}}$$

$$= 1.7 \times 10^{-10} \text{ m}.$$

When Davisson and Germer found that the scattered 54-eV electrons were concentrated in just that direction ($\theta = 50°$) predicted by the theory of interference for waves of $\lambda = 1.7 \times 10^{-10}$ m, instead of the more even distribution expected for purely billiard-ball scattering, they interpreted the result as support for de Broglie's hypothesis. **(1)**

On the other hand, a 1500-kg car whose velocity is 30 m/s has a de Broglie wavelength of

$$\lambda = \frac{h}{mv}$$

$$= \frac{6.63 \times 10^{-34} \text{ J-s}}{1.5 \times 10^3 \text{ kg} \times 30 \text{ m/s}}$$

$$= 1.5 \times 10^{-38} \text{ m}.$$

The car's wavelength is so small relative to its dimensions that no wave behavior can be detected.

The above examples are extreme ones. More ambiguous is the case of a moving atom or molecule. Earlier we saw how a model of a gas in which molecules are considered as particles was very successful in explaining the properties of gases. But at 0°C the average wavelength of helium atoms, to give a specific illustration, is 7.6×10^{-11} m, which is of the same order of magnitude as atomic dimensions. Such atoms may exhibit either particle or wave characteristics depending upon the situation.

WAVE FUNCTION AND PROBABILITY DENSITY

In water waves, the physical quantity that varies periodically is the height of the water surface. In sound waves, the variable quantity is pressure in the medium the waves travel through. In light waves, the variable quantities are the electric and magnetic field strengths. What is it that varies in the case of matter waves?

The quantity whose variations constitute the matter waves of a moving object is known as its *wave function*. The symbol for wave function is ψ, the Greek letter *psi*.

The value of ψ^2 for a particular object at a certain place and time is proportional to the probability of finding the object at that place at that time.

For this reason the quantity ψ^2 is called the *probability density* of the object. A large value of ψ^2 signifies that the body is likely to be found at the specified place and time; a small value of ψ^2 signifies that the body is unlikely to be found at that place and time. Thus matter waves may be regarded as waves of probability.

Why does the probability of finding an object depend upon ψ^2 and not upon the wave function ψ itself? The answer is subtle. The amplitude of every wave varies from $-A$ to $+A$ to $-A$ to $+A$ and so on, where A is the maximum absolute value of whatever the wave variable is. But a negative probability is meaningless: the probability that an object be at a given place at a given time must lie between 0 (the object is *not* there) and 1 (the object *is* there). An intermediate probability, say 0.4, means that there is a 40% chance of finding the object there at that time. A probability of -0.4, however, makes no sense at all. Procedures are known for calculating the wave function ψ for moving objects in a great many situations, and each value of ψ must be squared to obtain a positive quantity that can be compared with experiment. (Actually, ψ sometimes turns out to be an imaginary quantity that contains the factor $\sqrt{-1}$, and in such cases ψ^2 is obtained by another procedure.)

Even though we cannot visualize what is meant by ψ and so cannot form a mental image of matter waves, the agreement between theory and experiment signifies that the notion of matter waves is a meaningful way to describe moving objects.

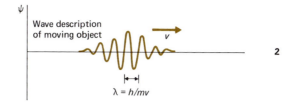

A group or packet of matter waves is associated with every moving object. The packet travels with the same velocity as the object does. The waves in the packet have the average wavelength $\lambda = h/mv$ given by de Broglie's formula. We shall now see why a single, precise wavelength cannot be associated with a packet of matter waves. (2)

THE UNCERTAINTY PRINCIPLE

To regard a moving object as a wave packet raises the problem of the ultimate accuracy with which such "particle" properties as its position and momentum can be determined. In the macroscopic world, where the wave aspects of matter are insignificant, the limit to how accurately position and momentum can be measured depends solely upon our instruments, and there is, in principle, no absolute limit at all. But in the microscopic world, where the wave aspects of matter are very significant indeed, these wave aspects set a fundamental limit to the accuracy of measurements of position and momentum. Of course, poor instruments will give poor results, but even if perfect instruments were available, they would not yield exact results. The *uncertainty principle* is the physical law which expresses quantitatively the basic indeterminacy which the wave nature of matter imposes on measurements of moving objects.

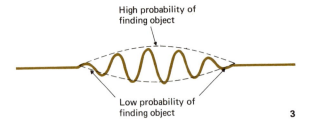

High probability of finding object

Low probability of finding object

3

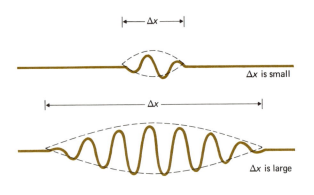

$\longmapsto \Delta x \longrightarrow$

Δx is small

$\longmapsto \Delta x \longrightarrow$

Δx is large

4

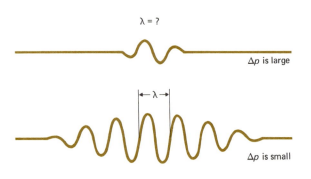

$\lambda = ?$

Δp is large

$\longmapsto \lambda \longrightarrow$

Δp is small

5

Let us see how the wave nature of matter leads to the uncertainty principle. In the wave packet that describes a moving object, the amplitude of the waves at any point is a measure of the probability of finding the object at that point. **(3)**

A narrow wave group means a small uncertainty Δx in the object's position. A wide wave group means a large uncertainty Δx in the object's position. **(4)**

On the other hand, the wavelength of a wide wave group is clearly defined, and the corresponding momentum of the object, which is given in terms of wavelength by

$$p = \frac{h}{\lambda},$$

has only a small uncertainty Δp. The wavelength of a narrow wave group is imprecise, and the corresponding momentum has a large uncertainty Δp. **(5)**

These considerations point to an inverse relationship between Δx and Δp, which a detailed analysis shows to be

$$\Delta x\, \Delta p \geqslant \frac{h}{2\pi}. \qquad \text{Uncertainty principle}$$

In words, this *uncertainty principle* states that

$$\begin{bmatrix} \text{Uncertainty} \\ \text{in position} \end{bmatrix} \times \begin{bmatrix} \text{uncertainty} \\ \text{in momentum} \end{bmatrix}$$

is equal to or greater than

$$\frac{\text{Planck's constant}}{2\pi}.$$

To repeat, the uncertainty principle follows from the wave character of moving objects, and has nothing to do with inaccuracies in making measurements.

The uncertainty principle thus has the following consequences:

1. If an object has a well-defined position at a certain time, its momentum must have a large uncertainty;

2. If an object has a well-defined momentum at a certain time, its position must have a large uncertainty.

On a macroscopic scale, since Planck's constant h is such a minute quantity, the limitation imposed upon measurements by the uncertainty principle is negligible, but on a microscopic scale the uncertainty principle dominates many phenomena.

ELECTRONS, ATOMS, AND NUCLEI

A significant application of the uncertainty principle is to the question of whether electrons are present within atomic nuclei. Experiments show that nuclei are about 10^{-14} m in radius. If an electron is to be confined within a nucleus, the uncertainty Δx in its position cannot exceed about 10^{-14} m. Hence the uncertainty in the electron's momentum must be given by

$$\Delta p \geqslant \frac{h}{2\pi \Delta x}$$

$$\geqslant \frac{6.63 \times 10^{-34} \text{ J-s}}{2\pi \times 10^{-14} \text{ m}}$$

$$\geqslant 1.1 \times 10^{-20} \frac{\text{kg-m}}{\text{s}}.$$

This momentum uncertainty corresponds to an energy uncertainty of over 20 MeV (a relativistic calculation is needed here). If this is the uncertainty in the electron's energy, the energy itself must be at least comparable in magnitude. However, the electrons associated with atoms, even unstable ones, never have much more than 10% of this amount of energy, and we therefore conclude that electrons are not present inside atomic nuclei.

How much energy must an electron have if it is confined to an atom? As we know, an atom is much larger in size than its nucleus, so the electron energy can be smaller than in the case of the nucleus. The hydrogen atom, to take a specific example, has a radius of about 5×10^{-11} m, which means that the uncertainty Δx in the location of its electron cannot exceed this figure. The momentum uncertainty of such an electron is

$$\Delta p \geqslant \frac{h}{2\pi \Delta x}$$

$$\geqslant \frac{6.63 \times 10^{-34} \text{ J-s}}{2\pi \times 5 \times 10^{-11} \text{ m}}$$

$$\geqslant 2.1 \times 10^{-24} \frac{\text{kg-m}}{\text{s}}.$$

This momentum uncertainty corresponds to an energy uncertainty of only 17 eV (the calculation here can be made using Newtonian mechanics, with KE $= p^2/2m$). Electrons in hydrogen atoms actually do have energies of about this amount.

AN EXPERIMENT WITH ELECTRONS

The uncertainty principle makes sense from the point of view of the particle properties of waves as well as from the point of view of the wave properties of particles. Suppose we wish to measure the position and momentum of something at a certain moment. To do so, we must touch it with something else that will carry the required information back to us;

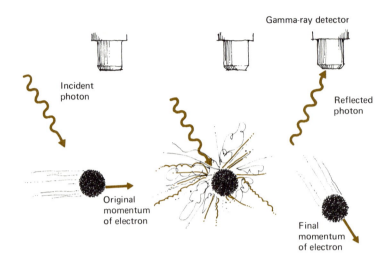

Incident photon

Gamma-ray detector

Reflected photon

Original momentum of electron

Final momentum of electron

7

that is, we must poke it with a stick, shine light on it, or perform some similar act. The measurement process itself thus requires that the body be interfered with in some way, and if we consider this interference in detail, we are led to the same uncertainty principle as before even without taking into account the wave nature of moving objects.

Let us imagine we are looking at an electron with the help of light whose wavelength is λ. We cannot use ordinary visible light, because the electron's diameter is perhaps a hundred million times smaller than the wavelengths found in visible light, but there is no reason in principle why gamma rays of ultrashort wavelength could not be employed instead. Each photon of this light has the momentum h/λ. (**6**)

$$\rightarrow| \; \lambda \; |\leftarrow$$

$$p = h/\lambda$$

6

When one of these photons bounces off the electron (which must occur if we are to "see" it), the electron's original momentum will be changed. The exact amount of the change Δp cannot be predicted, but it will be of the same order of magnitude as the photon momentum h/λ. Hence

$$\Delta p \approx \frac{h}{\lambda}.$$

The *larger* the wavelength, the smaller the uncertainty in momentum. (**7**)

Because light is a wave phenomenon as well as a particle phenomenon, we cannot expect to determine the electron's location with perfect accuracy even with the best of instruments. A reasonable estimate of the irreducible uncertainty Δx in the measurement might be one wavelength. That is,

$$\Delta x \geqslant \lambda.$$

The *smaller* the wavelength, the smaller the

uncertainty in location. Hence if we use light of short wavelength to increase the accuracy of our position measurement, there will be a corresponding decrease in the accuracy of our momentum measurement, whereas light of long wavelength will yield an accurate momentum but an inaccurate position. (8)

8

By combining the two formulas above, we find that

$$\Delta x \, \Delta p \geqslant h.$$

A more detailed calculation shows that somewhat better accuracy is actually possible, so that the limit of the product $\Delta x \, \Delta p$ is $h/2\pi$ instead of just h.

EXERCISES

1. How is the de Broglie wavelength of a particle related to its linear momentum? to its angular momentum?

2. A proton, an electron, and a car all have the same wavelength. Which of them has the most kinetic energy?

3. A photon and an electron have the same wavelength. How are their linear momenta related?

4. A proton and an electron have the same kinetic energies. How are their de Broglie wavelengths related?

5. Can the rest mass of a moving particle be determined by measuring its de Broglie wavelength?

6. What is the simplest experimental procedure that can distinguish between a gamma ray whose wavelength is 10^{-11} m and an electron whose de Broglie wavelength is also 10^{-11} m?

7. The uncertainty principle applies to all objects, yet its consequences are only significant for such minute particles as electrons, protons, and neutrons. Why?

8. The atoms in a solid would vibrate with a certain minimum "zero-point" energy even at $0°K$, whereas the molecules in an ideal gas would be at rest at $0°K$. Use the uncertainty principle to explain this apparent paradox.

9. Find the de Broglie wavelength of a 10^4 kg truck whose speed is 24 m/s.

10. Calculate the de Broglie wavelength of
    a) an electron whose speed is 1×10^8 m/s, and
    b) an electron whose speed is 2×10^8 m/s.
    Use relativistic formulas.

11. Calculate the de Broglie wavelength of a proton whose kinetic energy is 1 MeV. This calculation may be made nonrelativistically.

12. Nuclear dimensions are of the order of 10^{-14} m.
    a) Find the energy in eV of an electron whose de Broglie wavelength is 10^{-15} m and so is capable of revealing details of nuclear structure.
    b) Make the same calculation for a neutron.

13. Find the resolving power of an electron microscope that uses 40-keV electrons by assuming that this is equal to the electron wavelength.

14. a) An electron is confined in a box 10^{-9} m in length. What is the uncertainty in its velocity?
    b) A proton is confined in the same box. What is the uncertainty in its velocity?

15. The position and momentum of a 1000-eV electron are determined at the same time. If the position is found to within 10^{-10} m, what is the percentage of uncertainty in the momentum?

16. At a certain time a measurement establishes the position of an electron with an accuracy of $\pm 10^{-11}$ m. Calculate the uncertainty in the electron's momentum and, from this, the uncertainty in its position 1 s later.

PART 6

PHYSICS OF THE ATOM

34

THE HYDROGEN ATOM

"Rutherford's model of the atom puts before us a task reminiscent of the old dream of philosophers: to reduce the interpretation of the laws of nature to the consideration of pure numbers." Niels Bohr (1885–1962)

A hydrogen atom consists of a proton with an electron circling it. In the Bohr model of this atom the particle and wave properties of the electron are combined: its orbital motion is such that the centripetal force needed is provided by the electric force exerted by the proton, and the circumference of its orbit is exactly one de Broglie wavelength. This model predicts that the hydrogen atom can have certain specific energies only, and no others.

CLASSICAL MODEL OF THE HYDROGEN ATOM

The picture of the atom that emerged from Rutherford's work, as we learned in Section 23, consists of a tiny, massive, positively charged nucleus surrounded by enough negatively charged electrons to leave the atom as a whole electrically neutral. These electrons cannot be stationary, since the electric attraction of the nucleus would pull them in at once. If the electrons are in motion around the nucleus, however, stable orbits like those of the planets about the sun would seem to be possible.

A hydrogen atom, the simplest of all, has a single electron and a nucleus that consists of a single proton. Experiments indicate that 13.6 eV of work must be performed to break apart a hydrogen atom into a proton and electron that go their separate ways. This figure leads to an orbital radius of

$$r = 5.3 \times 10^{-11} \text{ m}$$

and an orbital velocity of

$$v = 2.2 \times 10^6 \frac{\text{m}}{\text{s}}$$

for the electron in the hydrogen atom. The proton mass is 1836 times the electron mass, so we can consider the proton as stationary

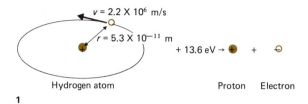

v = 2.2 X 10⁶ m/s

r = 5.3 X 10⁻¹¹ m

+ 13.6 eV → ⊕ + ⊝

Hydrogen atom   Proton   Electron

1

with the electron revolving around it in such a way that the required centripetal force is provided by the electric force exerted by the proton. **(1)**

What the preceding analysis overlooks is that, according to electromagnetic theory, all accelerated electric charges radiate electromagnetic waves—and an electron moving in a circular path is certainly accelerated. Thus an atomic electron circling its nucleus to keep from falling into it by electrostatic attraction cannot help radiating away energy, which means that it must spiral inward until it is swallowed up by the nucleus. **(2)**

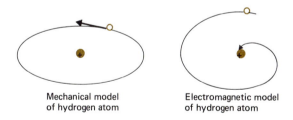

Mechanical model of hydrogen atom

Electromagnetic model of hydrogen atom

2

ELECTRON WAVES IN THE ATOM

As we have seen, the ordinary laws of physics cannot account for the stability of the hydrogen atom, the simplest atom of all, whose electron must be whirling around the nucleus to keep from being pulled into it and yet must be radiating electromagnetic energy continuously. However, since other phenomena similarly impossible to understand on a classical basis—the photoelectric effect, for instance—find complete explanation in terms of quantum concepts, it is appropriate for us to inquire whether this might also be true for the atom.

In the discussion that follows of the Bohr model of the atom, the initial argument is somewhat different from that of Bohr who did not, in 1913, have the notion of matter waves to guide his thinking. The results are exactly the same as Bohr obtained, though.

Let us begin by looking into the wave properties of the electron in the hydrogen atom. The de Broglie wavelength λ of an object of mass m and velocity v is

$$\lambda = \frac{h}{mv}$$

where h is Planck's constant. Now the velocity of the electron in a hydrogen atom is

$$v = 2.2 \times 10^6 \frac{m}{s},$$

and so the wavelength of its matter waves is

$$\lambda = \frac{h}{mv}$$

$$= \frac{6.63 \times 10^{-34} \text{ J-s}}{9.1 \times 10^{-31} \text{ kg} \times 2.2 \times 10^6 \text{ m/s}}$$

$$= 3.3 \times 10^{-10} \text{ m}.$$

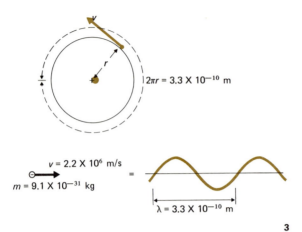

$$2\pi r = 3.3 \times 10^{-10}\ \text{m}$$

$v = 2.2 \times 10^{6}\ \text{m/s}$

$m = 9.1 \times 10^{-31}\ \text{kg}$

$$\lambda = 3.3 \times 10^{-10}\ \text{m}$$

3

This is a most exciting result, because the electron's orbit has a circumference of exactly

$$2\pi r = 2\pi \times 5.3 \times 10^{-11}\ \text{m}$$

$$= 3.3 \times 10^{-10}\ \text{m}. \quad (\textbf{3})$$

We therefore conclude that *the orbit of the electron in a hydrogen atom corresponds to one complete electron wave joined on itself.* (**4**)

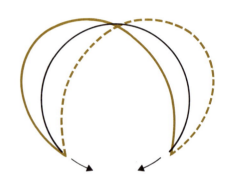

The fact that the electron orbit in a hydrogen atom is one electron wavelength in circumference is just the clue we need to construct a theory of the atom. If we examine the vibrations of a wire loop, we see that their wavelengths always fit an integral number of times into the loop's circumference, with each wave joining smoothly with the next. In the absence of dissipative effects, such vibrations would persist indefinitely. (**5**)

——— Electron path

——— De Broglie electron wave

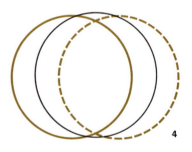

4

Circumference =
2 wavelengths

Circumference =
4 wavelengths

Circumference =
9 wavelengths

5

6

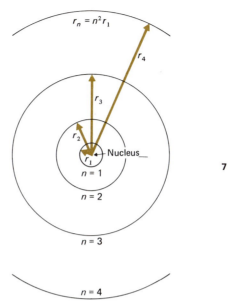

$$r_n = n^2 r_1$$

r_4

r_3

r_2

r_1 ← Nucleus

$n = 1$

$n = 2$

$n = 3$

$n = 4$

7

Why are these the only vibrations possible in a wire loop? A fractional number of wavelengths cannot be fitted into the loop and still allow each wave to join smoothly with the next; the result would be destructive interference as the waves travel around the loop, and the vibrations would die out rapidly. **(6)**

CONDITION FOR ORBITAL STABILITY

By considering the behavior of electron waves in the hydrogen atom as analogous to the vibrations of a wire loop, then, we may postulate that:

An electron can circle an atomic nucleus only if its orbit is an integral number of electron wavelengths in circumference.

This postulate is the decisive one in our understanding of the atom. We note that it combines both the particle and wave characters of the electron into a single statement; although we can never observe these antithetical characters at the same time in an experiment, they are inseparable in nature.

It is easy to express in a formula the condition that an integral number of electron wavelengths fits into the electron's "orbit". The cir-

cumference of a circular orbit of radius r is $2\pi r$, and so the condition for orbit stability is

$$n\lambda = 2\pi r_n \quad n = 1, 2, 3, \ldots \qquad \text{Condition for orbit stability}$$

where r_n designates the radius of the orbit that contains n wavelengths. The quantity n is called the *quantum number* of the orbit.

A straightforward calculation shows that the stable electron orbits are those whose radii are given by the formula

$$r_n = n^2 r_1 \quad n = 1, 2, 3, \ldots \qquad \text{Orbital radii in Bohr atom}$$

where

$$r_1 = 5.3 \times 10^{-11} \text{ m}$$

is the radius of the innermost orbit. **(7)**

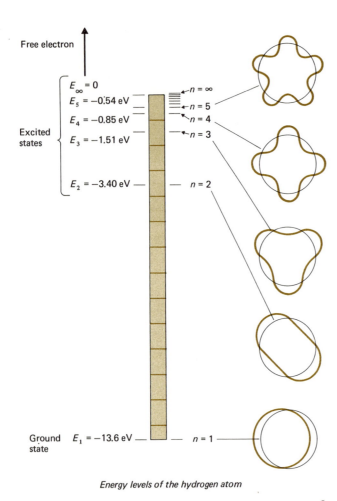

Energy levels of the hydrogen atom

8

where $E_1 = -13.6 \text{ eV} = -2.18 \times 10^{-18} \text{ J}$ is the energy corresponding to the innermost orbit. The energies specified by the above formula are called the *energy levels* of the hydrogen atom.

The energy levels of the hydrogen atom are all less than zero, which signifies that the electron does not have enough energy to escape from the atom. The lowest energy level E_1, corresponding to the quantum number $n = 1$, is called the *ground state* of the atom; the higher levels E_2, E_3, E_4, and so on are called *excited states.* (**8**)

As the quantum number n increases, the energy E_n approaches closer and closer to zero. In the limit of $n = \infty$, $E_\infty = 0$, and the electron is no longer bound to the proton to form an atom. An energy greater than zero signifies an unbound electron which, since it has no closed orbit that must satisfy quantum conditions, may have any positive energy whatever.

The work needed to remove an electron from an atom in its ground state is called its *ionization energy.* The ionization energy is therefore equal to the amount of energy that must be provided to raise an electron from its ground state to an energy of $E = 0$, when it is free. In the case of hydrogen, the ionization energy is 13.6 eV, since the ground-state energy of the hydrogen atom is -13.6 eV.

The presence of definite energy levels in an atom—which is true for all atoms, not just the hydrogen atom—is another example of the fundamental graininess of physical quantities on a microscopic scale. In the everyday world, matter, electric charge, energy, and so on seem continuous and capable of being cut up, so to speak, into parcels of any size we like. In the world of the atom, however,

ENERGY LEVELS

The total energy of a hydrogen atom is not the same in the various permitted orbits. The energy E_n of a hydrogen atom whose electron is in the nth orbit is given by

$$E_n = \frac{E_1}{n^2} \qquad n = 1, 2, 3, \ldots$$

Energy levels of hydrogen atom

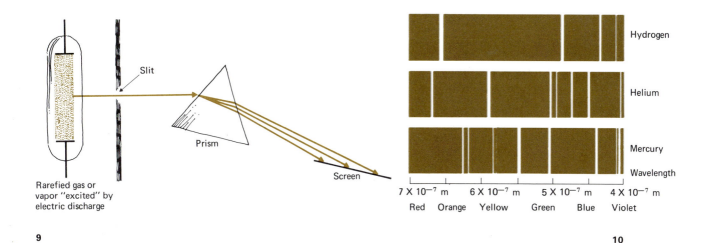

Slit

Prism

Screen

Rarefied gas or vapor "excited" by electric discharge

9

Hydrogen

Helium

Mercury

Wavelength

7×10^{-7} m 6×10^{-7} m 5×10^{-7} m 4×10^{-7} m

Red   Orange   Yellow   Green   Blue   Violet

10

matter consists of elementary particles of fixed masses which join together to form atoms of fixed masses; electric charge always comes in multiples of $+e$ and $-e$; energy in the form of electromagnetic waves of frequency f always comes in separate photons of energy hf; and stable systems of particles, such as atoms, can have only certain energies and no others.

Other quantities, too, are grainy, or *quantized*, and it has turned out that this graininess is the key to understanding how the properties of matter we are familiar with in everyday life originate in the interactions of elementary particles. In the case of the atom, the quantization of energy is a consequence of the wave nature of moving bodies: in an atom the electron wave functions can only occur in the form of standing waves, much as a violin string can vibrate only at those frequencies that give rise to standing waves.

ATOMIC SPECTRA

The Bohr theory is strikingly confirmed in its successful explanation of atomic spectra. When an atomic gas or vapor at somewhat less than atmospheric pressure is "excited" by the passage of an electric current through it, light whose spectrum consists of a limited number of discrete wavelengths is emitted. **(9)**

Atomic spectra are called *line spectra* from their appearance. Every element exhibits a unique line spectrum when a sample of it is suitably excited, and the presence of any element in a substance of unknown composition can be ascertained by the appearance of its characteristic wavelengths in the spectrum of the substance. Here are the most prominent lines in the spectra of hydrogen, helium, and mercury in the visible region. Spectral lines are also found in the infrared and ultraviolet regions. **(10)**

Absorption spectrum
of sodium vapor

Emission spectrum
of sodium vapor

11

It is worth noting that whereas unexcited gases and vapors do not radiate their characteristic spectral lines, they do *absorb* light of certain of those wavelengths when white light is passed through samples of them. In other words, the *absorption spectrum* of an element is closely related to its *emission spectrum*. Emission spectra consist of bright lines on a dark background; absorption spectra consist of dark lines on a bright background. Here are the absorption and emission spectra of sodium vapor. (**11**)

The wavelengths present in atomic spectra fall into definite series. The spectral series of hydrogen are shown here; the wavelength scale is not linear in order to cover the entire spectrum. (**12**)

ORIGIN OF SPECTRAL LINES

The presence of a sequence of definite, discrete energy levels in the hydrogen atom suggests a connection with line spectra. Let us assert that when an electron in an excited state drops to a lower state, the difference in energy between the states is emitted as a single photon of light. Because electrons cannot, according to our model, exist in an atom except in certain specific energy levels, a rapid "jump" from one level to the other, with the energy difference being given off all at once in a photon rather than in some gradual manner, fits in well with this model.

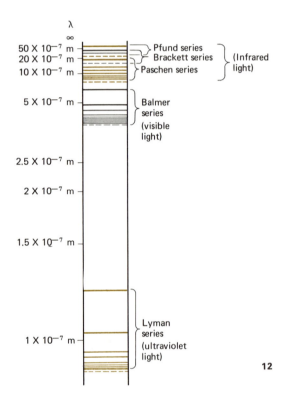

λ

∞

50×10^{-7} m — Pfund series
20×10^{-7} m — Brackett series — (Infrared light)
10×10^{-7} m — Paschen series

5×10^{-7} m — Balmer series (visible light)

2.5×10^{-7} m

2×10^{-7} m

1.5×10^{-7} m

Lyman series (ultraviolet light)

1×10^{-7} m

12

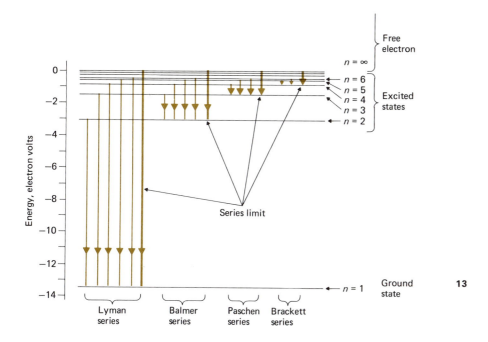

If the quantum number of the initial (higher energy) state is n_i and the quantum number of the final (lower energy) state is n_f, what we assert is that

Initial energy − final energy = quantum energy,

$$E_i - E_f = hf,$$

where f is the frequency of the emitted photon and h is Planck's constant.

Here is an energy-level diagram for the hydrogen atom showing the possible transitions from initial quantum states to final ones. Each transition—or "jump"—involves a characteristic amount of energy, and hence a photon of a certain characteristic frequency. The larger the energy difference between

initial and final energy levels, as indicated by the lengths of the arrows, the higher the frequency of the emitted photon. The origins of the various series of spectral lines are indicated on the diagram. (13)

ATOMIC EXCITATION

There are two principal mechanisms by which an atom may be excited to an energy level above that of its ground state and thereby become capable of radiating. One of these mechanisms is a collision with another atom during which part of their kinetic energy is transformed into electron energy within either or both of the participating atoms. An atom excited in this way will then lose its excitation energy by emitting one or more photons in the

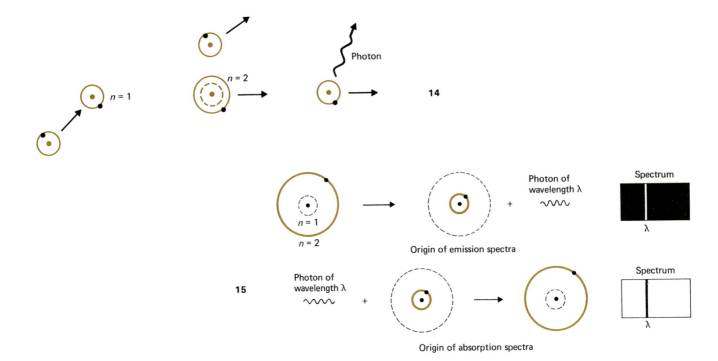

Photon

14

n = 2

n = 1

Photon of
wavelength λ

Spectrum

λ

n = 1

n = 2

Origin of emission spectra

Photon of
wavelength λ

15

+

Spectrum

λ

Origin of absorption spectra

course of returning to its ground state. In an electric discharge in a rarefied gas, an electric field accelerates electrons and charged atoms and molecules (whose charge arises from either an excess or a deficiency in the electrons required to neutralize the positive charge of their nuclei) until their kinetic energies are sufficient to excite atoms they happen to collide with. A neon sign is a familiar example of how applying a strong electric field between electrodes in a gas-filled tube leads to the emission of the characteristic spectral radiation of that gas, which happens to be orange light in the case of neon. (14)

Another excitation mechanism is the absorption by an atom of a photon of light whose energy is just the right amount to raise it to a higher energy level. A photon of wavelength 1.217×10^{-7} m is emitted when a hydrogen atom in the $n = 2$ state drops to the $n = 1$ state; hence the absorption of a photon of wavelength 1.217×10^{-7} m by a hydrogen atom initially in the $n = 1$ state will bring it up to the $n = 2$ state. This process explains the origin of absorption spectra. (15)

When white light (in which all wavelengths are present) is passed through hydrogen gas, photons of those wavelengths that correspond to transitions between hydrogen energy levels are absorbed. The resulting excited hydrogen atoms reradiate their excitation energy almost at once, but these photons come off in random directions, not all in the same direction as in the original beam of white light. The dark lines in an absorption spectrum are therefore never totally dark, but only appear so by contrast with the bright background of transmitted light. We would expect the lines in the absorption spectrum of a particular substance to be the same as lines in its emission spectrum, which is what is found. Only a few of the emission lines appear in absorption, however, since reradiation is so rapid that essentially all the atoms in an absorbing gas or vapor are initially in their lowest ($n = 1$) energy states and therefore take up and reemit only photons that represent transitions to and from the $n = 1$ state. (16)

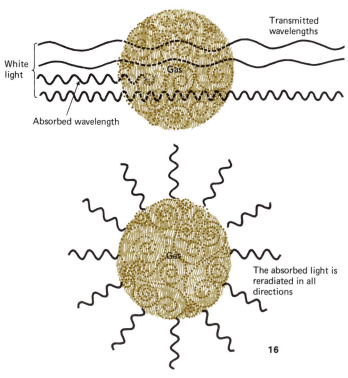

White light

Absorbed wavelength

Transmitted wavelengths

Gas

Gas

The absorbed light is reradiated in all directions

16

EXERCISES

1. How are the Bohr and Rutherford models of the hydrogen atom related?

2. In the Bohr model of the hydrogen atom, why is the electron pictured as revolving around the nucleus?

3. How does the energy difference between adjacent energy levels vary with increasing quantum number in the Bohr model of the hydrogen atom?

4. Is the Bohr theory of the atom compatible with the uncertainty principle?

5. Why does the spectrum of hydrogen have many lines while a hydrogen atom contains only a single electron?

6. Of the following transitions in a hydrogen atom, which leads to the emission of the photon of highest frequency? of lowest frequency? Which leads to the absorption of the photon of highest frequency?

   $n = 1 \rightarrow n = 2$ $n = 2 \rightarrow n = 6$
   $n = 2 \rightarrow n = 1$ $n = 6 \rightarrow n = 2$

7. A proton and an electron, both at rest initially, combine to form a hydrogen atom in the ground state. A single photon is emitted in this process. What is its energy?

8. To what temperature must a hydrogen gas be heated if the average molecular kinetic energy is to equal the ionization energy of the hydrogen atom?

9. Calculate the average kinetic energy per molecule in a gas at room temperature (20°C), and show that this is much less than the energy required to raise a hydrogen atom from its ground state ($n = 1$) to its first excited state ($n = 2$).

10. a) Calculate the de Broglie wavelength of the earth.
    b) What is the quantum number that characterizes the earth's orbit about the sun? (The earth's mass is 6.0×10^{24} kg, its orbital radius is 1.5×10^{11} m, and its orbital speed is 3×10^4 m/s.)

11. How much energy (in joules and in electron volts) is required to remove the electron from a hydrogen atom when it is in the $n = 5$ state?

35

QUANTUM THEORY OF THE HYDROGEN ATOM

"Now what *are* these particles, these atoms and molecules? . . . They can at the most perhaps be thought of as more or less temporary creations within the wave field, whose structure and structural variety, in the widest sense of the word, are so clearly and sharply determined by means of wave laws as they recur always in the same manner; that much takes place *as if* they were a permanent material reality." Erwin Schrödinger (1887–1961)

"I am a little world made cunningly." John Donne (1571–1631)

The Bohr model is quite successful in accounting for a number of the properties of the hydrogen atom, but for a complete understanding of not only this atom but more complex ones as well it is necessary to make use of the quantum theory of the atom. When this theory is applied to the hydrogen atom, it turns out that the electron is not confined to a specific orbit at all and four quantum numbers, not just one, are needed to describe its physical state.

BOHR THEORY VERSUS QUANTUM THEORY

The Bohr theory of the atom is indeed impressive in its agreement with experiment, but it has certain serious limitations. These limitations are absent from the *quantum theory of the atom*, which was developed a decade after Bohr's work and whose refinement and application to new problems continues to the present day. Not only has the quantum theory of the atom provided the theoretical framework for understanding the structure of the atom itself, but it has also furnished the key insights that explain how and why atoms join together to form molecules, solids, and liquids.

The traditional laws of mechanics and electromagnetism do not "work" when applied to the atomic world, as we have seen—these laws lead to predictions that do not agree with experimental findings. But quantum theory works on all scales of size, and it turns out that the traditional laws of physics are merely approximate versions of quantum theory that are successful for large-scale phenomena because on a large scale all that is apparent is the average behavior of a great many separate particles. In dealing with individual particles, however, no averaging is possible, and only quantum theory can be applied.

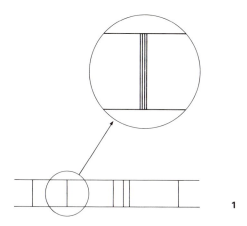

The Bohr theory is unable to account for many important atomic phenomena. While correctly predicting the wavelengths of the spectral lines in hydrogen, which has but a single atomic electron, the Bohr theory fails when attempts are made to apply it to more complex atoms. Even in hydrogen it is not possible to calculate from the Bohr theory the relative probabilities of the various transitions between energy levels, for instance, whether it is more likely that an atom in the $n = 3$ quantum state will go directly to the $n = 1$ state or instead first drop to the $n = 2$ state. (In other words, we cannot find from the Bohr theory which of the spectral lines of hydrogen will show up brightest in an emission spectrum and which will be faint.) For another thing, the careful study of spectral lines shows that many of them actually consist of two or more separate lines that are close together, something that the Bohr theory cannot account for. (1)

In cataloging these objections to the Bohr theory, the intent is not to detract from its eminence in the history of science, which is certainly secure, but instead to emphasize that a more general approach capable of wider application is necessary. Such an approach was developed in the 1920's by Schrödinger, Heisenberg, and others under the name of the quantum theory of the atom. Instead of trying to visualize an atomic electron as a kind of hybrid of a particle and a wave and thinking of it as occupying one of various possible orbits, the quantum theory of the atom avoids all reference to anything not capable of direct measurement and restricts itself only to such observable quantities as photon energies, the mass and charge of the electron, and so on.

In the Bohr theory we compute the radius of an electron orbit from a knowledge of its de Broglie wavelength, which depends upon the electron's momentum; however, according to the uncertainty principle, the position (and hence orbital radius) and momentum of an electron can never simultaneously have well-defined values, and so even in principle the Bohr theory cannot be subjected to an experimental test. The quantum theory sacrifices such easily pictured notions as that of electrons circling a nucleus like planets around the sun in favor of a wholly abstract mathematical formulation, each of whose statements can be verified. Freed from the necessity of working in terms of a model of any kind, the quantum theory is able to tackle successfully a broad range of atomic problems.

QUANTUM NUMBERS

Unfortunately the quantum theory of the hydrogen atom involves advanced mathematics and so cannot be developed here. The conclusions of the theory, however, are fairly straightforward.

In the Bohr model of the hydrogen atom, the motion of the orbital electron is essentially one-dimensional. There the electron is regarded as being confined to a definite circular orbit, and the only quantity that changes as it revolves around its nucleus is its position on this fixed circle. A single quantum number is all that is required to specify the physical state of such an electron.

In the more general quantum theory of the atom, the electron is not restricted to a specific orbit, and three quantum numbers turn out to be necessary instead of the single one of the Bohr theory. These are:

1. The principal quantum number n;
2. The orbital quantum number l;
3. The magnetic quantum number m_l.

The value of n is the chief factor that governs the total energy of an electron bound to a nucleus. The energy levels of a hydrogen atom are the same in the Bohr and quantum theories, and are specified by the formula

$$E = \frac{E_1}{n^2} \qquad n = 1, 2, 3, \ldots$$

where $E_1 = -13.6 \text{ eV}$ is the energy of the ground state.

ANGULAR MOMENTUM QUANTIZATION

The orbital quantum number l governs the magnitude L of the electron's angular momentum (Section 8). Even though the quantum theory discards the picture of an atomic electron as circling the nucleus in favor of a certain distribution of probability density ψ^2, angular momentum can be a property of the atom. The possible angular momenta are restricted to

$$L = \sqrt{l(l+1)} \, \frac{h}{2\pi} \qquad l = 0, 1, 2, \ldots, (n-1).$$

There are several interesting things about this result. First, there is the quantization of angular momentum itself, which is a universal phenomenon: like mass, electric charge, and the energy of a trapped particle, the angular momentum of a particle or system of particles can have only certain specific values. The quantity $h/2\pi$ is the natural unit of angular momentum. Because h is so small, the quantization of angular momentum is only perceptible in the physics of atoms and molecules, where it plays an important role.

The orbital quantum number l can be 0 or any integer up to $n - 1$, where n is the principal quantum number of the electron's quantum state. If $n = 3$, for instance, l can be 0, 1, or 2. When $l = 0$, the angular momentum $L = 0$ also. Here is a significant difference from the Bohr theory, since an electron revolving in a circle *must* possess angular momentum. No such requirement emerges from the quantum theory of the atom.

The magnetic quantum number m_l governs the *direction* of the electron's angular momentum L. Angular momentum is a vector quantity with both direction and magnitude, so it is natural (though perhaps surprising) that the direction of L should also be restricted in some way. Here is the right-hand rule for the direction of angular momentum. (2)

What can be the significance of the direction of L? And with respect to what is this direction reckoned? The answers to these questions become clear when we consider that an electron with angular momentum behaves like a loop of electric current and interacts with an external magnetic field in a manner similar to that of a bar magnet. The greater the angular momentum L the stronger the equivalent bar magnet. The potential energy of a bar magnet in a magnetic field B varies

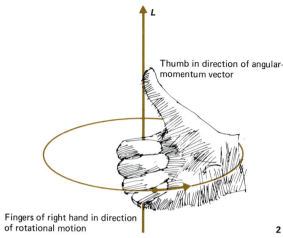

Thumb in direction of angular-momentum vector

Fingers of right hand in direction of rotational motion

L

2

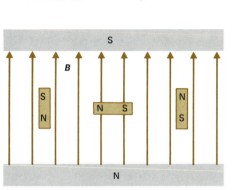

S

B

S
N

N   S

N
S

N

High
PE

Intermediate
PE

Low
PE

3

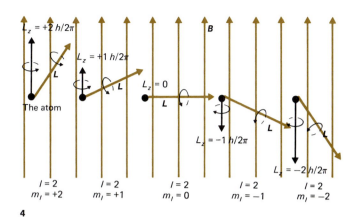

$L_z = +2\, h/2\pi$

$L_z = +1\, h/2\pi$

B

$L_z = 0$

The atom

L

L

L

$L_z = -1\, h/2\pi$

L

$L_z = -2\, h/2\pi$

L

| $l = 2$ | $l = 2$ | $l = 2$ | $l = 2$ | $l = 2$ |
| $m_l = +2$ | $m_l = +1$ | $m_l = 0$ | $m_l = -1$ | $m_l = -2$ |

4

with the strength of the magnet and with its direction relative to **B**; the energy is least when the magnet is parallel to the field, and is most when it is antiparallel. **(3)**

The magnetic quantum number m_l determines the angle between **L** and **B** and hence governs the extent of the magnetic contribution to the total energy of the atom when the atom is in a magnetic field. An atomic electron characterized by a certain value of m_l will assume a certain corresponding orientation of its angular momentum **L** relative to a magnetic field when placed in such a field.

The direction of **L** relative to **B** is specified in terms of the component of **L** parallel to the magnetic field **B**. By convention this component is called L_z, and its possible values are also in units of $h/2\pi$:

$$L_z = m_l \frac{h}{2\pi}.$$

The magnetic quantum number m_l can be any integer from $-l$ through 0 to $+l$. An atomic electron for which $l = 2$, for instance, could have a magnetic quantum number of -2, -1, 0, $+1$, or $+2$. The angular momentum vector **L** of the atom can never be exactly parallel to **B**, even when $m_l = l$, since L_z is always smaller than L. **(4)**

The uncertainty principle enables us to understand why the angular momentum **L** of an atomic electron can never be exactly parallel to **B**. If **L** were parallel to **B**—or if **L** pointed in any other definite direction in space—it would mean that the electron is always present in a specific plane perpendicular to **B**. But if there is such precision in the z coordinate of the electron, then the electron's momentum

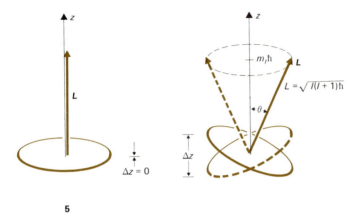

5

component in the z direction must be infinitely uncertain. An infinite uncertainty in the momenta of atomic electrons cannot be reconciled with the existence of stable atoms. However, the problem does not in fact arise because $L > L_z$ for all values of m_l. As the diagram shows, L may point anywhere along the circle at its tip and still have the component $m_l h/2\pi$ in the direction of B. Thus there is a built-in uncertainty in the electron's position, and its momentum uncertainty in consequence does not exceed an amount compatible with its presence in an atom. **(5)**

Although the interpretation of the magnetic quantum number m_l was discussed in terms of an external magnetic field, actually the component of L in *any* direction must be $m_l h/2\pi$. The point is that, for an isolated atom, an external magnetic field provides a definite, experimentally meaningful reference direction. But when the atom is part of a molecule, for instance, there are then other experimentally meaningful reference directions, namely those defined by the relative positions of the various atoms in the molecule. Thus a line joining the two H atoms in the hydrogen molecule H_2 is a quite specific reference

direction, and along this line the components of the angular momenta of the two H atoms are fixed by their m_l values.

PROBABILITY DENSITY

In place of the picture of a hydrogen atom as an electron circling a proton in a definite orbit, the quantum theory of the atom provides a description in terms of the probability density of the electron. We recall from Section 33 what is meant by the probability density ψ^2: the value of ψ^2 for a particular object at a certain place and time is proportional to the probability of finding the object at that place at that time. A large value of ψ^2 signifies that the object is likely to be found; a small value of ψ^2 signifies that the object is unlikely to be found.

The calculation of probability densities starts from *Schrödinger's equation*, an equation as central to atomic physics as Newton's laws of motion are to classical mechanics, but far more complicated. The results show that the electron probability density distribution is different for each quantum state of a hydrogen atom, that is, for each set of quantum numbers n, l, and m_l. We might call the ψ^2 distribution that corresponds to a particular quantum state a *probability cloud*. In the Bohr model, an atomic electron travels in a specific orbit; in the quantum theory, an atomic electron moves about within a probability cloud that forms a certain pattern in space.

THE $n = 1$ CLOUD

There is only a single probability cloud possible for a ground-state hydrogen atom, since when $n = 1$, $l = 0$ and $m_l = 0$. This cloud is spherically symmetric—that is, the likelihood of finding the electron at a given

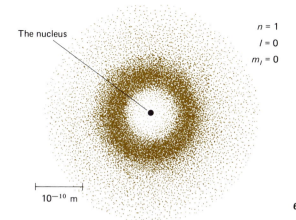

The nucleus

$n = 1$
$l = 0$
$m_l = 0$

$\vdash\!\!-\!\!\dashv$
10^{-10} m

6

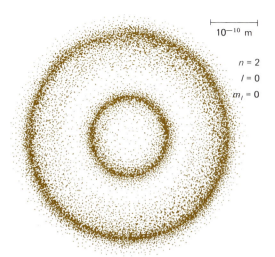

$\vdash\!\!-\!\!\dashv$
10^{-10} m

$n = 2$
$l = 0$
$m_l = 0$

7

distance from the nucleus is the same in all directions. We can picture ψ^2 for an $n = 1$ electron by a shaded drawing in which the darker the shading, the greater the value of ψ^2. As we can see, the probability cloud does not have a sharp boundary. The electron spends most of the time near the nucleus, but there is a certain probability, which becomes vanishingly small at large distances, that it be found anywhere. **(6)**

THE $n = 2$ CLOUDS

An electron with the principal quantum number $n = 2$ may exist in one of four quantum states:

$$n = 2 \begin{cases} l = 0 & m_l = 0 \\ l = 1 & \begin{cases} m_l = +1 \\ m_l = 0 \\ m_l = -1. \end{cases} \end{cases}$$

The $n = 2$, $l = 0$ probability cloud is spherically symmetric (indeed, *all* $l = 0$ clouds are spherically symmetric), but it does not fade out uniformly with distance as does the $n = 1$, $l = 0$ cloud. Instead there is a gap at a radius of 1.06×10^{-10} m; the electron is *never*

exactly this far from the nucleus. This gap corresponds to a node in a standing-wave pattern—in fact, we can think of the entire probability cloud as a kind of standing-wave pattern in three dimensions. The dense part of the cloud, which covers the region in which the electron is nearly always to be found, is four times as far across as the $n = 1$ cloud. This is in accord with the Bohr theory, in which an electron's orbital radius is proportional to n^2.

It is important to keep in mind that the cloud represented by the drawing is the same in all directions: we can think of it as representing a fuzzy sphere inside of a fuzzy spherical shell. **(7)**

The sizes of the $n = 2$, $l = 1$ probability clouds are roughly the same as that of the $n = 2$, $l = 0$ cloud, but they have different shapes. The $m_l = +1$ cloud has the form of a doughnut centered on the z axis. Because $m_l = +1$, the cloud possesses angular momentum whose component in the z direction is up (that is, in the $+z$ direction). This is indicated in the drawing by the arrow shown, in accord with the right-hand rule for angular momentum.

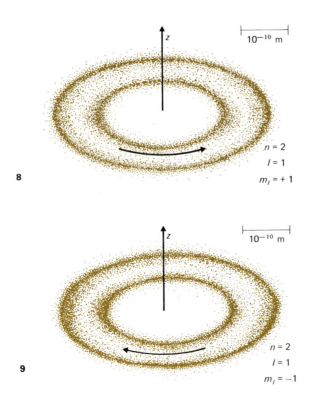

$n = 2$
$l = 1$
$m_l = +1$

8

$n = 2$
$l = 1$
$m_l = -1$

9

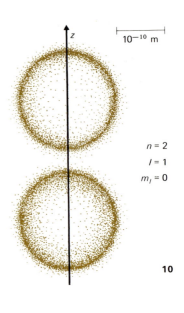

$n = 2$
$l = 1$
$m_l = 0$

10

If we like, we can think of the cloud as a rotating distribution of charge, with the rotation furnishing both the observed angular momentum and the observed magnetic behavior. In probability terms, if we were to determine the electron's direction of motion within the atom in a series of experiments, more often than not we would find it to be moving counterclockwise around the z axis. (8)

The $m_l = -1$ probability cloud is exactly the same as the $m_l = +1$ cloud except that the angular momentum is in the opposite direction. The rotational motion associated with

the cloud is now in the clockwise sense around the z axis. (9)

The $m_l = 0$ probability cloud has two concentrations around the z axis on either side of the nucleus. These concentrations are symmetric about the z axis—the pattern is the same on all sides. (10)

Because $m_l = 0$, there is no angular momentum about the z axis. A series of measurements of the electron's direction of motion would show it going as often clockwise as counterclockwise. But since $l = 1$, the atom nevertheless has angular momentum. Evidently the

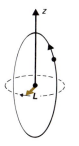

11

rotational motion that corresponds to this angular momentum occurs in such a way that L is always perpendicular to the z axis; then L has no component along this axis. In orbit terms, we can picture the electron as moving around in a circle about an axis perpendicular to the z axis, with the axis itself changing constantly in direction so that, on the average, the atom has no angular momentum in any particular direction. (11)

The probability clouds for quantum states whose principal quantum number n is greater than 2 are more complicated than those for $n = 1$ and $n = 2$. However, when $l = 0$, the clouds are always spherically symmetric, and when $l = 1$, the $m_l = 0, \pm 1$ clouds are similar in form to the corresponding ones for $n = 2$, $l = 1$ depicted above.

EXERCISES

1. The quantum theory of the atom is consistent with what fundamental principle that is violated by the Bohr theory?

2. What quantity is governed by each of the quantum numbers n, l, and m_l?

3. In the quantum theory of the atom, an atomic electron can have zero angular momentum. Is this possible in the Bohr theory?

4. What are the possible values for the orbital and magnetic quantum numbers of an atomic electron with the principal quantum number $n = 3$?

5. Why is there only one possible probability cloud for an atomic electron with no orbital angular momentum?

6. What is the condition for the probability cloud of an atomic electron to be spherically symmetric?

7. According to the Bohr model, the radius of an atomic electron's orbit is proportional to n^2. Is there any aspect of the probability-cloud model that corresponds to this relationship?

8. Verify that an atomic electron of principal quantum number n might have any one of n^2 possible probability clouds depending upon its l and m_l values.

36

COMPLEX ATOMS

"The periodic law, together with the revelations of spectrum analysis, have contributed to again revive an old but remarkably long-lived hope—that of discovering, if not by experiment, at least by a mental effort, the *primary matter*—which had its genesis in the minds of the Grecian philosophers." Dmitri Mendeleev (1834–1907)

The probability clouds of atoms that contain more than one electron are similar in shape to the corresponding ones in the hydrogen atom. Normally the electron in a hydrogen atom is in its ground state of $n = 1$. What about the electrons in a more complex atom? Are they *all* in the same $n = 1$ state, jammed together in a single probability cloud, or are they distributed in some special way among the various other quantum states? To answer this natural question, we must first look into electron spin and the exclusion principle.

ELECTRON SPIN

We mentioned earlier that one of the many problems facing the atomic theorist is the observed splitting of many spectral lines into several components close together. For example, the first line of the Balmer series of hydrogen, which both theory and experiment place at a wavelength of 6.56280×10^{-7} m, actually consists of a pair of lines 0.00014×10^{-7} m apart.

In 1925 it was pointed out that the fine structure of spectral lines would occur if the electron behaves like a charged sphere spinning on its axis rather than like a single point charge; several years later the British physicist P. A. M. Dirac was able to show on the basis of a relativistic version of quantum theory that the electron *must* have the spin attributed to it. A spinning electron is in effect a tiny bar magnet, and it interacts with the magnetic field produced by its own motion in an atom as well as with any magnetic fields originating outside the atom.

Electron spin is described by the *spin magnetic quantum number*, m_s, whose values are $+\frac{1}{2}$ and $-\frac{1}{2}$. The magnitude of the angular momentum associated with its spin is the same for every

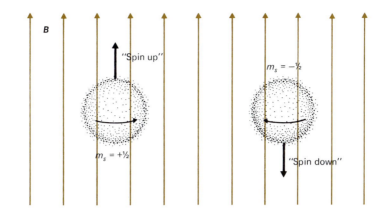

electron; the quantum number m_s refers to the *direction* of the spin. Here is how electrons with $m_s = +\frac{1}{2}$ and $m_s = -\frac{1}{2}$ align themselves in an external magnetic field. The component L_{sz} of spin angular momentum in the direction of $\mathbf{B}$ is $m_s h/2\pi$. **(1)**

The notion of an electron as a spinning charged sphere is simply a model to make it easier to understand what is going on. No one knows what an electron really looks like, and in fact it is doubtful whether any picture based on our perceptions of the macroscopic world can be relevant to elementary particles. But we *can* say that the possession of angular momentum and the magnetic behavior that goes with it is as fundamental a property of an electron as its rest mass and charge.

In its own frame of reference an atomic electron "sees" the positively charged nucleus moving around *it*, and the electron accordingly experiences a magnetic field. Because the energy of an electron is different in each orientation of its spin relative to the magnetic field it experiences, the various energy levels of an atom are split into sublevels. The presence of these sublevels of energy is responsible for the observed fine structure of spectral lines.

QUANTUM NUMBERS OF AN ATOMIC ELECTRON

The introduction of electron spin into the theory of the atom means that a total of four quantum numbers, n, l, m_l, and m_s, is required to describe each possible state of an atomic electron.

Table 36.1

Name	Symbol	Possible values	Quantity determined
Principal	n	1, 2, 3, ...	Electron energy
Orbital	l	0, 1, 2, . . . , $n-1$	Magnitude of angular momentum
Magnetic	m_l	$-l, \ldots, 0, \ldots, +l$	Direction of angular momentum
Spin magnetic	m_s	$-\frac{1}{2}, +\frac{1}{2}$	Direction of electron spin

There are $2n^2$ possible quantum states for each value of the principal quantum number n. For example, when $n = 3$ there are 18 states.

Table 36.2

	$m_l=0$	$m_l=-1$	$m_l=+1$	$m_l=-2$	$m_l=+2$
	↓↑				
$l=1$:	↓↑	↓↑	↓↑		
$l=2$:	↓↑	↓↑	↓↑	↓↑	↓↑

↑ $m_s=+\frac{1}{2}$
↓ $m_s=-\frac{1}{2}$

THE EXCLUSION PRINCIPLE

Wolfgang Pauli discovered the key to understanding the ground-state configurations of many-electron atoms. His *exclusion principle* states:

No two electrons in an atom can exist in the same quantum state.

That is, each electron in a complex atom must have a different set of the quantum numbers n, l, m_s, and m_l. The exclusion principle can be generalized to refer to the electrons in *any* small region of space, regardless of whether or not they constitute an atom.

Pauli was led to the exclusion principle by a study of atomic spectra. It is possible to determine the quantum states of an atom empirically by analyzing its spectrum, since states with different quantum numbers differ in energy (even if only slightly) and the various wavelengths present correspond to transitions between these states. Pauli found several lines missing from the spectra of every element except hydrogen; the missing lines correspond to transitions to and from atomic states having certain sets of quantum numbers. Every one of these absent states has two or more electrons with identical sets of quantum numbers, a result that is expressed in the exclusion principle. Hydrogen, with a single electron, naturally has no absent states.

The Pauli exclusion principle is the final piece of information we require in order to understand atomic structure.

FAMILIES OF ELEMENTS

Certain elements resemble one another so closely in their physical and chemical properties that it is natural to think of them as forming a "family". Three particularly striking examples of such families are the *halogens*, the *inert gases*, and the *alkali metals*. The members of these groups of elements are listed here together with the number of electrons the respective atoms contain. This number is called the *atomic number* of the element.

Table 36.3

Halogens	Inert gases	Alkali metals
	(2) Helium	(3) Lithium
(9) Fluorine	(10) Neon	(11) Sodium
(17) Chlorine	(18) Argon	(19) Potassium
(35) Bromine	(36) Krypton	(37) Rubidium
(53) Iodine	(54) Xenon	(55) Cesium
(85) Astatine	(86) Radon	(87) Francium

The halogens are nonmetallic elements with a high degree of chemical activity. At room temperature fluorine and chlorine are gases, bromine a liquid, and iodine and astatine solids. The halogens have valences of -1, and form diatomic molecules in the vapor state.

The inert gases, as their name suggests, are inactive chemically: they form virtually no compounds with other elements, and their atoms do not join together into molecules. The inert gases have valences of 0.

The alkali metals, like the halogens, are chemically very active, but they are active as reducing agents rather than as oxidizing agents. They are soft, not very dense, and have low melting points (all but lithium are liquid above 100°C). The alkali metals have valences of + 1.

THE PERIODIC LAW

A curious feature of the three groups listed above is that, while the atomic numbers of the member elements of each group bear no obvious relation to one another, each inert gas is preceded in atomic number by a halogen (except in the case of helium) and followed by an alkali metal. Thus fluorine, neon, and sodium have the atomic numbers 9, 10, and 11, respectively, a sequence that persists through astatine (85), radon (86), and francium (87). If we list *all* of the elements in order of their

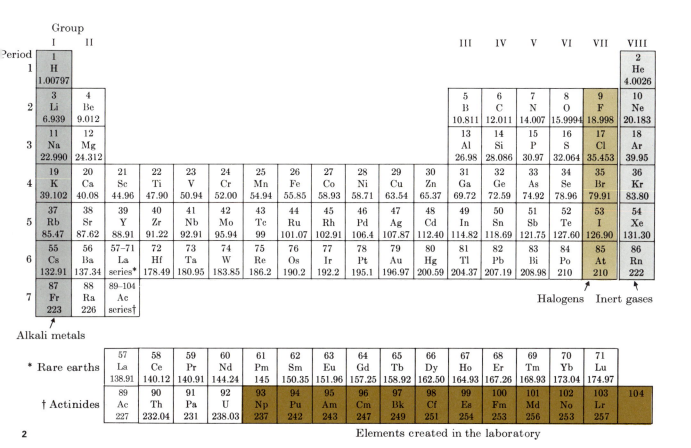

Elements created in the laboratory

atomic number, *elements with similar properties recur at regular intervals*. This observation, first formulated in detail by Dmitri Mendeleev about 1869, is known as the *periodic law*.

A periodic table is a listing of the elements according to atomic number in a series of rows such that elements with similar properties form vertical columns. The accompanying table is perhaps the most common form of periodic table; the number above the symbol of each element is its atomic number, and the number below the symbol is its atomic mass. The elements whose atomic masses are given as whole numbers are radioactive and are not found in nature, but have been prepared in nuclear reactions. The atomic mass in each such case is the mass number of the longest-lived isotope of the element. (2)

The columns in the periodic table are called *groups*. We recognize group I as the alkali metals plus hydrogen, group VII as the halogens, and group VIII as the inert gases. In addition to the elements forming the eight principal groups there are a number of *transition elements* falling between groups II and III. The transition elements are metals which share certain general properties: most are hard and brittle, have high melting points, exhibit several different valences, and form compounds that are paramagnetic. The rows in the periodic table are called *periods*. Each period starts with an active alkali metal and proceeds through less active metals to weakly active nonmetals to an active halogen and an inactive inert gas. The transition elements in each period may be very much alike; the rare earths and actinides are so much alike that they are usually considered as separate categories.

For nearly a century the periodic law has been a mainstay of the chemist by permitting him to predict the behavior of undiscovered elements and by providing a framework for organizing his knowledge. It is one of the triumphs of the quantum theory of the atom that it enables us to account for the periodic law in complete detail without invoking any new assumptions or postulates.

SHELLS AND SUBSHELLS

There are two basic principles that govern the structures of atoms with more than one electron:

1. An atom is stable when its total energy is a minimum. This means that in the normal configuration of an atom its various electrons are present in the lowest energy states available to them.

2. Only one electron can exist in each quantum state of an atom. This is the exclusion principle.

The notion of shells and subshells of electrons in an atom is a useful one. The electrons in an atom that share the same principal quantum number n are said to occupy the same *shell*. These electrons average about the same distance from the nucleus and have comparable, though not identical, energies as a rule. Electrons in a given shell that also have the same orbital quantum number l are said to occupy the same *subshell*. In complex atoms the various subshells of the same shell vary in energy because electrons with different angular momenta have different probability density distributions and hence different average distances from the nucleus. The higher the value of l, the higher the energy, in the same shell.

STRUCTURES OF COMPLEX ATOMS

We can construct the periodic table of the elements with the help of the above considerations. Our procedure will be to investigate the status of a new electron added to an existing electronic structure. (Of course, the nuclear charge must also increase by $+e$ each time this is done.) In the simplest case we add an electron to a hydrogen atom ($Z = 1$) to give a helium atom ($Z = 2$). Both electrons in helium fall into the same $n = 1$ shell. Since $l = 0$ is the only value l can have when $n = 1$, both electrons have $l = m_l = 0$. The exclusion principle is not violated here since one electron can have the spin magnetic quantum number $m_s = +\frac{1}{2}$ while the other has $m_s = -\frac{1}{2}$. It is customary to describe this situation by saying that the electrons have *opposite spins*, that is, that they behave as though they rotate in opposite directions.

Because no more than two electrons can occupy the $n = 1$ shell, helium atoms have *closed shells*. The characteristic properties of closed shells and subshells are that the orbital and spin angular momenta of their constituent electrons cancel out independently and that their effective electric charge distributions are perfectly symmetrical. As a result atoms with closed shells do not tend to interact with other atoms, which we know to be true of helium.

Lithium, with $Z = 3$, has one more electron than helium; it is the lightest of the alkali metals. There is no room left in the $n = 1$ shell, and so the additional electron goes into the $l = 0$ subshell of the $n = 2$ shell. The outer electron is relatively far from the nucleus in this shell, and is much less tightly bound. The chemical activity of lithium is a consequence of the low binding energy of this electron, which is readily lost to other atoms.

The next element, beryllium, has two electrons of opposite spin in the $l = 0$ subshell of the $n = 2$ shell. The nuclear charge is $+ 4e$, and so these outer electrons are more tightly held than the single outer electron in lithium; beryllium is accordingly less reactive than lithium.

Boron, with $Z = 5$, has an electron in the $l = 1$ subshell as well as two in the $l = 0$ one. The $l = 1$ subshell can contain a total of six electrons, corresponding to two electrons of opposite spin in the $m_l = + 1$, $m_l = 0$, and $m_l = - 1$ states. This subshell is closed (and the $n = 2$ shell is also closed) in neon, whose atomic number is 10. We therefore expect neon to be chemically inert, as indeed it is.

Fluorine, the element just before neon in the periodic table and the lightest of the halogens, lacks but one electron of having a closed outer shell. Just as lithium tends to lose its single outermost electron in interacting with other elements, thereby leaving it with a closed shell configuration, fluorine tends to gain a single electron to close its outer shell. The very different behaviors of the alkali metals, the inert gases, and the halogens thus find explanation in terms of their respective atomic structures.

Here is a highly schematic representation of the electron structures of the elements in the first three periods of the periodic table. Each dashed circle represents an electron shell. (**3**)

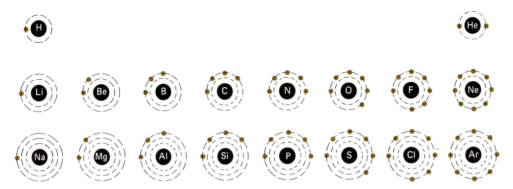

3

EXERCISES

1. Which of the various quantum numbers that describe an atomic electron is always the largest?

2. Can you think of any property that can distinguish one atomic electron from another when both have the same n, l, m_l, and m_s values?

3. Do electrons in inner shells spin faster than those in outer shells?

4. What is true in general of the chemical characteristics of elements in each horizontal row of the periodic table?

5. What is true in general of the chemical characteristics of elements in each vertical column of the periodic table?

6. How many elements would there be if atoms with occupied electron shells up through $n = 6$ could exist?

7. What property of their electron structures is responsible for the characteristic chemical behavior of
   a) the halogens?
   b) the inert gases?
   c) the alkali metals?

37

MOLECULES, LIQUIDS, AND SOLIDS

"Atoms by themselves have only a few interesting properties: their spectra, their dielectric and magnetic behavior, hardly any others. It is when they come into combination with each other that problems of real physical and chemical interest arise." John C. Slater (1900–)

Under certain circumstances, individual atoms interact with one another to form molecules, liquids, and solids. The quantum theory of the atom is able to account for these interactions in a natural way, with no additional assumptions, which is further testimony to the power of this approach.

WHAT IS A MOLECULE?

A molecule is a group of atoms that stick together strongly enough to act as a single particle. A molecule always has a certain definite structure; hydrogen molecules always consist of two hydrogen atoms each, for instance, and water molecules always consist of one oxygen atom and two hydrogen atoms each. A piece of iron is not a molecule because, even though its atoms stay together, any number of them do so to form an object of any size or shape.

A molecule of a given kind is complete in itself with little tendency to gain or lose atoms. If one of its atoms is somehow removed or another atom is somehow attached, the result is a molecule of a different kind with different properties. A liquid or a solid, on the other hand, can gain or lose additional atoms of the kinds already present without changing its character. (1)

THE HYDROGEN MOLECULE

The chief mechanism that bonds atoms together to form molecules involves the sharing of electrons by the atoms involved. As the shared electrons circulate around the atoms, they spend more time between the atoms than they do on the outside, which produces an attractive force. To see why this force should occur, we may think of the atoms that are sharing the electrons as positive ions, so the

Hydrogen molecule | Water molecule | **1**

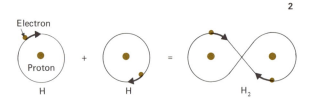

Electron
Proton
H + H = H$_2$

2

High probability of finding electrons

H + H = H$_2$

3

Low probability of finding electrons

presence of the shared electrons between them means a negative charge that holds the positive ions together. This is especially easy to see in the case of the hydrogen molecule, H$_2$, which consists of two hydrogen atoms.

In the hydrogen molecule, the two protons are 7.42×10^{-11} m apart and the two electrons, one contributed by each atom, belong to the entire molecule rather than to their parent nuclei. Here is a rather crude way to picture the H$_2$ molecule. (**2**)

A more realistic picture of the hydrogen molecule is provided by the quantum theory, in which the notion of electrons with definite, predictable positions and velocities at every moment is replaced by the notion of probability density distributions. Here is how the electron probability clouds of two hydrogen atoms join to form the probability cloud of a hydrogen molecule. (**3**)

Because the electrons spend more time on the average between the protons than they do on the outside, there is effectively a net negative charge between the protons. The attractive force this charge exerts on the protons is enough to counterbalance the direct repulsion between them. If the protons are too close together, however, their repulsion becomes dominant and the molecule is not stable. The balance between attractive and repulsive forces occurs at a separation of 7.42×10^{-11} m, where the total energy of the H$_2$ molecule is -4.5 eV. Hence 4.5 eV of work must be done to break a H$_2$ molecule into two H atoms:

$$H_2 + 4.5\,\text{eV} \rightarrow H + H.$$

By comparison, the binding energy of the hydrogen atom is 13.6 eV:

$$H + 13.6\,\text{eV} \rightarrow p^+ + e^-.$$

This is an example of the general rule that it is easier to break up a molecule than to break up an atom.

WHY THE HYDROGEN MOLECULE IS STABLE

There is a very interesting argument based on the uncertainty principle that makes it possible, even without a detailed calculation starting from Schrödinger's equation, to understand why a hydrogen molecule should have less energy (and hence more stability) than two separate hydrogen atoms. According to the uncertainty principle, the smaller the uncertainty in a particle's position, the greater the uncertainty in its momentum. Therefore the smaller the region of space in which an electron is confined, the greater its momentum and hence energy must be. An electron shared by two protons has more room in which to move about than an electron bound to a single proton to form a hydrogen atom, so its energy is less.

According to the Pauli exclusion principle, two electrons can jointly occupy the same region while in their ground states only if their spin magnetic quantum numbers are different, with one having $m_s = +\frac{1}{2}$ and the other $m_s = -\frac{1}{2}$. That is, the two electrons must have opposite spins. If the spins were the same, the electrons would have to spend most of their time near the far ends of the H_2 molecule where they interact least. The result would be both a direct repulsion between the nuclei and a repulsion induced by the outward attractions of the electrons, making a stable molecule impossible. The graph shows the total energy of two hydrogen atoms plotted versus their separation. Bonding is evidently possible only when the electron spins are antiparallel. The minimum of the lower curve corresponds to the H_2 molecule, which is a stable configuration since work must be done to move the H atoms closer together or farther apart than the distance $a = 7.42 \times 10^{-11}$ m. (4)

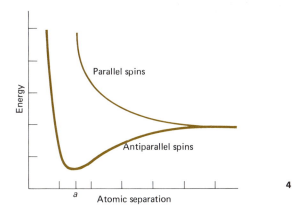

SATURATION

Why do only two hydrogen atoms join together to form a molecule? Why not three, or four, or a hundred?

The basic reason for the limited size of molecules is the exclusion principle, which prohibits more than one electron in an atomic system from having the same set of quantum numbers. If a third H atom were to be brought up to an H_2 molecule, its electron would have to leave the $n = 1$ shell and go to the $n = 2$ shell in order for an H_3 molecule to be formed, since only two electrons can occupy the $n = 1$ shell. But an $n = 2$ electron in the hydrogen atom has over 10 eV more energy than an $n = 1$ electron, and the binding energy in H_2 is only 4.5 eV. Hence an H_3 molecule, if it could somehow be put together, would immediately break apart into $H_2 + H$ with the release of energy. Similar considerations hold for other molecules.

The ability of atoms to join with only a limited number of others to form molecules is referred to as *saturation*.

The exclusion principle is also responsible for the fact that the inert gases—helium, neon, argon, and so on—do not form molecules. The electron shells of the inert gas atoms are all filled to capacity, so in order for two of them to share an electron pair, one of the electrons would have to go into an empty shell of higher energy. The increase in energy involved would be more than the energy decrease produced by sharing the electrons, and therefore no such molecules as He_2 or Ne_2 occur.

COVALENT BONDING

The mechanism by which electron sharing holds atoms together to form molecules is known as *covalent bonding*. Often it is convenient to think of the atoms as being held together by *covalent bonds*, with each shared pair of electrons constituting a bond.

More complex atoms than hydrogen also join together to form molecules by sharing electrons. Depending upon the electron structures of the atoms involved, there may be one, two, or three covalent bonds—shared electron pairs—between the atoms. For example, in the oxygen molecule O_2 there are two bonds between the O atoms, and in the nitrogen molecule N_2 there are three bonds between the N atoms. Covalent bonds are represented either by a pair of dots or a single dash for each shared pair of electrons. Thus the H_2, O_2, and N_2 molecules can be represented as follows:

H:H or H—H

O::O or O=O

N:::N or N≡N.

Covalent bonds are not limited to atoms of the same element nor to only two atoms per molecule. Here are two examples of more complicated molecules, with a line representing each covalent bond:

Water, H_2O

H
|
O—H

Ammonia, NH_3

H
|
N—H
|
H

We notice that oxygen participates in two bonds in H_2O and nitrogen in three bonds in NH_3, while H participates in only one bond in both cases. This behavior is consistent with the fact that the hydrogen, oxygen, and nitrogen molecules respectively have one, two, and three bonds between their atoms.

Carbon atoms tend to form four covalent bonds at the same time, since they have four electrons in their outer shells and these shells lack four electrons for completion. Various distributions of these bonds are possible, including bonds between adjacent carbon atoms in a complex molecule. The structures of the common covalent molecules methane, carbon dioxide, and acetylene illustrate the different bonds in which carbon atoms can participate to form molecules:

H
|
H—C—H   O=C=O   H—C≡C—H
|
H

Methane   Carbon dioxide   Acetylene

Carbon atoms are so versatile in forming covalent bonds with each other as well as with other atoms that literally millions of carbon compounds are known, some whose molecules contain tens of thousands of atoms. Such compounds were once thought to originate only in living things, and their study is accordingly known even today as organic chemistry.

LIQUIDS AND SOLIDS

Most solids are crystalline in nature, with their constituent atoms or molecules arranged in regular, repeated patterns. A crystal is thus characterized by the presence of *long-range order* in its structure.

Other solids lack the definite arrangements of atoms and molecules so conspicuous in crystals. They can be thought of as liquids whose stiffness is due to an exaggerated viscosity. (*Viscosity* is the term used to describe internal friction in a liquid; water has a low viscosity, honey has a high viscosity.) Examples of such *amorphous* ("without form") solids are pitch, glass, and many plastics. The structures of amorphous solids exhibit *short-range order* only.

Some substances, for instance B_2O_3, can exist in either crystalline or amorphous forms. In both cases each boron atom is surrounded by three larger oxygen atoms, which represents a short-range order. In a B_2O_3 crystal a long-range order is also present, as shown here in a two-dimensional representation (bottom), whereas amorphous B_2O_3, a glassy material, lacks this additional regularity (top). (5)

The lack of long-range order in amorphous solids means that the various bonds vary in strength. When an amorphous solid is heated, the weakest bonds are ruptured at lower temperatures than the others and the solid softens gradually, while in a crystalline solid the bonds break simultaneously, and melting is a sudden process.

Liquids have more in common with solids than with gases, even though liquids share with gases the ability to flow from place to place. Because the density of a given liquid is usually approximately the same as that of the same substance in solid form, we infer that the bonding mechanism is similar in both cases. When a solid is heated to its melting point, its atoms or molecules acquire enough energy to shift the bonds holding them together so that they form different groupings, but not until the liquid is heated to its vaporization point are the atoms or molecules able to break loose completely and form a gas. This interpretation of the liquid state is confirmed by x-ray studies that reveal definite groupings of atoms or molecules in a liquid (that is, short-range order like that in amorphous

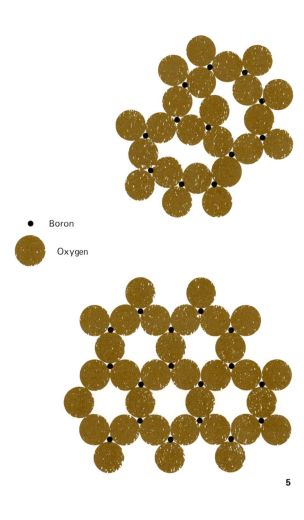

- Boron

Oxygen

5

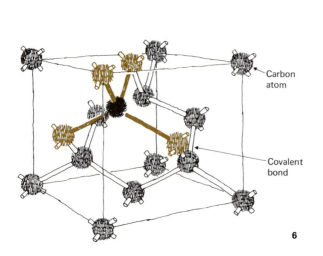

Carbon atom

Covalent bond

6

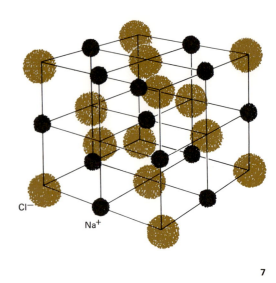

Cl^-

Na^+

7

solids), but the groupings are constantly shifting their arrangements unlike the permanent arrangements in a solid.

COVALENT BONDING
IN SOLIDS

The covalent bonds between atoms that are responsible for the formation of molecules also act to hold certain crystalline solids together. The accompanying figure shows the array of carbon atoms in a diamond crystal, with each carbon atom sharing electron pairs with the four other carbon atoms adjacent to it. All the electrons in the outer shells of the carbon atoms participate in the bonding, and it is therefore not surprising that diamond is extremely hard and must be heated to over 3500°C before its crystal structure is disrupted and it melts. Other covalent crystals are those of silicon, silicon carbide ("Carborundum"), and germanium. **(6)**

IONIC BONDING

In covalent bonding, two atoms share one or more pairs of electrons. In *ionic bonding*, one or more electrons from one atom transfer to another atom, producing a positive ion and a negative ion that attract each other.

The structure of an NaCl crystal is shown here. Each ion behaves essentially like a point charge, and thus tends to attract to itself as many ions of opposite sign as can fit around it. The latter ions, of course, repel one another, and so the resulting crystal is an equilibrium configuration in which the various attractions and repulsions balance out. In a NaCl crystal each Na^+ ion is surrounded by six Cl^- ions and vice versa. In crystals having different structures the number of "nearest neighbors" around each ion may be 3, 4, 6, 8, or 12. Ionic bonds are usually fairly strong, and consequently ionic crystals are strong, hard, and have high melting points. **(7)**

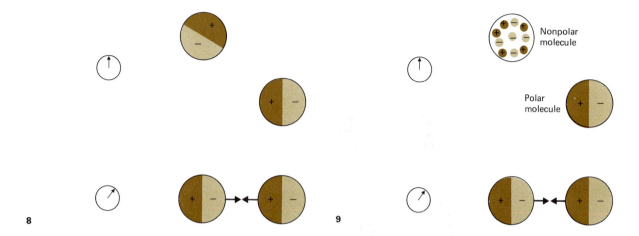

8

9

Many crystalline bonds are partly ionic and partly covalent in origin. An example of such mixed bonding is quartz, SiO_2.

VAN DER WAALS' BONDING

A number of molecules and nonmetallic atoms exist whose electronic structures do not lend themselves to either of the above kinds of binding. The inert gas atoms, which have filled outer electron shells, and organic molecules such as methane,

$$\begin{array}{c} H \\ | \\ H—C—H, \\ | \\ H \end{array}$$

whose valence electrons are fully involved in the molecular bond itself, fall into this category. However, even these virtually non-interacting substances condense into solids and liquids at low enough temperatures through the action of what are known collectively as *van der Waals' forces.*

The electrostatic attraction between polar molecules that was discussed on page 208 is an example of a van der Waals force. A polar molecule has a nonsymmetric distribution of charge, with one end positive and the other end negative. When one polar molecule is near another, the ends of opposite sign attract each other to hold the molecules together. (8)

A nonpolar molecule has a symmetric distribution of charge. When a nonpolar molecule is near a polar one, the latter's electric field distorts the charge distribution of the nonpolar molecule. The two molecules then attract each other in the same way as any other pair of polar molecules. (9)

Nonpolar molecules attract one another in much the same way that polar molecules

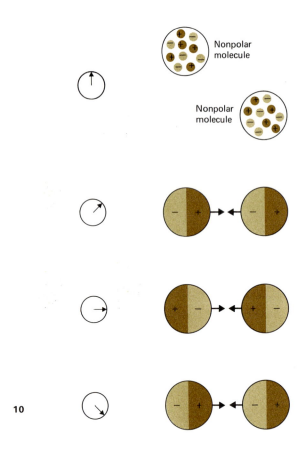

In general van der Waals' bonds are considerably weaker than ionic, covalent, and metallic bonds; only about 1% as much energy is needed to remove an atom or molecule from a van der Waals solid as is required in the case of ionic or covalent crystals. As a result the inert gases and compounds with symmetric molecules liquify and vaporize at rather low temperatures. Thus the boiling point of argon is $-186°C$ and the boiling point of methane is $-161°C$. Molecular crystals, whose lattices consist of individual molecules held together by van der Waals' forces, generally lack the mechanical strength of other kinds of crystals.

METALLIC BONDING

A fourth important type of cohesive force in crystalline solids is the *metallic bond*, which has no molecular counterpart.

A characteristic property of all metal atoms is the presence of only a few electrons in their outer shells, and these electrons can be detached relatively easily to leave behind positive ions. According to the theory of the metallic bond, a metal in the solid state consists of an assembly of atoms that have given up their outermost electrons to a common "gas" of freely moving electrons that pervades the entire metal. The electrostatic interaction between the positive ions and the negative electron gas holds the metal together.

This theory has many attractive features. The high electrical and thermal conductivity of metals follows from the ability of the free electrons to migrate through their crystal structures, while all of the electrons in ionic and covalent crystals are bound to particular atoms or pairs of atoms. Unlike other crystals, metals may be considerably deformed without fracture, because the electron gas permits the

attract nonpolar ones. In a nonpolar molecule, the electrons are distributed symmetrically *on the average*, but *at any moment* one part of the molecule contains more electrons than usual and the rest of the molecule contains fewer. Thus *every* molecule (and atom) behaves as though it is polar, though the direction and magnitude of the polarization vary constantly. The fluctuations in the charge distributions of nearby nonpolar molecules keep in step through the action of electric forces, and these forces also hold the molecules together. (10)

Type	Covalent	Ionic	Molecular	Metallic
	Shared electrons	Negative ion / Positive ion	Instantaneous charge separation in molecule	Metal ion / Electron gas
Bond	Shared electrons	Electrostatic attraction	Van der Waals forces	Electron gas
Properties	Very hard; high melting point; soluble in very few liquids	Hard; high melting point; may be soluble in polar liquid such as water	Soft; low melting and boiling points; soluble in covalent liquids	Ductile; metallic luster; ability to conduct heat and electric current readily
Example	Diamond (C)	Sodium Chloride (NaCl)	Methane (CH_4)	Iron (Fe)

11

atoms to slide past one another in the absence of directional bonds between specific atoms. While certain solids, such as rubber or asphalt, that are amorphous rather than crystalline in structure can also be deformed readily, they lack the strength and hardness conferred by the metallic bond. Furthermore, since the atoms in a metal interact through the medium of a common electron gas, the properties of mixtures of different metal atoms should not depend critically on the relative proportions of each kind of atom provided their sizes are similar. This prediction is fulfilled in the observed behavior of alloys, in contrast to the specific atomic proportions characteristic of ionic and covalent solids.

The opacity and metallic luster exhibited by metals may also be traced to the gas of free electrons that pervades them. When light of any frequency shines on a metal, the free electrons are set in vibration by the oscillating electromagnetic fields and thereby absorb the light. The oscillating electrons themselves then act as sources of light, sending out electromagnetic waves of the original frequency in all directions. Those waves that happen to be directed back toward the metal surface are able to escape, and their emergence gives the metal its lustrous appearance. If the metal surface is smooth, the reradiated waves appear to us as a reflection of the original incident light. (**11**)

EXERCISES

1. What is wrong with the model of a hydrogen molecule in which the two electrons follow figure-eight orbits that encircle the two protons?

2. What is the chief cause of the repulsive force that keeps atoms from meshing together despite any attractive forces that may be present?

3. At what temperature would the average energy of the hydrogen molecules in a gas sample be equal to their binding energy?

4. Under what circumstances can two electrons have the same probability cloud in a molecule?

5. In what state of matter is short-range order never found?

6. From the point of view of its structure, does a liquid resemble a gas or a solid more closely?

7. What are the individual particles in an ionic crystal?

8. What are the individual particles in a covalent crystal?

9. What are the individual particles in the van der Waals crystal of a compound?

10. What kind of solids contain a "gas" of freely moving electrons? Does this gas include all the electrons present?

11. What class of solids has the lowest melting points?

12. Van der Waals' forces can hold inert gas atoms together to form solids, but they cannot hold such atoms together to form molecules in the gaseous state. Why not?

13. Which of the four fundamental interactions is responsible for each of the principal bonding mechanisms in solids?

14. The temperature of a gas falls when it passes slowly from a full container to an empty one through a porous plug. Since the expansion is into a rigid container, no mechanical work is done. What is the origin of the fall in temperature?

38

THE ATOMIC NUCLEUS

"The complicated aspect of the nuclear force suggests to many physicists that it is an involved manifestation of something deeper that is actually simpler and more basic in character. One hopes that the nuclear force will one day be explained by some fundamental phenomenon in the internal dynamics of the nucleon, just as the complicated chemical forces today can be traced back to the simple electrical attraction between nuclei and electrons." Victor F. Weisskopf (1908–   )

Until now we have not had to regard the nucleus of an atom as anything but a tiny positively charged lump whose sole functions are to provide the atom with most of its mass and to hold its several electrons in place. Since the behavior of atomic electrons is responsible for the behavior of matter in bulk, the properties of matter we have been exploring, save for mass, have nothing directly to do with atomic nuclei. Nevertheless, the nucleus turns out to be of supreme importance in the universe: the elements exist by virtue of the ability of nuclei to hold multiple electric charges, and the energy involved in nearly all natural processes has its ultimate origin in nuclear reactions and transformations.

NUCLEONS

The nature and behavior of the electron structure of the atom was understood before even the composition of its nucleus was known. The reason is that the nucleus is held together as a unit by forces vastly stronger than the electric forces that hold the electrons to the nucleus, and it is correspondingly harder to break apart a nucleus to find out what is inside. Changes in the electron structure of an atom, such as those that occur in the emission of photons or in the formation of chemical bonds, involve energies of only several eV, whereas changes in nuclear structure involve energies of several MeV, a million times more. Let us first inquire into the composition of the nucleus: What is it that gives a nucleus its characteristic mass and charge?

The nucleus of the hydrogen atom consists of a single proton, whose charge is $+e$ and whose mass is

$$m_{proton} = 1.673 \times 10^{-27} \text{ kg.}$$

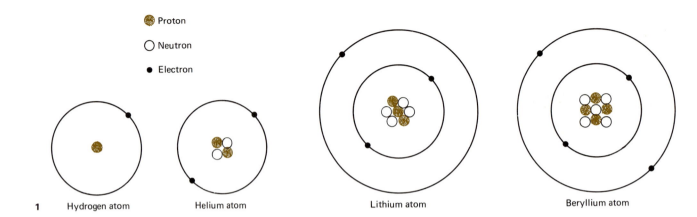

Proton
Neutron
Electron

1   Hydrogen atom   Helium atom   Lithium atom   Beryllium atom

The proton mass is 1836 times that of the electron, so by far the major part of the hydrogen atom's mass resides in its nucleus. This is true of all other atoms as well.

Elements more complex than hydrogen have nuclei that contain *neutrons* as well as protons. The neutron, as its name suggests, is uncharged. The neutron mass is slightly more than that of the proton:

$$m_{\text{neutron}} = 1.675 \times 10^{-27} \text{ kg.}$$

Neutrons and protons are jointly called *nucleons*.

Every atom contains the same number of protons and electrons; this number is the *atomic number* of the element involved. Except in the case of ordinary hydrogen atoms, the number of neutrons in a nucleus equals or exceeds the number of protons. The compositions of atoms of the four lightest elements are illustrated here. (1)

ISOTOPES

The mass of an atom can be determined with the help of a mass spectrometer, such as the one described in Section 26. Atomic masses are conventionally expressed in terms of the *atomic mass unit*, abbreviated amu, whose value is

$$1 \text{ amu} = 1.660 \times 10^{-27} \text{ kg.}$$

The proton and neutron masses in amu are respectively

$$m_{\text{p}} = 1.00728 \text{ amu,}$$
$$m_{\text{n}} = 1.00867 \text{ amu.}$$

Because m_{p} and m_{n} are so close to 1 amu, we would expect atomic masses to be very nearly whole numbers of amu. This is often true. For example, the atomic mass of helium is 4.003 amu, that of lithium is 6.939, and that of beryllium is 9.012. However, the chlorine found in nature has an atomic mass of 35.45, which does not fit in with this picture.

Chlorine is an example of an element whose nuclei do not all have the same composition. The several varieties of an element are called its *isotopes*. The number of protons is always equal to the atomic number Z of the element, of course, but the number of neutrons may be different. Thus chlorine consists of two isotopes, one whose nuclei contain 17 protons and 18 neutrons and another whose nuclei contain 17 protons and 20 neutrons. There are about three times as many nuclei of the former type as there are of the latter, which yields an average atomic mass of 35.45.

All elements have isotopes, even hydrogen. The most abundant hydrogen isotope has nuclei that each consist of a single proton. Less common is the isotope deuterium, whose nuclei each consist of a proton and a neutron, and the isotope *tritium*, whose nuclei each consist of a proton and two neutrons. Deuterium is stable, but tritium is radioactive and a sample of it gradually changes to an isotope of helium. The flux of cosmic rays from space continually replenishes the earth's tritium by nuclear reactions in the atmosphere; there is only about 2 kg of tritium of natural origin on the earth's surface, nearly all of it in the oceans. (See table below.)

Because the chemical properties of an element depend upon the distribution of the electrons in its atoms, which in turn depends upon the nuclear charge, nuclear structure beyond the number of protons present has little significance for the chemist. The physical properties of an element, however, depend strongly on the nuclear structures of its isotopes, whose behavior may be very different from one another although chemically they are indistinguishable.

The conventional symbols for isotopes all follow the pattern

$$^A_Z X,$$

where

X = chemical symbol of the element;
Z = atomic number of the element,
   = number of protons in the nucleus;
A = mass number of the isotope,
   = number of protons and neutrons in the nucleus.

Hence ordinary hydrogen is designated $^1_1 H$, since its atomic number and mass number are both 1, while tritium is designated $^3_1 H$. The

Table 38.1

| Element | Properties of element | | Properties of isotope | | | |
	Atomic number	Atomic mass, amu	Protons in nucleus	Neutrons in nucleus	Atomic mass, amu	Relative abundance
Hydrogen	1	1.008	1	0	1.008	99.985%
			1	1	2.014	0.015%
			1	2	3.016	very small
Chlorine	17	35.45	17	18	34.97	75.53%
			17	20	36.97	24.47%

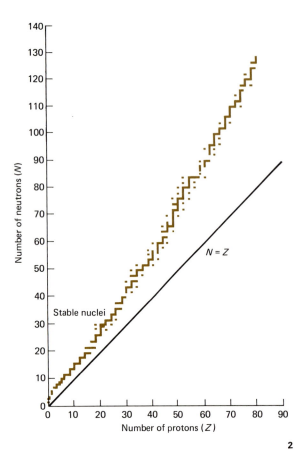

2

neutrons than protons. The accompanying graph is a plot of N versus Z for stable nuclei. Evidently the number of neutrons in the nuclei of a given element is a fairly critical quantity. Let us see why. (2)

There are two opposing tendencies in a nucleus. The first is a tendency for N to equal Z, which arises from the existence of quantum states of different energy in a nucleus. Just like an electron in an atom, a nucleon in a nucleus can possess only certain specific energies. Because neutrons and protons obey the exclusion principle, at most two of each kind of nucleon (one whose spin is "up" and one whose spin is "down") can occupy each quantum state. Again as in the case of atomic energy levels, nuclear energy levels are filled in sequence to achieve nuclei of minimum energy and hence maximum stability. Thus the boron isotope $^{12}_{5}B$ has more energy than the carbon isotope $^{12}_{6}C$ because one of its neutrons is in a higher energy level, and it is accordingly unstable. If created in a nuclear reaction, a $^{12}_{5}B$ nucleus changes by beta decay (described in Section 40) into a $^{12}_{6}C$ nucleus in a fraction of a second. (3)

The other tendency in a nucleus is for the number of neutrons to exceed the number of protons. This is a consequence of the strong electric repulsion exerted by the protons upon one another, which must be balanced by the attractive nuclear forces that act between nucleons. The repulsive electric forces increase more rapidly with Z than the attractive nuclear forces increase with A, the total number of nucleons, and so a greater proportion of neutrons, which produce only attractive forces owing to their electrical neutrality, is necessary for stability in large nuclei.

two isotopes of chlorine mentioned above are designated $^{35}_{17}Cl$ and $^{37}_{17}Cl$ respectively.

The number of neutrons in a nucleus is its *neutron number N*. Thus the mass number of a nucleus is equal to $Z + N$.

STABLE NUCLEI

Only certain combinations of protons and neutrons form stable nuclei. In light nuclei there are about as many neutrons as protons, while in heavier ones there are somewhat more

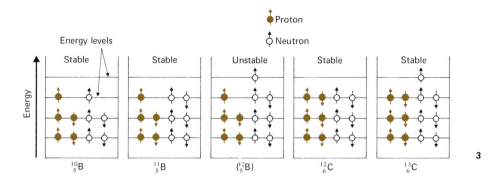

Proton
Neutron

Energy levels

Stable | Stable | Unstable | Stable | Stable

Energy

$^{10}_{5}B$ | $^{11}_{5}B$ | $(^{12}_{5}B)$ | $^{12}_{6}C$ | $^{13}_{6}C$

3

NUCLEAR SIZES

The Rutherford scattering experiment (Section 23) provides information on nuclear dimensions as well as on atomic structure. The observed distribution of scattering angles is consistent with a nucleus of infinitely small size provided the alpha particles are not too energetic. That is, below a certain particle energy the size of the nucleus is small compared with the minimum distance to which the incident alpha particles approach it. At higher energies discrepancies occur between theory and data which suggest that the particles have come so close to the nucleus that it no longer can be thought of as a point charge, and from the energy at which these discrepancies appear an estimate can be made of nuclear dimensions.

More recent experiments that employ high-energy electrons, protons, and neutrons yield more precise figures for nuclear sizes. The gold nucleus, for instance, turns out to be about 5×10^{-15} m in radius. The radii of other nuclei range from about 1.1×10^{-15} m for the proton to about 7.4×10^{-15} m for the $^{235}_{92}U$ nucleus. These experiments are, in essence, observations of the diffraction of the matter waves of the fast particles by the nuclei of the target atoms; the result is a diffraction pattern formed by the scattered particles, from which the size of the "obstacle"—the nucleus—can be inferred. The observations show that the diffraction patterns are almost, but not quite, the same as those that would be produced by a black disc. The nucleus does not have a sharp boundary, but (like the atom itself) a fuzzy one.

The density of nuclear matter is about 2×10^{17} kg/m^2, which is 3 billion tons per cubic inch. "White dwarf" stars consist of atoms whose electron structures have collapsed because of immense pressures, and the densities of these stars approach the density of nuclear matter.

THE STRONG NUCLEAR INTERACTION

The existence of stable nuclei can only be explained on the basis of a special interaction between nucleons. This interaction cannot be electrical, because neutrons are uncharged and the positive charges of protons lead to repulsive forces only. It cannot be gravitational, because gravitational forces are far too weak to be able to counterbalance the repulsive electric forces between protons. Thus we have a third fundamental interaction, the

The strong nuclear interaction 353

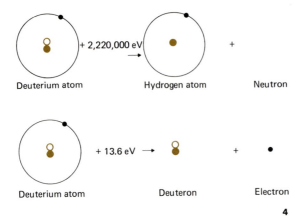

+ 2,220,000 eV

Deuterium atom   Hydrogen atom   Neutron

+ 13.6 eV

Deuterium atom   Deuteron   Electron

4

strong nuclear interaction, which is responsible for the existence of atomic nuclei more complex than that of $_1^1$H, which is a single proton.

Two properties of the strong interaction stand out. First, it is by far the strongest of all the fundamental interactions, as we can tell from the magnitude of nuclear binding energies. To pull apart the neutron and proton in a deuterium nucleus takes 2.22 MeV, but to pull the electron in a deuterium atom away from the nucleus takes only 13.6 eV, over a hundred thousand times less work. **(4)**

The second noteworthy aspect of the strong interaction is its short range. Electric and gravitational forces fall off with distance as $1/r^2$, and are effective at considerable separations between the interacting objects—the planet Pluto is 6×10^{12} m from the sun, yet is kept in orbit by the gravitational attraction of the sun. But nuclear forces are effective only over a range of a few nucleon diameters. Up to a separation of about 3×10^{-15} m, the nuclear attraction between two protons is about one hundred times stronger than the electric repulsion between them, but beyond this distance the nuclear force dies out rapidly. The nuclear interactions between protons and protons, between protons and neutrons, and between neutrons and neutrons appear to be identical.

The short range of nuclear forces is responsible for the restricted number of stable elements. The larger a nucleus, the stronger the electric repulsive forces that act on each of its protons, but the attractive nuclear forces on each nucleon cannot increase indefinitely because only a limited number of other nucleons are close enough to interact with it. The largest stable nucleus is the bismuth isotope $_{83}^{209}$Bi, and nuclei larger than the uranium isotope $_{91}^{238}$U are too unstable to have survived on earth since its formation.

EXERCISES

1. What property of its nucleus governs the chemical behavior of an atom?

2. The concept that atomic nuclei consist of neutrons and protons cannot account for the existence of atomic masses such as that of chlorine, 35.45 amu, which is nowhere near being a multiple of nucleon masses. The notion of isotopes, however, rescues the concept. How does this situation differ from that of the caloric theory of heat (see Section 16), which was simply abandoned rather than modified when phenomena were observed that could not be accounted for in terms of it?

3. State the number of neutrons and protons in each of the following nuclei: $_3^6$Li, $_6^{13}$C, $_{15}^{31}$P, $_{40}^{94}$Zr, $_{56}^{137}$Ba.

4. Which nucleus would you expect to be more stable, $_3^7$Li or $_3^8$Li? $_4^9$Be or $_4^{10}$Be?

5. Compare the density of the $_1^1$H atom—assuming it to be a sphere whose radius is equal to that of the first Bohr orbit—with the density of its nucleus. The volume of a sphere of radius r is $\frac{4}{3}\pi r^3$.

39

NUCLEAR ENERGY

"It seems therefore possible that the uranium nucleus has only small stability of form and may, after neutron capture, divide itself into two nuclei of roughly equal size." Lise Meitner (1878–1968)

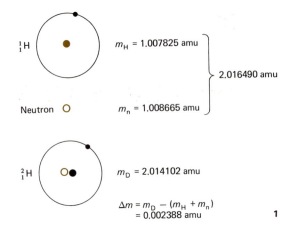

$m_H = 1.007825$ amu

Neutron $m_n = 1.008665$ amu

2.016490 amu

$m_D = 2.014102$ amu

$$\Delta m = m_D - (m_H + m_n)$$
$$= 0.002388 \text{ amu}$$

1

The mass of every atomic nucleus is less than the total of the masses of its constituent neutrons and protons—the whole is less than the sum of its parts. The "missing" mass represents energy liberated during the processes which led to the formation of the nucleus. The formation of atomic nuclei from individual nucleons is the ultimate source of nearly all the energy of the universe.

BINDING ENERGY

It was stated earlier that the nucleus of a deuterium atom consists of a proton and a neutron. Thus we would expect that the mass of a deuterium atom, ^2_1H, should be equal to the mass of an ordinary hydrogen atom, ^1_1H, plus the mass of a neutron. However, it turns out that the mass of ^2_1H is 0.002388 amu *less* than the combined masses of a ^1_1H atom and a neutron. **(1)**

The above result is an example of a general observation: stable atoms always have less mass than the combined masses of their constituent particles. The energy equivalent of the "missing" mass of a nucleus is called its *binding energy*. In order to break a nucleus

apart into its constituent nucleons, an amount of energy equal to its binding energy must be supplied either in a collision with another particle or by the absorption of a sufficiently energetic photon. Binding energies are due to the action of the nuclear forces that hold nuclei together, just as ionization energies of atoms, which must be supplied to remove electrons from them, are due to the action of the electric forces that hold them together.

It is useful to know the equivalent in MeV of the mass unit. Since 1 amu = 1.66×10^{-27} kg,

$$E = m_0 c^2$$
$$= 1.66 \times 10^{-27} \text{ kg} \times (3.00 \times 10^8 \text{ m/s})^2$$
$$= 1.49 \times 10^{-10} \text{ J}$$

is the equivalent in joules of 1 amu, and so

$$1 \text{ amu} = \frac{1.49 \times 10^{-10} \text{ J}}{1.60 \times 10^{-19} \text{ J/eV}}$$
$$= 9.31 \times 10^8 \text{ eV}$$
$$= 931 \text{ MeV}.$$

The difference between the ^2_1H atomic mass and the combined masses of ^1_1H and a neutron is 0.002388 amu. The binding energy of the deuteron (as the deuterium nucleus is called) is therefore

$$0.002388 \text{ amu} \times 931 \frac{\text{MeV}}{\text{amu}} = 2.22 \text{ MeV}.$$

This figure is confirmed by experiments that show that the minimum energy a photon must have in order to disrupt a deuteron is 2.22 MeV. **(2)**

Deuteron   Proton

2.22 MeV

Photon   Neutron

2

BINDING ENERGY PER NUCLEON

The binding energy per nucleon in a nucleus is equal to the total binding energy (calculated from the mass deficiency of the nucleus) divided by the number of neutrons and protons it contains. A very interesting curve results when we plot binding energy per nucleon versus mass number. Except for an anomalously high peak for $^4_2$He, whose nucleus is an *alpha particle*, the curve is a quite regular one. Nuclei of intermediate size have the highest binding energies per nucleon, which means that their nucleons are held together more securely than the nucleons in both heavier and lighter nuclei. The maximum in the curve is 8.8 MeV/nucleon at $A = 56$, which corresponds to the iron nucleus $^{56}_{26}$Fe. **(3)**

A significant feature of nuclear structure is illustrated by this curve. Suppose that we split the nucleus $^{235}_{92}$U, whose binding energy is 7.6 MeV/nucleon, into two fragments. Each fragment will be the nucleus of a much lighter element, and therefore will have a higher binding energy per nucleon than the uranium nucleus. The difference is about 0.8 MeV/nucleon, and so, if such *nuclear fission* were to take place, an energy of

$$0.8 \frac{\text{MeV}}{\text{nucleon}} \times 235 \text{ nucleons} \approx 190 \text{ MeV}$$

would be given off per splitting. This is a truly immense amount of energy to be produced in a single atomic event. As a comparison, chemical processes involve energies of the order of magnitude of one electron volt per reacting atom, 10^{-8} the energy involved in fission. **(4)**

The graph of binding energy per nucleon versus mass number also shows that if two of the very light nuclei are combined to form a

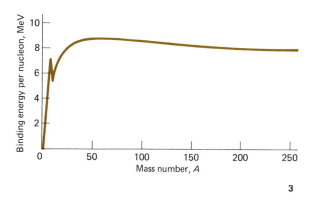

3

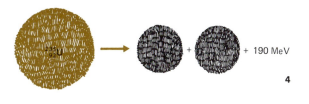

4

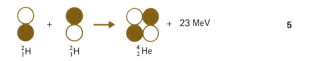

5

heavier one, the higher binding energy of the latter will also result in the evolution of energy. For instance, if two deuterons were to join to make up a $^4_2$He nucleus, over 23 MeV would be released. This process is known as *nuclear fusion*. **(5)**

NUCLEAR REACTIONS

When two nuclei approach close enough together, it is possible for a rearrangement of their constituent nucleons to occur with one or more new nuclei formed. Such a process is called a nuclear reaction, by analogy with chemical reactions in which two or more compounds may combine to form new ones.

Atoms and molecules are neutral, so it is easy for them to come together and react. Nuclei all have positive charges, as much as $+92e$ in nuclei found in nature, and the electric repulsion between them is sufficient to keep them beyond the range where they can interact unless they are moving very fast to begin with. In the sun and other stars, whose interiors are at temperatures of many millions of $°K$, the nuclei present are moving fast enough on the average for nuclear reactions to be frequent, and indeed nuclear reactions provide the energy that maintains these temperatures.

In the laboratory it is easy enough to produce nuclear reactions on a very small scale, either with alpha particles from radioactive substances or with protons, deuterons, or even heavier nuclei accelerated in various devices. But only one type of nuclear reaction has as yet proved to be a practical source of energy on the earth, namely the fission that occurs when neutrons strike the nuclei of certain very heavy nuclei.

NUCLEAR FUSION

Virtually all the energy in the universe originates in the fusion of hydrogen nuclei into helium nuclei in stellar interiors, where hydrogen is the most abundant element. Two different reaction sequences are possible, with the likelihood of each depending upon the properties of the star involved. The *proton-proton cycle* is the predominant energy source of stars whose interiors are cooler than that of the sun, perhaps $1.5 \times 10^{7}°K$. The proton-proton cycle proceeds by means of the following reactions:

$$_1^1H + _1^1H \rightarrow _1^2H + e^+ + 0.4 \text{ MeV,}$$

$$_1^1H + _1^2H \rightarrow _2^3He + 5.5 \text{ MeV,}$$

$$_2^3He + _2^3He \rightarrow _2^4He + 2\,_1^1H + 12.9 \text{ MeV.}$$

Proton-proton cycle

The first two of these reactions must each occur twice for every synthesis of $_2^4He$, so that the total energy produced per cycle is 24.7 MeV.

The symbol e^+ represents a *positron* (positively charged electron), whose role in nuclear transformations is discussed in the next section. (6)

Stars hotter than the sun obtain most of their energy from the *carbon cycle*. This cycle requires a $_6^{12}C$ nucleus for its first step and in its last step regenerates a $_6^{12}C$ nucleus, so that this isotope may be thought of as a catalyst for the process. The carbon cycle proceeds as follows:

$$_1^1H + _6^{12}C \rightarrow _7^{13}N + 2.0 \text{ MeV,}$$

$$_7^{13}N \rightarrow _6^{13}C + e^+ + 1.2 \text{ MeV,}$$

$$_6^{13}C + _1^1H \rightarrow _7^{14}N + 7.6 \text{ MeV,}$$

$$_7^{14}N + _1^1H \rightarrow _8^{15}O + 7.3 \text{ MeV,}$$

$$_8^{15}O \rightarrow _7^{15}N + e^+ + 1.7 \text{ MeV,}$$

$$_7^{15}N + _1^1H \rightarrow _6^{12}C + _2^4He + 4.9 \text{ MeV.}$$

Carbon cycle

Here again the net result is the formation of an alpha particle and two positrons from four protons with a total of 24.7 MeV of energy evolved. In the sun both the proton-proton and carbon cycles take place with comparable probabilities. (7)

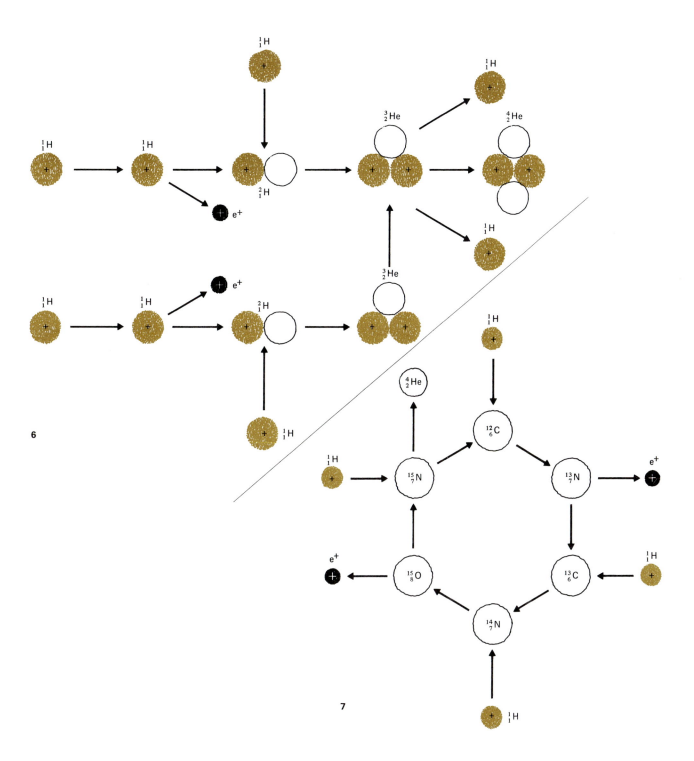

6

7

The energy liberated by nuclear fusion is often called *thermonuclear energy*. High temperatures and densities are necessary for fusion reactions to occur in such quantity that a substantial amount of thermonuclear energy is produced: the high temperature assures that the initial light nuclei have enough thermal energy to overcome their mutual electrostatic repulsion and come close enough together to react, while the high density assures that such collisions are frequent. A further condition for the proton-proton and carbon cycles is a large reacting mass, such as that of a star, since a number of separate steps is involved in each cycle and much time may elapse between the initial fusion of a particular proton and its ultimate incorporation in an alpha particle.

On the earth, where any reacting mass must be very limited in size, an efficient thermonuclear process cannot involve more than a single step. Two reactions that appear promising as sources of commercial power involve the combination of two deuterons to form a triton and a proton,

$$\mathrm{^2_1H + {^2_1}H \rightarrow {^3_1}H + {^1_1}H + 4.0\ MeV,}$$

or their combination to form a $^3_2$He nucleus and a neutron,

$$\mathrm{^2_1H + {^2_1}H \rightarrow {^3_2}He + {^1_0}n + 3.3\ MeV.}$$

Both reactions have about equal probabilities. A major advantage of these reactions is that deuterium is relatively abundant on the earth, so that there should be no fuel problems in power plants operating on deuteron fusion. While there are many difficulties to surmount in achieving practical thermonuclear power, it will almost certainly become an eventual reality. The use of "magnetic bottles" to contain thermonuclear reactions in the laboratory is discussed in Section 26.

NUCLEAR FISSION

A particularly significant nuclear reaction that requires a neutron to initiate it is *fission*. In this process, which can take place only in certain very heavy nuclei such as $^{235}_{92}$U, the absorption of an incoming neutron causes the target nucleus to split into two smaller nuclei called *fission fragments*. Fission evolves a great deal of energy because the lighter nuclei that result have higher binding energies per nucleon than the original heavy one did.

Because stable light nuclei have proportionately fewer neutrons than do heavy nuclei (see the graph on page 352), the fragments are unbalanced when they are formed and at once release one or two neutrons each. Usually the fragments are still somewhat unstable, and may undergo radioactive decay to achieve appropriate neutron: proton ratios. The products of fission, such as the fallout from a nuclear bomb burst, are accordingly highly radioactive. Although a variety of nuclear species may appear as fission fragments, we might cite as a typical fission reaction

$$\mathrm{^{235}_{92}U + {^1_0}n \rightarrow {^{236}_{92}}U \rightarrow {^{140}_{54}}Xe + {^{94}_{38}}Sr}$$
$$\mathrm{+ {^1_0}n + {^1_0}n + 200\ MeV.}$$

The $^{236}_{92}$U which is first formed is the nucleus that actually splits in two. (8)

The fission fragments $^{140}_{54}$Xe and $^{94}_{38}$Sr are both beta radioactive; the former decays four successive times until it becomes the stable isotope $^{140}_{58}$Ce, and the latter decays twice in becoming the stable isotope $^{94}_{40}$Zr. About 84% of the total energy liberated during fission appears as kinetic energy of the fission fragments, about 2.5% as kinetic energy of the neutrons, and about 2.5% in the form of instantaneously emitted gamma rays, with

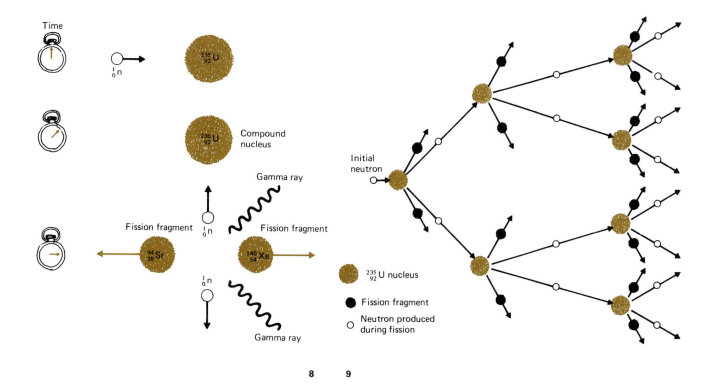

Time

$^{235}_{92}$U

$^{236}_{92}$U Compound nucleus

Gamma ray

Fission fragment $^{1}_{0}$n Fission fragment

$^{94}_{38}$Sr $^{140}_{54}$Xe

$^{1}_{0}$n $^{1}_{0}$n

Gamma ray

8

Initial neutron

$^{235}_{92}$U nucleus

● Fission fragment

○ Neutron produced during fission

9

the remaining 11% being given off in the decay of the fission fragments. (Radioactive decay is discussed in the next section.)

CHAIN REACTION

Because each fission event liberates two or three neutrons whereas only one neutron is required to initiate it, a rapidly multiplying sequence of fissions can occur in a lump of suitable material. Here is a sketch of such a *chain reaction* in $^{235}_{92}$U. (**9**)

When uncontrolled, a chain reaction evolves an immense amount of energy in a short time. If we assume that two neutrons emitted in each fission are able to induce further fissions (the average figure is lower in practice) and that 10^{-8} s elapses between the emission of a

neutron and its subsequent absorption, a chain reaction starting with a single fission will release 2×10^{13} J of energy in less than 10^{-6} s! An uncontrolled chain reaction evidently can cause an explosion of exceptional magnitude.

When properly controlled so as to assure that exactly one neutron per fission causes another fission, a chain reaction occurs at a constant level of power output. A reaction of this kind makes a very efficient source of power: an output of about 1000 kW is produced by the fission of 1 g of a suitable isotope per day, as compared with the consumption of over 3 tons of coal per day per 1000 kW in a conventional power plant. A device in which a chain reaction can be initiated and controlled is called a *nuclear reactor*.

EXERCISES

1. How does the ionization energy of an atom compare with the binding energy per nucleon of its nucleus?

2. A nucleus of $^{15}_{7}N$ is struck by a proton. A nuclear reaction takes place with the emission of (a) a neutron, or (b) an alpha particle. Give the atomic number, mass number, and chemical name of the remaining nucleus in each of the above cases.

3. A nucleus of $^{9}_{4}Be$ is struck by an alpha particle. A nuclear reaction takes place with the emission of a neutron. Give the atomic number, mass number, and chemical name of the resulting nucleus.

4. A reaction often used to detect neutrons occurs when a neutron strikes a $^{10}_{5}B$ nucleus, with the subsequent emission of an alpha particle. What is the atomic number, mass number, and chemical name of the remaining nucleus?

5. Complete the following nuclear reactions:

$$^{6}_{3}Li + ? \rightarrow {}^{7}_{4}Be + {}^{1}_{0}n$$

$$^{10}_{5}B + ? \rightarrow {}^{7}_{3}Li + {}^{4}_{2}He$$

$$^{35}_{17}Cl + ? \rightarrow {}^{32}_{16}S + {}^{4}_{2}He$$

6. What are the differences and similarities between nuclear fusion and nuclear fission?

7. What advantage is there in using neutrons as bombarding particles to investigate nuclear reactions?

8. The mass of $^{20}_{10}Ne$ is 19.99244 amu. What is its binding energy? What is its binding energy per nucleon?

9. The mass of $^{35}_{17}Cl$ is 34.96885 amu. What is its binding energy? What is its binding energy per nucleon?

10. The atomic masses of $^{15}_{7}N$, $^{15}_{8}O$, and $^{16}_{8}O$ are respectively 15.0001, 15.0030, and 15.9949 amu.
    a) Find the average binding energy per nucleon in $^{16}_{8}O$.
    b) How much energy is needed to remove one proton from $^{16}_{8}O$?
    c) How much energy is needed to remove one neutron from $^{16}_{8}O$?
    d) Why are these figures different from one another?

11. What is the minimum energy a gamma-ray photon must have if it is to split an alpha particle into
    a) a triton and a proton?
    b) a $^{3}_{2}He$ nucleus and a neutron?
    The atomic masses involved are $^{1}_{0}n$, 1.008665 amu; $^{1}_{1}H$, 1.007825 amu; $^{3}_{1}H$, 3.016050 amu; $^{3}_{2}He$, 3.016030 amu; and $^{4}_{3}He$, 4.002603 amu.

12. How much energy is evolved in the nuclear reaction

$$^{2}_{1}H + {}^{3}_{1}H \rightarrow {}^{4}_{2}He + {}^{1}_{0}n?$$

13. The nuclear reaction

$$^{6}_{3}Li + {}^{2}_{1}H \rightarrow 2 \, {}^{4}_{2}He$$

evolves 22.4 MeV. Calculate the mass of $^{6}_{3}Li$ in amu.

14. In certain stars three alpha particles join in succession to form a $^{12}_{6}C$ nucleus. The mass of $^{12}_{6}C$ is 12.0000 amu. How much energy is evolved in this reaction?

15. How much mass is lost per day by a nuclear reactor operated at a 10 MW (10×10^{6} W) power level?

16. If each fission in $^{235}_{92}U$ releases 200 MeV, how many fissions must occur per second to produce a power of 1 kW?

17. $^{235}_{92}U$ loses about 0.1% of its mass when it undergoes fission.
    a) How much energy is released when 1 kg of $^{235}_{92}U$ undergoes fission?
    b) One ton of TNT releases about 10^{6} kcal when it is detonated. How many tons of TNT are equivalent in destructive power to a bomb that contains 1 kg of $^{235}_{92}U$?

40

RADIOACTIVITY

"One may also imagine that in criminal hands radium might become very dangerous, and here we may ask ourselves if humanity has anything to gain by learning the secrets of nature, if it is ripe to profit by them, or if this knowledge is not harmful . . . I am among those who think that humanity will obtain more good than evil from the new discoveries."
Pierre Curie (1859–1906)

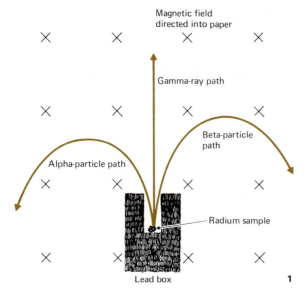

Magnetic field directed into paper

Gamma-ray path

Beta-particle path

Alpha-particle path

Radium sample

Lead box

1

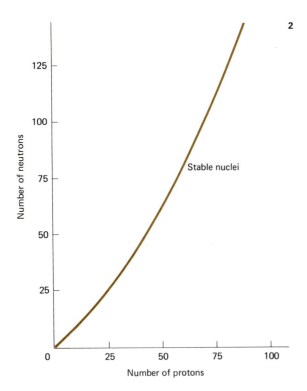

Stable nuclei

Number of neutrons

Number of protons

2

Not all atomic nuclei are stable. At the beginning of the twentieth century it became known, as the result of research by Becquerel, the Curies, and others, that some nuclei exist which spontaneously transform themselves into other nuclear species with the emission of radiation. Such nuclei are said to be *radioactive*.

ALPHA, BETA, AND GAMMA RADIATION

Unstable nuclei in nature emit three kinds of radiation:

1. *Alpha particles*, which are the nuclei of $^4_2$He atoms;

2. *Beta particles*, which are electrons;

3. *Gamma rays*, which are photons of high energy.

The early experimenters identified these radiations with the help of a magnetic field. The diagram shows a radium sample in a magnetic field directed into the paper: the positively charged alpha particles are deflected to the left and the negatively charged beta particles are deflected to the right. Gamma rays carry no charge and are not affected by the magnetic field. (1)

ORIGIN OF RADIOACTIVE DECAY

In Section 38 we saw that a nucleus with a given number of protons has only a very narrow range of possible numbers of neutrons if it is to be stable. Here is a sketch of the stability curve shown in more detail earlier. (2)

Suppose now that a nucleus exists which has too many neutrons for stability relative to the number of protons present. If one of the excess neutrons transforms itself into a proton, this will simultaneously reduce the number of neutrons while increasing the number of protons. To conserve electric charge, such a transformation requires the emission of a negative electron, and we may write it in equation form as

$$n^0 \rightarrow p^+ + e^-.$$

The electron leaves the nucleus, and is detectable as a "beta particle." The residual nucleus may be left with some extra energy as a consequence of its shifted binding energy, and this energy is given off in the form of gamma rays. Sometimes more than one such *beta decay* is required for a particular unstable nucleus to reach a stable configuration.

Should the nucleus have too few neutrons, the inverse reaction

$$p^+ \rightarrow n^0 + e^+,$$

in which a proton becomes a neutron with the emission of a *positive* electron (known as a *positron*), may take place. This is also called beta decay, since it resembles the emission of negative electrons from an unstable nucleus in every way save for the difference in charge.

Another way of altering its structure to achieve stability may involve a nucleus in *alpha decay*, in which an alpha particle consisting of two neutrons and two protons is emitted. Thus negative beta decay increases the number of protons by one and decreases the number of neutrons by one; positive beta decay decreases the number of protons by one and increases the number of neutrons by one; and alpha decay decreases both the number of protons and the number of neutrons by two. These processes are shown schematically in the diagram. Very often a succession of alpha and beta decays, with accompanying gamma decays to carry off excess energy, is required before a nucleus reaches stability. **(3)**

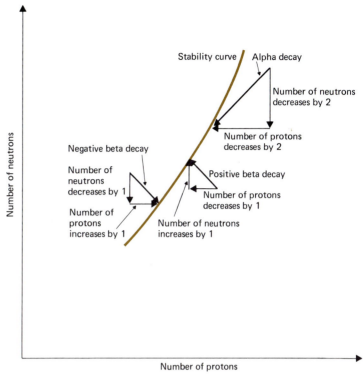

3

ALPHA DECAY

Nuclei that contain more than about 210 nucleons are so large that the short-range forces holding them together are barely able to counterbalance the long-range electrostatic repulsive forces of their protons. Such a nucleus can reduce its bulk and thereby achieve greater stability by emitting an alpha particle, which decreases its mass number A by 4.

It is appropriate to ask why it is that only alpha particles are given off by excessively heavy nuclei, and not, for example, individual protons or ^3_2He nuclei. The reason is a consequence of the high binding energy of the alpha particle, which means that it has

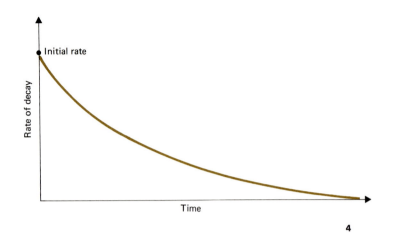

Initial rate

Rate of decay

Time

4

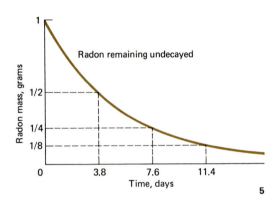

Radon remaining undecayed

Radon mass, grams

1

1/2

1/4

1/8

0   3.8   7.6   11.4

Time, days

5

significantly less mass than four individual nucleons. Because of this small mass, an alpha particle can be ejected by a heavy nucleus with energy to spare. Thus the alpha particle released in the decay of $^{232}_{92}U$ has a kinetic energy of 5.4 MeV, while 6.1 MeV would have to be supplied from the outside to this nucleus if it is to release a proton, and 9.6 MeV supplied if it is to release a ^3_2He nucleus.

HALF-LIFE

One of the characteristics of all types of radioactivity is that the rate at which the nuclei in a given sample decay always follows a curve whose shape is like that shown here. If we start with a sample whose rate of decay is, say, 100 disintegrations/s, it will not continue to decay at that rate but instead fewer and fewer disintegrations will occur in each successive second. **(4)**

Some isotopes decay faster than others, but in each case a certain definite time is required for half of an original sample to decay. This

time is called the *half-life* of the isotope. The element radon has a half-life of 3.8 days, for instance. Should we start with 1 g of radon in a closed container (since it is a gas), $\frac{1}{2}$ g will remain undecayed after 3.8 days; $\frac{1}{4}$ g will remain undecayed after 7.6 days; $\frac{1}{8}$ g will remain undecayed after 11.4 days; and so on. **(5)**

Half-lives range from billionths of a second to billions of years. Samples of radioactive isotopes decay in the manner illustrated because a great many individual nuclei are involved, each having a certain probability of decaying. The fact that radon has a half-life of 3.8 days signifies that every radon nucleus has a 50% chance of decaying in any 3.8-day period. Because a nucleus does not have a memory, this does *not* mean that a radon nucleus has a 100% chance of decaying in 7.6 days: the likelihood of decay of a given nucleus stays the same until it actually does decay. Thus a half-life of 3.8 days means a 75% probability of decay in 7.6 days, an 87.5% probability of decay in 11.4 days, and so on, because in each interval of 3.8 days the probability is 50%.

The above discussion suggests that radioactive decay involves individual events that take place within individual nuclei, rather than collective processes that involve more than one nucleus in interaction. This idea is confirmed by experiments which show that the half-life of a particular isotope is invariant under changes of pressure, temperature, electric and magnetic fields, and so on, which might, if strong enough, influence internuclear phenomena.

Problem. The hydrogen isotope tritium, $_1^3H$, is radioactive and emits an electron with a half-life of 12.5 yr. (a) What does $_1^3H$ become after beta decay? (b) What percentage of an original sample of tritium will remain 25 yr after its preparation?

Solution. (a) When a nucleus emits an electron, its atomic number increases by 1 unit (corresponding to an increase in nuclear charge of $+e$) and its mass number is unchanged. Helium has the atomic number 2, and so

$$_1^3H \rightarrow \; _2^3He + e^-.$$

(b) Twenty-five years represents two half-lives of tritium; hence $\frac{1}{2} \times \frac{1}{2} = \frac{1}{4}$ of the original sample remains undecayed.

Three aspects of radioactivity are wholly remarkable from the point of view of classical physics. First, the atomic number of a nucleus that undergoes alpha or beta decay changes, so that the nucleus becomes one characteristic of a different element. Elements *can* be transmuted into other elements, though hardly in a manner anticipated by alchemy.

Second, radioactive decay liberates energy that can only come from *within* individual atoms. One gram of radium in a sealed container (to prevent the escape of radon, a gaseous product of its decay that is itself radioactive) evolves energy at the rate of 0.14 kcal/hr; this rate decreases so slowly that after 1600 yr it has dropped by only 50%. Where does all this energy come from? Not until 1905, when Einstein proposed the equivalence of mass and energy, was this puzzle understood.

Third, radioactivity is a statistical process. Every nucleus of a radioisotope has a certain likelihood of decaying, but, because the decay obeys the laws of chance, we have no way of predicting *which* nuclei will actually decay at a particular time. There is no cause-effect relationship in radioactivity as there is in all classical physics. The revolution in physics that finally overthrew determinism in 1925 with the advent of quantum theory thus had its beginnings several decades earlier in the discovery of radioactive decay.

RADIOCARBON DATING

Among the other effects they produce, cosmic-ray particles from space often disrupt the nuclei of atoms in the earth's atmosphere. The protons set free in this way eventually pick up electrons and become hydrogen atoms. Some of the neutrons leave the atmosphere and beta-decay into protons and electrons near the earth, where they are trapped by the earth's magnetic field and become part of the magnetosphere. Other neutrons are "captured" by the nuclei of nitrogen atoms in the atmosphere, which undergo the nuclear reaction

$$_7^{14}N + \; _0^1n \rightarrow \; _6^{14}C + \; _1^1H.$$

The result is a free proton and the carbon isotope $_6^{14}C$. This isotope has too many neutrons for stability and ultimately it undergoes

Time after death of animal or plant	Radiocarbon ($^{14}_{6}$C) content of specimen, which decreases with time	Ordinary carbon ($^{12}_{6}$C) content of specimen, which remains unchanged
0 yr		
5600 yr (1/2 has decayed)		
11,200 yr (3/4 has decayed)		
16,800 yr (7/8 has decayed)		

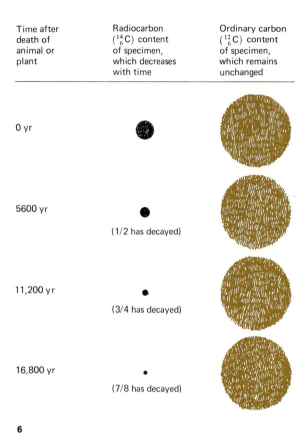

6

Thus every living thing on the earth has a very small proportion of radiocarbon in its tissues. Because the mixing of radiocarbon is relatively efficient on the earth, living plants and animals all have the same ratio of radiocarbon to ordinary carbon $^{12}_{6}$C.

When a plant or animal dies, however, it no longer takes in radiocarbon atoms, while the radiocarbon atoms it already has continually decay to nitrogen. After 5600 yr a dead plant or animal has left only half as much radiocarbon as it had while alive, after 11,200 yr only a quarter as much, and so on. It is therefore possible to determine the time that has elapsed since the death of a plant or animal by measuring the ratio between its radiocarbon and ordinary carbon contents. This technique, called *radiocarbon dating*, has made it possible to find the ages of such archeological specimens as wooden tools, mummies, pieces of cloth, and charcoal from cooking fires. Specimens up to about 40,000 yr old can be dated in this way. **(6)**

EXERCISES

1. Radium undergoes spontaneous decay into helium and radon. Why is radium regarded as an element rather than as a chemical compound of helium and radon?

2. What happens to the atomic number and mass number of a nucleus that emits a gamma ray? What happens to the actual mass of the nucleus?

3. The nuclei $^{14}_{8}$O and $^{19}_{8}$O both undergo beta decay in order to become stable nuclei. Which would you expect to emit a positron and which an electron?

4. The neutron decays in free space into a proton and an electron. What must be the minimum binding energy contributed by a neutron to a nucleus in order that the neutron not decay

beta decay to become $^{14}_{7}$N. The half-life of such *radiocarbon* is 5600 yr. At the present time there are perhaps 90 tons of radiocarbon on the earth's surface and in the atmosphere.

Not long after their formation, radiocarbon atoms combine with oxygen molecules to form carbon dioxide (CO_2) molecules. Green plants convert water and carbon dioxide into carbohydrates in the process of photosynthesis, which means that every plant contains a certain amount of radiocarbon. Animals eat plants and so they too contain radiocarbon.

inside the nucleus? How does this figure compare with the observed binding energies per nucleon in stable nuclei?

5. It has recently been discovered that certain artificially prepared isotopes decay by means of proton emission. What property do you suppose is characteristic of such nuclei?

6. The nucleus $^{233}_{90}$Th undergoes two negative beta decays in becoming an isotope of uranium. What is the symbol of the isotope?

7. The nucleus $^{238}_{92}$U decays into a lead isotope through the successive emissions of eight alpha particles and six electrons. What is the symbol of the resulting lead nucleus?

8. After two hours have elapsed, $\frac{1}{16}$ of the initial amount of a certain radioactive isotope remains undecayed. What is the half-life of this isotope?

9. The half-life of radium is 1600 yr. How long will it take for $\frac{15}{16}$ of a given sample of radium to decay?

10. Sixty hours after a sample of the beta emitter $^{24}_{11}$Na has been prepared, only $\frac{1}{8}$ of it remains undecayed. What is the half-life of this isotope?

11. $^{232}_{92}$U (mass 232.1095 amu) alpha decays into $^{228}_{90}$Th (mass 228.0998 amu).
    a) Find the amount of energy released in the decay.
    b) The alpha particle emitted in the decay of $^{232}_{92}$U is observed to have a kinetic energy of 5.3 MeV. If this is not the same as the answer to (a), account for the difference.
    c) Is it possible for $^{232}_{92}$U to decay into $^{231}_{92}$U (mass 231.1082 amu) by emitting a neutron? Why?
    d) Is it possible for $^{232}_{92}$U to decay into $^{231}_{91}$U (mass 231.1078) by emitting a proton? Why? ($m_\alpha = 4.0039$ amu, $m_n = 1.0087$ amu, $m_H = 1.0078$ amu)

41

NEUTRINOS AND ANTIPARTICLES

"The explanation of the process of beta decay presents very serious difficulties for the modern quantum theory. The chief puzzle of beta disintegration is the fact that the magnetic spectrum of the emitted beta particles is always continuous, which shows that the energy of the electrons emitted from the same substance varies within very wide limits . . . There is no doubt that the question is of fundamental importance and will lead to revolutionary changes in our present picture of the physical world." (Written in 1931.) George Gamow (1904–1968)

Ordinary matter is composed of neutrons, protons, and electrons. Another important elementary particle is the *neutrino*. All these particles have *antiparticle* counterparts whose mass is the same but whose charge (and other properties) are different. When a particle and its antiparticle come together, they both disappear as their mass is turned into energy.

THE NEUTRINO

In radioactive decay, as in all other natural processes, energy (including mass energy) is conserved. For this reason the total mass of the products of a particular decay must be less than the mass of the initial nucleus, with the missing mass appearing as photon energy in the case of gamma decay and as kinetic energy in the cases of alpha and beta decay. In gamma and alpha decay the liberated energy is indeed precisely equal to the energy equivalent of the lost mass, but in beta decay a strange effect occurs: instead of all having the same energy, the emitted electrons from a particular isotope exhibit a variety of energies. These energies range from zero up to a maximum figure equal to the energy equivalent of the missing mass in the transformation. This effect is illustrated in the graph, which shows the spread in electron energy in the decay of $^{210}_{83}$Bi. (**1**)

Momentum as well as energy is apparently not conserved in beta decay. When an object at rest disintegrates into two parts, they must move apart in opposite directions in order that the total momentum of the system remain zero. Experiments show, however, that the emitted electron and the residual nucleus do *not* in general travel in opposite directions after beta decay occurs, so that their momenta cannot cancel out to equal the initial momentum of zero.

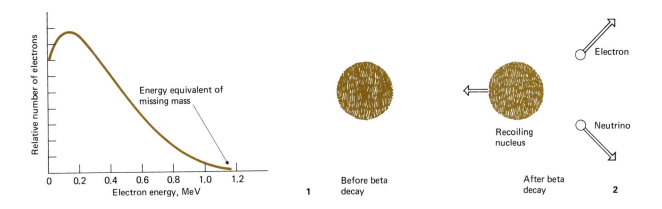

Energy equivalent of missing mass

Relative number of electrons

Electron energy, MeV

1

Before beta decay

Recoiling nucleus

After beta decay

Electron

Neutrino

2

A third difficulty concerns angular momentum. The spin angular momentum component in a given direction of the neutron, the proton, the electron, and the positron is in every case $\pm\frac{1}{2}h/2\pi$. The conversion of a neutron into a proton and an electron or of a proton into a neutron and a positron therefore leaves an angular momentum discrepancy of $\frac{1}{2}h/2\pi$.

The process of beta decay is the first one we have encountered in which the conservation laws of energy, linear momentum, and angular momentum do not seem to hold. To account for the above discrepancies without abandoning three of the most fundamental and otherwise well-established physical principles, the existence of a new particle was postulated, the *neutrino*, symbol ν_e (Greek letter "nu"). The neutrino has no electric charge and no mass, but is able to possess both energy and momentum and has an intrinsic spin. (Lest this seem unlikely, we might reflect that the photon, also massless, has energy, momentum, and angular momentum. The neutrino is *not* a photon, however, but is an entirely different entity.) According to the neutrino theory, an electron and neutrino are simultaneously emitted in beta decay, which permits energy and momentum to be conserved. (**2**)

For a quarter of a century the neutrino hypothesis was accepted despite the absence of any direct evidence in its support. This was a very unusual situation: here was a particle of a rather odd kind, which nobody had ever detected experimentally, yet practically no physicists doubted that it really existed. It was far from being a matter of blind faith, however. There were no theoretical objections to the neutrino hypothesis, whereas there were strong theoretical and experimental objections to dropping the principles of conservation of energy, linear momentum, and angular momentum. Furthermore, the neutrino, which lacks mass and charge and is not electromagnetic in nature as is the photon, interacts only feebly with matter, so it was not easy to think of a way to detect it.

Finally, in 1956, an experiment was performed in which a nuclear reaction that, in theory, could only be caused by a neutrino, was actually found to take place. In the operation of a nuclear reactor a great many beta decays take place, and as a result more than 10^{16} neutrinos per second may emerge from each square meter of the shielding around a reactor. A neutrino striking a proton has a small probability of inducing the reaction

$$\nu_e + p \rightarrow n + e^+$$

in which a neutron and positron are created. By placing a sensitive detecting chamber containing hydrogen near a nuclear reactor, the simultaneous appearance of a neutron and a positron could be registered each time the above reaction occurred. Calculations were made initially of how many such reactions per second should occur based on the known properties of the detecting apparatus and on the theoretical properties of the neutrino. When this reaction rate was actually found, there was no doubt that neutrinos indeed exist.

THE WEAK INTERACTION

We have now come to the last of the four fundamental interactions between elementary particles. This is the *weak interaction* that leads to beta decay. The range of the weak interaction is so short that it operates only *within* certain elementary particles and leads to their transformation into other particles with the simultaneous creation of neutrinos. Insofar as the structure of matter is concerned, the role of the weak interaction seems to be confined to causing beta decays in nuclei whose neutron:proton ratios are not appropriate for stability. This interaction also affects elementary particles that are not part of a nucleus. The name "weak interaction" arose because the other short-range force acting upon nucleons is extremely strong, as the high binding energies of nuclei attest. Actually the gravitational interaction is weaker than the weak interaction at distances where the latter is a factor.

Neutrinos are able to travel unimpeded through vast amounts of matter because they are limited to the weak interaction. On the average, a neutrino must traverse 130 light-years of solid iron before being absorbed—and a light-year, the distance light goes in a year, is 9.5×10^{15} m.

In both the proton-proton and carbon cycles, two beta decays occur for each helium nucleus formed. A vast number of neutrinos is therefore produced in the sun at all times. Because neutrinos travel freely through matter, almost all of these neutrinos escape into space and take with them 6 to 8% of the total energy generated by the sun. The flux of neutrinos from the sun is such that every cubic inch on the earth contains perhaps 100 neutrinos at any moment! The energy carried by the neutrinos created in the sun and the other stars is apparently lost forever from the universe in the sense that it is no longer available for conversion into other forms of energy, such as matter.

ANTIPARTICLES

Another recent experimental discovery of a particle whose existence had been predicted theoretically decades ago is that of the negative proton, or *antiproton*, whose symbol is $\bar{p}$. This is a particle with the same properties as the proton except that it has a negative electric charge. The existence of antiprotons was predicted largely on the basis of symmetry arguments: since the electron has a positive counterpart in the positron, why

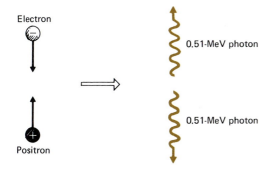

Electron

0.51-MeV photon

0.51-MeV photon

Positron

3

should the proton not have a negative counterpart as well? Actually, as sophisticated theories show, this is an excellent argument, and few physicists were surprised when the antiproton was actually found.

The reason positrons and antiprotons are so difficult to find is that they are readily *annihilated* upon contact with ordinary matter. When a positron is in the vicinity of an electron, they attract one another electrostatically, come together, and then both vanish simultaneously, with their missing mass appearing in the form of two gamma-ray photons:

$$e^+ + e^- \rightarrow \gamma + \gamma.$$

The total mass of the two particles is the equivalent of 1.02 MeV, and so each photon has an energy of 0.51 MeV. (Their energies must be equal and they must be emitted in opposite directions in order that momentum be conserved.) **(3)**

While the similar reaction

$$p + \bar{p} \rightarrow \gamma + \gamma$$

can occur when a proton and antiproton undergo annihilation, it is more usual for the vanished mass to reappear in the form of several mesons, particles which we shall consider in the next section.

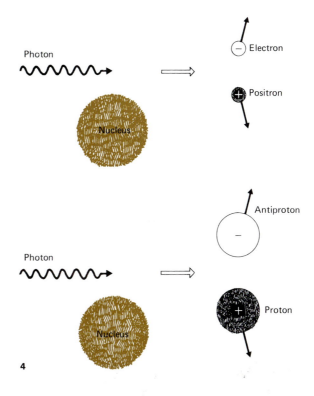

Photon

Nucleus

Electron

Positron

Photon

Nucleus

Antiproton

Proton

4

The reverse of annihilation can also take place, with the electromagnetic energy of a photon materializing into a positron and an electron or, if it is energetic enough, into a proton and an antiproton. This phenomenon is known as *pair production*, and requires the presence of a nucleus in order that momentum as well as energy be conserved. Any photon energy in excess of the amount required to provide the mass of the created particles (1.02 MeV for a positron-electron pair, 1872 MeV for a proton-antiproton pair) appears as kinetic energy. **(4)**

Antineutrons (symbol $\bar{n}$) and antineutrinos symbol $\bar{\nu}_e$) have also been identified. Antineutrons can be detected through their mutual annihilation with neutrons, while more indirect, though equally definite, evidence supports the existence of antineutrinos. The antineutrino differs from the neutrino in that, while the spin axes of both are parallel to their directions of motion, the spin of the former is clockwise and that of the latter is counterclockwise when viewed from behind. A moving neutrino may be thought of as resembling a left-handed screw, and a moving antineutrino as resembling a right-handed screw. An antineutrino is released during a beta decay in which an electron is emitted, and a neutrino is released during a beta decay in which a positron is emitted. Thus the fundamental equations of beta decay are

$$p \rightarrow n + e^+ + \nu_e,$$

$$n \rightarrow p + e^- + \bar{\nu}_e. \quad (5)$$

Neutrino   Antineutrino

Annihilation and pair production are consequences of the facts that matter is a form of energy and that conversions from matter to energy and from energy to matter are no more improbable than conversions from, say, gravitational potential energy to kinetic energy.

ANTIMATTER

Ordinary atoms are composed of neutrons, protons, and electrons. There is apparently no reason why atoms composed of antineutrons, antiprotons, and positrons should not be stable and behave in every way like ordinary atoms. It is an attractive notion that equal amounts of matter and *antimatter* came into being at the origin of the universe which became segregated into separate galaxies. The spectra of the light emitted by the members of antimatter galaxies would be exactly the same as the spectra of the light emitted by the members of galaxies of ordinary matter, which gives us no way to distinguish between the two. Of course, if antimatter comes in contact with ordinary matter, their mutual annihilation occurs with the release of a great deal of energy, and it is possible that certain astrophysical phenomena derive their energy from this source.

EXERCISES

1. Discuss the similarities and differences between the photon and the neutrino.

2. How much energy must a gamma-ray photon have if it is to materialize into a proton-antiproton pair with each particle having a kinetic energy of 10 MeV?

3. How much energy must a gamma-ray photon have if it is to materialize into a neutron-antineutron pair? Is this more or less than that required to form a proton-antiproton pair?

4. A 1-MeV positron collides head on with a 1-MeV electron, and the two are annihilated. What is the energy and wavelength of each of the resulting gamma-ray photons?

42

ELEMENTARY PARTICLES

"This is not the end. It is not even the beginning of the end. But it is, perhaps, the end of the beginning." Winston Churchill (1874–1964)

A great many elementary particles are known besides the ones we have already studied. All these other particles are extremely unstable and decay rapidly after being created in high-energy collisions between other elementary particles. The true significance of these short-lived particles is not fully understood at present, though it seems probable that most of them should not be considered as true "elementary particles" but rather as different excited states of a smaller number of basic entities. The study of elementary particles is one of the most active areas in current physics research.

Attractive force

MESON THEORY OF NUCLEAR FORCES

In 1935 the Japanese physicist Yukawa suggested that the strong nuclear interaction could be regarded as the result of an interchange of certain particles between nucleons. Today these particles are called *pions*. Pions may be charged or neutral; those with charges of $+e$ or $-e$ have rest masses of 273 times the electron mass, while neutral pions have rest masses of 264 times the electron mass. Pions are members of a class of elementary particles collectively called *mesons*—the word pion is a contraction of the original name π-meson.

Repulsive force

1

The crude analogy illustrated here may help in understanding how meson exchange can lead to both attractive and repulsive forces between nuclei. Each child in the figure has a pillow. When the children exchange pillows by snatching them from each other's grasp, the effect is like that of a mutually attractive force. On the other hand, the children may also exchange pillows by throwing them at each other. Here conservation of momentum requires that the children move apart, just as if a repulsive force were present between them. (1)

According to Yukawa's theory, nearby nucleons constantly exchange mesons without themselves being altered. We note that the emission of a meson by a nucleon at rest which does not lose a corresponding amount of mass violates the law of conservation of energy. However, the law of conservation of energy, like all physical laws, deals only with measurable quantities. Because the uncertainty principle restricts the accuracy with which we can perform certain measurements, it limits the range of application of physical laws such as that of energy conservation.

In Section 33 the origin of the uncertainty principle in the form

$$\Delta x \, \Delta mv \geqslant \frac{h}{2\pi}$$

was discussed. This formula states that the product of the uncertainty in the position of a body and the uncertainty in its momentum cannot be less than $h/2\pi$, where h is Planck's constant. There is another form of the uncertainty principle that relates the uncertainty ΔE in an energy measurement with the uncertainty Δt in the time when the energy measurement is made. This form of the uncertainty principle states that

$$\Delta E \, \Delta t \geqslant \frac{h}{2\pi}.$$

We conclude that a process can take place in which energy is *not* conserved by an amount ΔE *provided* that the time interval Δt in which the process takes place is not more than $h/2\pi\Delta E$. Thus the creation, transfer, and disappearance of a meson do not conflict with the conservation of energy provided the sequence takes place fast enough. The latter condition provides a way to estimate the mass of the pion.

Let us assume that the temporary energy discrepancy ΔE is of the same magnitude as the rest energy mc^2 of the pion, and that the pion travels at very nearly the speed of light c as it goes from one nucleon to another. (These assumptions are crude because the kinetic energy of the pion is ignored, but all we are after is an approximate figure for m.) The time Δt the pion spends between its creation in one nucleon and its absorption in another cannot be greater than R/c, where R is the maximum distance that can separate interacting nucleons. **(2)**

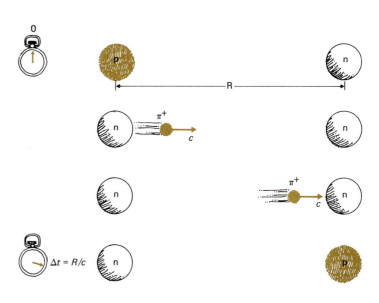

2

We therefore have, using the symbol $\approx$ to indicate that the result is only a rough approximation,

$$\Delta E \Delta t \approx \frac{h}{2\pi},$$

$$mc^2 \times \frac{R}{c} \approx \frac{h}{2\pi},$$

$$m \approx \frac{h}{2\pi Rc}.$$

The strong nuclear interaction responsible for the attractive forces between nucleons has a range of about 1.7×10^{-15} m. When we substitute

$$R = 1.7 \times 10^{-15} \text{ m}$$

in the above formula, we obtain $m \approx 2.1 \times 10^{-28}$ kg for the pion mass, which is about 230 electron masses. Of course, the preceding calculation is hardly a rigorous one, but if Yukawa's theory has any validity, the pions he postulated should have masses somewhere in this vicinity—as they do.

PIONS AND MUONS

Not long after Yukawa's work, charged particles of about the right mass were experimentally discovered in the cosmic radiation. Their discovery was not unexpected because a sufficiently energetic nuclear collision should be able to liberate mesons by providing enough energy to create them without violating conservation of energy, and nuclear collisions between fast cosmic-ray protons from space and oxygen and nitrogen nuclei occur constantly in the atmosphere. However, these particles did not behave at all in the way they were expected to behave. Far from strongly interacting with nuclei, as Yukawa's mesons were supposed to do, they barely interacted at all. Instead of being absorbed in at most a meter of earth, they penetrated thousands of meters into the ground.

Finally, in 1947, the explanation for their peculiar behavior was found. The weakly interacting mesons, known as *muons* (a contraction of μ-meson, where μ is the Greek letter "mu") and found profusely in cosmic rays at sea level, are not the direct products of nuclear collisions, but are secondary particles that result from the decay of pions.

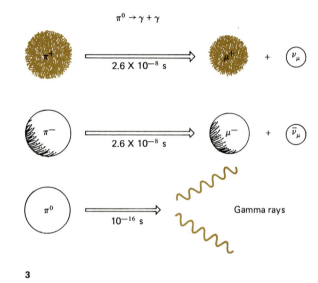

3

Outside a nucleus a charged pion decays in an average of 2.6×10^{-8} s into a muon and a neutrino (or antineutrino) plus kinetic energy. That is,

$$\pi^+ \rightarrow \mu^+ + \nu_\mu,$$
$$\pi^- \rightarrow \mu^- + \bar{\nu}_\mu.$$

A small proportion of charged pions also decay directly into an electron or positron plus a neutrino or antineutrino. Neutral pions, whose masses of $264\, m_e$ are a little less than the charged pion mass of $273\, m_e$, decay in about 10^{-16} s into a pair of gamma-ray photons:

$$\pi^0 \rightarrow \gamma + \gamma. \quad \textbf{(3)}$$

The neutrinos involved in pion decay have been denoted ν_μ, whereas those involved in beta decay have been denoted ν_e. Until 1962 it had been thought that there is only a single kind of neutrino. In that year an experiment was performed in which pions were produced by bombarding a metal target with high-energy protons from an accelerator. The pion decays liberated neutrinos, and the interactions of these neutrinos with matter was studied. The only inverse reactions found led to the production of muons; no electrons whatever were created. Hence the neutrinos set free in pion decay are different from those set free in beta decay.

The muon, whose mass is $207\, m_e$, is less well understood than the pion. The relatively long muon lifetime of 2.2×10^{-6} s and their feeble interaction with matter account for the penetrating ability of these particles. In contrast to the pion, which fits in well with the theory of nuclear forces, the muon seems to have no particular function in the scheme of things; why such a particle should exist is an unsolved problem. The positive and negative muons decay into positrons and electrons respectively together with a neutrino-antineutrino pair in each case. **(4)**

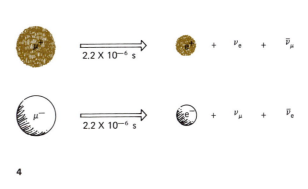

4

CATEGORIES OF ELEMENTARY PARTICLES

Several hundred elementary particles have been discovered in recent years in addition to the sixteen we have thus far discussed. Such abundance where scarcity had been expected stimulated an intense research effort, with the result that a number of suggestive regularities have been found in the properties of the various particles. But there is as yet no theory of elementary particles even remotely comparable in an ability to account for their existence and behavior to, say, the quantum theory of the atom. The search for such a theory continues, perhaps more difficult, frustrating, and exciting than any other in the history of science.

The following table is a list of the 37 longest-lived elementary particles, together with some of their properties. The other known particles have much briefer lifetimes, in most cases so short (10^{-22}–10^{-23} s) that their existence is inferred on the basis of indirect evidence. The particles in the table, together with the unlisted ones, fall into four categories: *photons, leptons, mesons,* and *baryons.* The photon, π^0 meson, and η^0 meson are their own antiparticles. Only the principal mode or modes of decay are given; some particles may decay in a variety of ways.

Table 42.1

Name	Particle	Anti-particle	Mass, in electron masses	Stability and chief decay mode	Average lifetime, s	Category
Photon	γ	(γ)	0	Stable		Photon
Neutrino	v_e	$\bar{v}_e$	0	Stable		Leptons
	v_μ	$\bar{v}_\mu$	0	Stable		
Electron	e^-	e^+	1	Stable		
Muon	μ^-	μ^+	207	Unstable; decays into electron plus two neutrinos	2.2×10^{-6}	
Pion	π^+	π^-	273	Unstable; decays into muon plus neutrino	2.6×10^{-8}	Mesons
	π^0	(π^0)	264	Unstable; decays into two gamma rays	8.9×10^{-17}	
K meson	K^+	K^-	966	Unstable; decays into muon plus neutrino or into two or three pions	1.2×10^{-3}	
	K_1^0	$\bar{K}_1^0$	974	Unstable; decays into two pions	8.7×10^{-11}	
	K_2^0	$\bar{K}_2^0$	974	Unstable; decays into three pions or into pion and neutrino plus muon or electron	5.3×10^{-8}	
Eta meson	η^0	(η^0)	1073	Unstable; decays into three pions or two gamma rays	10^{-18}	
Proton	p^+	p^-	1836	Stable		Baryons
Neutron	n^0	$\bar{n}^0$	1839	Unstable in free space; decays into electron, proton, and neutrino	1×10^3	
Lambda hyperon	Λ^0	$\bar{\Lambda}^0$	2182	Unstable; decays into pion plus neutron or proton	2.5×10^{-10}	
Sigma hyperon	Σ^+	$\bar{\Sigma}^+$	2328	Unstable; decays into pion plus neutron or proton	8×10^{-11}	
	Σ^-	$\bar{\Sigma}^-$	2341	Unstable; decays into neutron plus pion	1.6×10^{-10}	
	Σ^0	$\bar{\Sigma}^0$	2332	Unstable; decays into lambda hyperon plus gamma ray	10^{-14}	
Xi hyperon	Ξ^-	$\bar{\Xi}^-$	2583	Unstable; decays into lambda hyperon plus pion	1.7×10^{-10}	
	Ξ^0	$\bar{\Xi}^0$	2571	Unstable; decays into lambda hyperon plus pion	2.9×10^{-10}	
Omega hyperon	Ω^-	$\bar{\Omega}^-$	3290	Unstable; decays into lambda hyperon plus K meson or into xi hyperon plus pion	10^{-10}	

PHOTONS

As we know, photons are quanta of electromagnetic energy that are stable in space and have zero rest mass. It is possible to think of the electromagnetic interaction as being "carried" from one charge to another by the circulation of photons between them, just as the strong nuclear interaction is carried from one nucleon to another by the circulation of mesons between them.

Since both mesons and photons actually exist, it is tempting to ask whether the gravitational and weak interactions also might have particles associated with them. Since gravitational forces are unlimited in range, the hypothetical *graviton* should have zero rest mass to avoid violating the conservation of energy. (The same argument accounts for the zero rest mass of the photon.) Also like the photon, the graviton should be stable, electrically neutral, and travel with the speed of light. Unlike the photon, the graviton should interact only feebly with matter, and so would be extremely hard to detect. There is no definite experimental evidence either for or against the existence of the graviton.

The quantum of the weak interaction is called the *intermediate boson*. Because the weak interaction has a shorter range than the strong nuclear interaction, the rest mass of the intermediate boson should exceed meson masses. It should be charged and decay rapidly. A number of careful searches has turned up no traces of the intermediate boson.

LEPTONS

Leptons include electrons, muons, and neutrinos. There are two families of leptons, one consisting of the electron, the positron, the v_e neutrino, and the $\bar{v}_e$ antineutrino, and the other consisting of the positive and negative muon, the v_μ neutrino, and the $\bar{v}_\mu$ antineutrino. A conservation law applies to each lepton family: in every known process, the number of leptons of each kind remains constant, reckoning a lepton as $+$ and an antilepton as $-$.

An easy way to apply these conservation laws is to assign the quantum numbers L and M to the various leptons as follows:

$L = +1$	$L = -1$	$M = +1$	$M = -1$
e^-	e^+	μ^-	μ^+
v_e	$\bar{v}_e$	v_μ	$\bar{v}_\mu$

All other particles have $L = 0$ and $M = 0$. Here are a few processes that involve leptons, showing that L and M have the same values before and after:

Pion decay
$$\pi^- \rightarrow \mu^- + \bar{v}_\mu$$
$$M = 0 \qquad +1 \quad -1$$

Muon decay
$$\mu^- \rightarrow e^- + v_\mu + \bar{v}_e$$
$$L = 0 \qquad +1 \quad 0 \quad -1$$
$$M = +1 \qquad 0 \quad +1 \quad 0$$

Pair production
$$\gamma \rightarrow e^- + e^+$$
$$L = 0 \quad +1 \quad -1$$

Other conservation principles, such as those of energy, linear momentum, angular momentum, and electric charge, can be traced to certain symmetries in the natural world. What symmetries the conservation of L and M are associated with, if any, remain unknown.

MESONS

The role of mesons as carriers of the strong interaction that binds nucleons together into nuclei has been examined earlier. There seems to be no conservation principle associated specifically with the number of mesons and antimesons involved in a process, although a number of other conservation principles must be obeyed.

BARYONS

Nucleons and heavier particles belong to the category of baryons. As in the case of the lepton families, a quantum number B can be assigned that is $+1$ for a baryon and -1 for an antibaryon, and it is found that in every process the value of B is conserved. An important example is the decay of the neutron, which also illustrates the conservation of L:

Neutron decay $\quad \mathrm{n}^0 \to \mathrm{p}^+ + \mathrm{e}^- + \bar{v}_\mathrm{e}$

$$L = \quad 0 \qquad 0 \qquad +1 \quad -1$$
$$B = +1 \quad +1 \qquad 0 \qquad 0$$

This is the only possible way in which the neutron can decay and still conserve both energy (since the proton is the only lighter baryon) and quantum number B. The same reasoning explains why the proton is a stable particle: there is no way in which it can decay without violating either conservation of energy or conservation of B. As in the case of L and M, the ultimate significance of B is not known.

It seems possible to consider the various baryons—not far from 100 in all, including the short-lived ones—as excited states of the proton, by analogy with the excited states of an atom. In the case of an atom, the quantum theory is able to explain precisely why there should be excited states of certain specific energies, and is also able to account for the

existence of different lifetimes against transitions between different pairs of states. No such detailed theory exists for the baryons, but there are a number of suggestive findings. For example, *all* baryons eventually decay into protons or antiprotons, just as all excited atoms eventually decay into atoms in their ground states. Other, more subtle evidence also supports the idea of the baryons as excited states of the proton. The same train of thought has led some physicists to consider the muon to be an excited state of the electron. Perhaps all elementary particles can be similarly regarded as excited states of just a few truly elementary particles . . . perhaps not.

A recent theory suggests that three particles called *quarks*, plus three antiquarks, are the ultimate constituents of all other elementary particles. The most novel thing about quarks is that two of them are predicted to have charges of $-\frac{1}{3}e$ and the third to have a charge of $+\frac{2}{3}e$. Quarks have the baryon number $B = \frac{1}{3}$; combinations of three quarks are supposed to make up the various baryons, and quark-antiquark pairs are supposed to make up the various mesons. For all the apparently bizarre properties of quarks, the theory behind them has proved quite successful in accounting for many aspects of elementary particle behavior. But no particles of fractional electric charge have ever been found, despite an intensive search. The physics of elementary particles remains on the frontier of science.

EXERCISES

1. Show that the laws of conservation of linear momentum and of energy make it impossible for a particle to decay into a single lighter particle.

2. Compare the distance traveled by a particle whose lifetime is 10^{-23} s which is moving at nearly the speed of light with nuclear diameters.

3. A negative muon collides with a proton, and a neutron plus another particle are created. What is the other particle?

4. Can a neutron decay into a positive muon and an electron?

5. One way to account for the decay of the π^0 meson into two photons, which are electromagnetic quanta, despite the absence of both charge and magnetic moment is to assume that the π^0 first becomes a nucleon-antinucleon pair which can interact electromagnetically to produce two photon whose energies are consistent with the π^0 mass.

   a) How long does the uncertainty principle permit such a nucleon-antinucleon pair to exist?

   b) Do any conservation principles other than conservation of energy prohibit the transformation of a π^0 meson into a nucleon-antinucleon pair?

6. According to the theory of the continuous creation of matter, the evolution of the universe can be traced to the spontaneous appearance of neutrons in free space. Which conservation laws would this process violate?

7. One proton strikes another, and the reaction

$$p + p \rightarrow n + p + \pi^+$$

takes place. What is the minimum energy the incident proton must have had?

8. Find the energy of each of the gamma-ray photons produced in the decay of the neutral pion. Why must their energies be the same?

9. Why does the Λ-hyperon not decay into a π^+- and a π^--meson?

APPENDIX A

USEFUL MATHEMATICS

The findings of physics about the natural world are almost always best expressed in mathematical statements. These statements summarize the results of experiment and observation in a compact way and often make it possible to draw general conclusions that are not evident from the raw data. In order to understand and use physics, then, a knowledge of basic mathematics is necessary. In this section those aspects of algebra employed in the text are reviewed briefly.

ALGEBRA

Algebra may be thought of as a generalized arithmetic in which symbols are used in place of specified numbers. The advantage of algebra is that with its help we can perform calculations without knowing in advance the numerical values of all the quantities involved. Sometimes algebra is no more than a help, perhaps by showing us how to simplify complex calculations, and sometimes it is the only way in which we can solve a problem.

The symbols of algebra are normally letters of the alphabet. If we have two quantities a and b and add them to give the sum c, we would write

$$a + b = c.$$

If we subtract b from a to give the difference d, we would write

$$a - b = d.$$

Multiplying a and b together to give e may be written $a \times b = e$, or more simply, as just

$$ab = e.$$

Whenever two algebraic quantities are written together with nothing between them, it is understood that they are to be multiplied together.

Dividing a by b to give the quotient f is usually written

$$\frac{a}{b} = f,$$

but it may sometimes be more convenient to write

$$a/b = f,$$

which has the same meaning.

Parentheses and brackets are used to show the order in which various operations are to be performed. Thus

$$\frac{(a + b)c}{d} - e = f$$

means that we are first to add a and b together, multiply their sum by c, then divide by d, and finally subtract e.

EQUATIONS

An *equation* is simply a statement that a certain quantity is equal to another one. Thus

$$7 + 2 = 9,$$

which contains only numbers, is an arithmetical equation, and

$$3x + 12 = 27,$$

which contains a symbol as well, is an algebraic equation. The symbols in an algebraic equation usually cannot have any arbitrary values if the equality is to hold. Finding the possible values of these symbols is called *solving* the equation. The *solution* of the latter equation above is

$$x = 5,$$

since only when x is 5 is it true that $3x + 12 = 27$.

In order to solve an equation a basic principle must be kept in mind.

Any operation performed on one side of an equation must be performed on the other.

An equation therefore remains valid when the same quantity, numerical or otherwise, is added to or subtracted from both sides, or when the same quantity is used to multiply or divide both sides. Other operations, for instance squaring or taking the square root, also do not alter the equality if the same thing is done to both sides. As a simple example, to solve $3x + 12 = 27$, we first subtract 12 from both sides:

$$3x + 12 - 12 = 27 - 12,$$
$$3x = 15.$$

To complete the solution we divide both sides by 3:

$$\frac{3x}{3} = \frac{15}{3},$$
$$x = 5.$$

To verify the correctness of a solution, we substitute it back in the original equation and see whether the equality is still true. Thus we can check that $x = 5$ by reducing the original algebraic equation to an arithmetical one:

$$3x + 12 = 27,$$
$$3 \cdot 5 + 12 = 27,$$
$$15 + 12 = 27,$$
$$27 = 27.$$

Problem. Solve the following equation for x:

$$\frac{3x - 42}{9} = 2(7 - x).$$

Solution. Our strategy is to assemble all the terms involving the unknown x on the left-hand side and all the purely numerical terms on the right-hand side. We begin by removing the parentheses on the right-hand side by actually multiplying $(7 - x)$ by 2. This yields

$$\frac{3x - 42}{9} = 2 \cdot 7 - 2x = 14 - 2x.$$

Next we multiply both sides by 9 to give

$$3x - 42 = 9(14 - 2x) = 126 - 18x.$$

We now add $42 + 18x$ to both sides:

$$3x - 42 + 42 + 18x = 126 - 18x + 42 + 18x,$$
$$3x + 18x = 126 + 42,$$
$$21x = 168.$$

The final step is to divide both sides by 21:

$$\frac{21x}{21} = \frac{168}{21},$$

$$x = 8.$$

To check this solution we substitute $x = 8$ in the original equation:

$$\frac{3 \cdot 8 - 42}{9} = 2(7 - 8),$$

$$\frac{24 - 42}{9} = 14 - 16,$$

$$\frac{-18}{9} = -2,$$

$$-2 = -2.$$

Two simple rules follow directly from the principle stated above. The first is,

Any term on one side of an equation may be transposed to the other side by changing its sign.

To verify this rule, we subtract b from each side of the equation

$$a + b = c$$

to obtain

$$a + b - b = c - b,$$
$$a = c - b.$$

We see that b has disappeared from the left-hand side and $-b$ is now on the right-hand side.

The second rule is,

A quantity which multiplies one side of an equation may be transposed so as to divide the other side, and vice versa.

To verify this rule, we divide both sides of the equation

$$ab = c$$

by b. The result is

$$\frac{ab}{b} = \frac{c}{b},$$

$$a = \frac{c}{b}.$$

We see that b, a multiplier on the left-hand side, is now a divisor on the right-hand side.

Problem. Solve the following equation for x:

$$4(x - 3) = 7.$$

Solution. The above rules make this problem a simple one to deal with. We proceed as follows:

$$4(x - 3) = 7,$$
$$x - 3 = \tfrac{7}{4},$$
$$x = \tfrac{7}{4} + 3 = 1.75 + 3 = 4.75.$$

EXPONENTS

It is often necessary to multiply a quantity by itself a number of times. This process is indicated by a superscript number called the exponent, according to the following scheme:

$$A = A^1,$$
$$A \times A = A^2,$$
$$A \times A \times A = A^3,$$
$$A \times A \times A \times A = A^4,$$
$$A \times A \times A \times A \times A = A^5.$$

We read A^2 as "A squared" because it is the area of a square of length A on a side; similarly A^3 is called "A cubed" because it is the volume of a cube each of whose sides is A long. More generally we speak of A^n as "A to the nth power." Thus A^5 is read as "A to the fifth power."

When we multiply a quantity raised to some particular power (say A^n) by the same quantity raised to another power (say A^m), the result is that quantity raised to a power equal to the sum of the original exponents. That is,

$$A^n A^m = A^{(n+m)}.$$

For example,

$$A^2 A^5 = A^7,$$

which we can verify directly by writing out the terms:

$$\underbrace{(A \times A)}_{A^2} \times \underbrace{(A \times A \times A \times A \times A)}_{A^5} =$$

$$\underbrace{A \times A \times A \times A \times A \times A \times A}_{A^7}.$$

From the above result we see that when a quantity raised to a particular power (say A^n)

is to be multiplied by itself a total of m times, we have

$$(A^n)^m = A^{nm}.$$

For example,

$$(A^2)^3 = A^6,$$

since

$$(A^2)^3 = A^2 \times A^2 \times A^2 = A^{(2+2+2)} = A^6.$$

Reciprocal quantities are expressed in a similar way with the addition of a minus sign in the exponent, as follows:

$$\frac{1}{A} = A^{-1}, \qquad \frac{1}{A^2} = A^{-2},$$

$$\frac{1}{A^3} = A^{-3}, \qquad \frac{1}{A^4} = A^{-4}.$$

Exactly the same rules as before are used in combining quantities raised to negative powers with one another and with some quantity raised to a positive power. Thus

$$A^5 A^{-2} = A^{(5-2)} = A^3,$$
$$(A^{-1})^{-2} = A^{-1(-2)} = A^2,$$
$$(A^3)^{-4} = A^{-4 \times 3} = A^{-12},$$
$$A A^{-7} = A^{(1-7)} = A^{-6}.$$

It is important to remember that any quantity raised to the zeroth power, say A^0, is equal to 1. Hence

$$A^2 A^{-2} = A^{(2-2)} = A^0 = 1.$$

This is more easily seen if we write A^{-2} as $1/A^2$:

$$A^2 A^{-2} = A^2 \times \frac{1}{A^2} = \frac{A^2}{A^2} = 1.$$

Fractional exponents are frequently useful. The simplest case is that of the *square root* of a quantity A, commonly written $\sqrt{A}$, which when multiplied by itself equals the quantity:

$$\sqrt{A} \times \sqrt{A} = A.$$

Using exponentials we see that, because

$$(A^{1/2})^2 = A^{2 \times (1/2)} = A^1 = A,$$

we can express square roots by the exponent $\frac{1}{2}$:

$$\sqrt{A} = A^{1/2}.$$

Other roots may be expressed similarly. The *cube root* of a quantity A, written $\sqrt[3]{A}$, when multiplied by itself twice equals A. That is,

$$\sqrt[3]{A} \times \sqrt[3]{A} \times \sqrt[3]{A} = A,$$

which may be more conveniently written

$$(A^{1/3})^3 = A,$$

where $\sqrt[3]{A} = A^{1/3}$.

In general the nth root of a quantity, $\sqrt[n]{A}$, may be written $A^{1/n}$, which is a more convenient form for most purposes. Some examples may be helpful:

$$(A^4)^{1/2} = A^{(1/2) \times 4} = A^2,$$
$$(A^{1/4})^{-7} = A^{-7 \times (1/4)} = A^{-7/4},$$
$$(A^3)^{-1/3} = A^{-(1/3) \times 3} = A^{-1},$$
$$(A^{1/2})^{1/2} = A^{(1/2) \times (1/2)} = A^{1/4}.$$

EQUATIONS WITH POWERS AND ROOTS

An equation that involves powers and/or roots is subject to the same basic principle that governs the manipulation of simpler equations: whatever is done to one side must be done to the other. Hence the following rules:

An equation remains valid when both sides are raised to the same power, that is, when each side is multiplied by itself the same number of times as the other side.

An equation remains valid when the same root is taken of both sides.

Problem. The mass m of an object whose speed is v is given by

$$m = \frac{m_0}{\sqrt{1 - v^2/c^2}}$$

where m_0 is the object's mass when it is at rest and c is the speed of light. Solve this equation for v.

Solution. The first step is to multiply both sides by $\sqrt{1 - v^2/c^2}$ to give

$$m\sqrt{1 - v^2/c^2} = m_0.$$

Next the square root is removed by squaring both sides, which yields

$$m^2(1 - v^2/c^2) = m_0^2.$$

Now we proceed in the usual way to find v^2:

$$1 - \frac{v^2}{c^2} = \frac{m_0^2}{m^2},$$

$$\frac{v^2}{c^2} = 1 - \frac{m_0^2}{m^2},$$

$$v^2 = c^2(1 - m_0^2/m^2).$$

The final step is to take the square root of both sides:

$$v = c\sqrt{1 - m_0^2/m^2}.$$

APPENDIX B

POWERS OF TEN

In physics, as in all the other sciences, numbers are usually expressed in powers-of-ten notation instead of in ordinary decimal notation. Powers-of-ten notation has important advantages, among them that large and small numbers can be written in more compact form, calculations are much easier to make, and the degree of accuracy with which a number is known is immediately apparent.

POWERS OF TEN

Very small and very large numbers are common in physics. For example, the mass of an electron is 0.000,000,000,000,000,000,000,000,-000,000,910,9 kg, and the mass of the earth is 5,983,000,000,000,000,000,000,000 kg. Such numbers in ordinary decimal form are clumsy to write and to make calculations with, and it is hard to appreciate their precise magnitudes because of the sea of zeros.

A better method for expressing numbers makes use of powers-of-ten notation. This method is based upon the fact that all numbers may be represented by a number between 1 and 10 multiplied by a power of 10. In powers-of-ten notation the mass of an electron is written simply as 9.109×10^{-31} kg and the mass of the earth is written as 5.983×10^{24} kg.

$10^{-10} = 0.000,000,000,1$	$10^0 = 1$
$10^{-9} = 0.000,000,001$	$10^1 = 10$
$10^{-8} = 0.000,000,01$	$10^2 = 100$
$10^{-7} = 0.000,000,1$	$10^3 = 1000$
$10^{-6} = 0.000,001$	$10^4 = 10,000$
$10^{-5} = 0.000,01$	$10^5 = 100,000$
$10^{-4} = 0.000,1$	$10^6 = 1,000,000$
$10^{-3} = 0.001$	$10^7 = 10,000,000$
$10^{-2} = 0.01$	$10^8 = 100,000,000$
$10^{-1} = 0.1$	$10^9 = 1,000,000,000$
$10^0 = 1$	$10^{10} = 10,000,000,000$

The adjacent table contains powers of 10 from 10^{-10} to 10^{10}. Evidently positive powers of 10 (which cover numbers greater than 1) follow this pattern:

$10^0 = 1$ = 1 with decimal point moved 0 places
$10^1 = 10$ = 1 with decimal point moved 1 place to the right
$10^2 = 100$ = 1 with decimal point moved 2 places to the right
$10^3 = 1,000$ = 1 with decimal point moved 3 places to the right
$10^4 = 10,000$ = 1 with decimal point moved 4 places to the right
$10^5 = 100,000$ = 1 with decimal point moved 5 places to the right
$10^6 = 1,000,000$ = 1 with decimal point moved 6 places to the right

and so on. The exponent of the 10 indicates how many places the decimal point is moved *to the right* from 1.000 . . . .

A similar pattern is followed by negative powers of 10, whose values always lie between 0 and 1:

$10^0 = 1$ = 1 with decimal point moved 0 places
$10^{-1} = 0.1$ = 1 with decimal point moved 1 place to the left
$10^{-2} = 0.01$ = 1 with decimal point moved 2 places to the left
$10^{-3} = 0.001$ = 1 with decimal point moved 3 places to the left
$10^{-4} = 0.000,1$ = 1 with decimal point moved 4 places to the left
$10^{-5} = 0.000,01$ = 1 with decimal point moved 5 places to the left
$10^{-6} = 0.000,001$ = 1 with decimal point moved 6 places to the left

and so on. The exponent of the 10 now indicates how many places the decimal point is moved *to the left* from 1.

Here are a few examples of powers-of-ten notation:

$$600 = 6 \times 100 = 6 \times 10^2$$
$$7940 = 7.94 \times 1000 = 7.94 \times 10^3$$
$$93{,}000{,}000 = 9.3 \times 10{,}000{,}000 = 9.3 \times 10^7$$
$$0.023 = 2.3 \times 0.01 = 2.3 \times 10^{-2}$$
$$0.000{,}035 = 3.5 \times 0.000{,}01 = 3.5 \times 10^{-5}$$

USING POWERS OF TEN

Let us see how to make calculations using numbers written in powers-of-ten notation.

To add or subtract numbers written in powers-of-ten notation, they must be expressed in terms of the *same* power of ten.

$$7 \times 10^4 + 2 \times 10^5 = 0.7 \times 10^5 + 2 \times 10^5 = 2.7 \times 10^5$$

$$5 \times 10^{-2} + 3 \times 10^{-4} = 5 \times 10^{-2} + 0.03 \times 10^{-2} = 5.03 \times 10^{-2}$$

$$8 \times 10^{-3} - 7 \times 10^{-4} = 8 \times 10^{-3} - 0.7 \times 10^{-3} = 7.3 \times 10^{-3}$$

$$4 \times 10^5 - 1 \times 10^6 = 4 \times 10^5 - 10 \times 10^5 = -6 \times 10^5$$

To multiply powers of ten together, add their exponents:

$$10^n \times 10^m = 10^{n+m}.$$

Be sure to take the sign of each exponent into account.

$$10^2 \times 10^3 = 10^{2+3} = 10^5$$
$$10^7 \times 10^{-3} = 10^{7-3} = 10^4$$
$$10^{-2} \times 10^{-4} = 10^{-2-4} = 10^{-6}$$

To multiply numbers written in powers-of-ten notation, multiply the decimal parts of the numbers together and add the exponents to find the power of ten of the product:

$$A \times 10^n \times B \times 10^m = AB \times 10^{n+m}.$$

If necessary, rewrite the result so the decimal part is a number between 1 and 10.

$$3 \times 10^2 \times 2 \times 10^5 = 3 \times 2 \times 10^{2+5} = 6 \times 10^7$$

$$8 \times 10^{-5} \times 3 \times 10^7 = 8 \times 3 \times 10^{-5+7} = 24 \times 10^2 = 2.4 \times 10^3$$

$$1.3 \times 10^{-3} \times 4 \times 10^{-5} = 1.3 \times 4 \times 10^{-3-5} = 5.2 \times 10^{-8}$$

$$-9 \times 10^{17} \times 6 \times 10^{-18} = -9 \times 6 \times 10^{17-18} = -54 \times 10^{-1} = -5.4$$

To divide one power of ten by another, subtract the exponent of the denominator from the exponent of the numerator:

$$\frac{10^n}{10^m} = 10^{n-m}.$$

Again, be sure to take the sign of each exponent into account.

$$\frac{10^5}{10^3} = 10^{5-3} = 10^2$$

$$\frac{10^{-2}}{10^4} = 10^{-2-4} = 10^{-6}$$

$$\frac{10^{-3}}{10^{-7}} = 10^{-3-(-7)} = 10^{-3+7} = 10^4$$

To divide a number written in powers-of-ten notation by another number written that way, divide the decimal parts of the numbers in the usual way and use the above rule to find the exponent of the power of ten of the quotient:

$$\frac{A \times 10^n}{B \times 10^m} = \frac{A}{B} \times 10^{n-m}.$$

If necessary, rewrite the result so the decimal part is a number between 1 and 10.

$$\frac{6 \times 10^5}{3 \times 10^2} = \frac{6}{3} \times 10^{5-2} = 2 \times 10^3$$

$$\frac{2 \times 10^{-7}}{8 \times 10^4} = \frac{2}{8} \times 10^{-7-4} = \frac{1}{4} \times 10^{-11}$$
$$= 0.25 \times 10^{-11}$$
$$= 2.5 \times 10^{-12}$$

$$\frac{-7 \times 10^5}{10^{-2}} = -7 \times 10^{5-(-2)}$$
$$= -7 \times 10^{5+2} = -7 \times 10^7$$

$$\frac{5 \times 10^{-2}}{-2 \times 10^{-9}} = -\frac{5}{2} \times 10^{-2(-9)}$$
$$= -2.5 \times 10^{-2+9}$$
$$= -2.5 \times 10^7.$$

To find the reciprocal of a power of ten, change the sign of the exponent:

$$\frac{1}{10^n} = 10^{-n},$$

$$\frac{1}{10^{-m}} = 10^m.$$

$$\frac{1}{10^5} = 10^{-5}$$

$$\frac{1}{10^{-3}} = 10^3.$$

Hence the prescription for finding the reciprocal of a number written in powers-of-ten notation is

$$\frac{1}{A \times 10^n} = \frac{1}{A} \times 10^{-n}$$

$$\frac{1}{2 \times 10^3} = \frac{1}{2} \times 10^{-3} = 0.5 \times 10^{-3} = 5 \times 10^{-4}$$

$$\frac{1}{4 \times 10^{-8}} = \frac{1}{4} \times 10^8 = 0.25 \times 10^8 = 2.5 \times 10^7.$$

The powers-of-ten method of writing large and small numbers makes arithmetic involving such numbers relatively easy to carry out. Here is a calculation that would be very tedious if each number were kept in decimal form.

$$\frac{3800 \times 0.0054 \times 0.000,001}{430,000,000 \times 73}$$

$$= \frac{3.8 \times 10^3 \times 5.4 \times 10^{-3} \times 10^{-6}}{4.3 \times 10^8 \times 7.3 \times 10^1}$$

$$= \frac{3.8 \times 5.4}{4.3 \times 7.3} \times \frac{10^3 \times 10^{-3} \times 10^{-6}}{10^8 \times 10^1}$$

$$= 0.65 \times 10^{(3-3-6-8-1)}$$

$$= 0.65 \times 10^{-15}$$

$$= 6.5 \times 10^{-16}.$$

POWERS AND ROOTS

To square a power of ten, multiply the exponent by 2; to cube a power of ten, multiply the exponent by 3:

$$(10^n)^2 = 10^{2n},$$

$$(10^n)^3 = 10^{3n}.$$

In general, to raise a power of ten to the mth power, multiply the exponent by m:

$$(10^n)^m = 10^{m \times n}.$$

Be sure to take the sign of each exponent into account.

$(10^3)^2 = 10^{2 \times 3} = 10^6$
$(10^{-2})^5 = 10^{5 \times -2} = 10^{-10}$
$(10^{-4})^{-2} = 10^{-2 \times -4} = 10^8.$

To raise a number written in powers-of-ten notation to the mth power, multiply the decimal part of the number by itself m times and multiply the exponent of the power of ten by m:

$$(A \times 10^n)^m = A^m \times 10^{m \times n}.$$

If necessary, rewrite the result so the decimal part is a number between 1 and 10. A table of squares and cubes of numbers from 1 to 100 is given in Appendix E.

$$(2 \times 10^5)^2 = 2^2 \times 10^{2 \times 5} = 4 \times 10^{10}$$

$$(3 \times 10^{-3})^3 = 3^3 \times 10^{3 \times -3} = 27 \times 10^{-9}$$
$$= 2.7 \times 10^{-8}$$

$$(5 \times 10^{-2})^{-4} = 5^{-4} \times 10^{-4 \times -2} = \frac{1}{5^4} \times 10^8$$

$$= \frac{1}{625} \times 10^8 = 0.0016 \times 10^8 = 1.6 \times 10^5.$$

A fractional exponent signifies a root. The *square root* of a quantity A, written $\sqrt{A}$, when multiplied by itself equals the quantity:

$$\sqrt{A} \times \sqrt{A} = A.$$

Because

$$(A^{\frac{1}{2}})^2 = A^{2 \times \frac{1}{2}} = A^1 = A,$$

we see that we can express a square root by using the exponent $\frac{1}{2}$:

$$\sqrt{A} = A^{\frac{1}{2}}.$$

$$\sqrt{A^4} = (A^4)^{\frac{1}{2}} = A^{\frac{1}{2} \times 4} = A^2$$

$$(\sqrt{A})^4 = (A^{\frac{1}{2}})^4 = A^{4 \times \frac{1}{2}} = A^2$$

$$\sqrt{\sqrt{A}} = (A^{\frac{1}{2}})^{\frac{1}{2}} = A^{\frac{1}{2} \times \frac{1}{2}} = A^{\frac{1}{4}}.$$

The *cube root* of A, written $\sqrt[3]{A}$, multiplied by itself twice equals A:

$$\sqrt[3]{A} \times \sqrt[3]{A} \times \sqrt[3]{A} = A.$$

Because

$$(A^{1/3})^3 = A^{3 \times 1/3} = A^1 = A,$$

we have

$$\sqrt[3]{A} = A^{1/3}.$$

In general, the mth root of a quantity may be written

$$\sqrt[m]{A} = A^{1/m}.$$

To take the mth root of a power of ten, divide the exponent by m:

$$\sqrt[m]{10^n} = (10^n)^{1/m} = 10^{n/m}.$$

$$\sqrt{10^4} = (10^4)^{\frac{1}{2}} = 10^{4/2} = 10^2$$

$$\sqrt[3]{10^9} = (10^9)^{1/3} = 10^{9/3} = 10^3$$

$$\sqrt[3]{10^{-9}} = (10^{-9})^{1/3} = 10^{-9/3} = 10^{-3}.$$

In powers-of-ten notation, the exponent of the 10 must be an integer. Hence in taking the mth root of a power of ten, the exponent should be an integral multiple of m. Instead of, for example,

$$\sqrt{10^5} = (10^5)^{\frac{1}{2}} = 10^{2\frac{1}{2}}$$

which, while correct, is hardly useful, we would write

$$\sqrt{10^5} = \sqrt{10^1 \times 10^4}$$
$$= \sqrt{10} \times \sqrt{10^4}$$
$$= 3.16 \times 10^2.$$

$$\sqrt{10^{-3}} = \sqrt{10^1 \times 10^{-4}} = \sqrt{10} \times \sqrt{10^{-4}}$$
$$= 3.16 \times 10^{-2}$$

$$\sqrt[3]{10^8} = \sqrt[3]{10^2 \times 10^6} = \sqrt[3]{100} \times \sqrt[3]{10^6}$$
$$= 4.64 \times 10^2.$$

To take the mth root of a number expressed in powers-of-ten notation, first write the number so the exponent of the 10 is an integral multiple of m. Then take the mth root of the decimal part of the number and divide

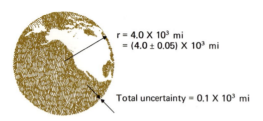

r = 4.0 X 10³ mi
= (4.0 ± 0.05) X 10³ mi

Total uncertainty = 0.1 X 10³ mi

r = 3.96 X 10³ mi
= (3.96 ± 0.005) X 10³ mi

Total uncertainty = 0.01 X 10³ mi

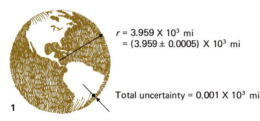

r = 3.959 X 10³ mi
= (3.959 ± 0.0005) X 10³ mi

Total uncertainty = 0.001 X 10³ mi

1

SIGNIFICANT FIGURES

An advantage of powers-of-ten notation is that it gives no false impression of the degree of accuracy with which a number is stated. For instance, the average radius of the earth is 3959 mi, but it is often taken as 4000 mi for convenience in making rough calculations. To indicate the approximate character of the latter figure, all we have to do is write

$$r = 4.0 \times 10^3 \text{ mi.}$$

Now how large the number is and how accurate it is are both clear. The accurately known digits, plus one uncertain digit, are called *significant figures;* in the above case r has two significant figures, 4 and 0. If we require more accuracy, we would write

$$r = 3.96 \times 10^3 \text{ mi,}$$

which contains three significant figures, or

$$r = 3.959 \times 10^3 \text{ mi,}$$

which contains four significant figures. **(1)**

Sometimes one or more zeros in a number are significant figures, and it is proper to retain them when expressing the number in exponential notation. Thus the number 1 represents any number between 0.5 and 1.5 expressed in one significant figure, whereas 1.0 narrows the range to any number between 0.95 and 1.05 and 1.00 further narrows it to any number between 0.995 and 1.005:

$$1 = 1 \pm 0.5,$$
$$1.0 = 1.0 \pm 0.05,$$
$$1.00 = 1.00 \pm 0.005. \textbf{ (2)}$$

When quantities are combined arithmetically, the result is no more accurate than the quantity with the largest uncertainty. Suppose a man weighing 162 lb picks up an 0.34-lb apple. The total weight of man + apple is still 162 lb

the exponent by m to find the power of ten of the result:

$$(A \times 10^n)^{1/m} = \sqrt[m]{A} \times 10^{n/m}.$$

$$\sqrt{9 \times 10^4} = \sqrt{9} \times 10^{4/2} = 3 \times 10^2$$

$$\sqrt{9 \times 10^{-6}} = \sqrt{9} \times 10^{-6/2} = 3 \times 10^{-3}$$

$$\sqrt{4 \times 10^7} = \sqrt{40 \times 10^6} = \sqrt{40} \times 10^{6/2}$$
$$= 6.32 \times 10^3$$

$$\sqrt[3]{3 \times 10^{-5}} = \sqrt[3]{30 \times 10^{-6}} = \sqrt[3]{30} \times 10^{-6/3}$$
$$= 3.11 \times 10^{-2}$$

A table of square and cube roots of numbers from 1 to 100 is given in Appendix E.

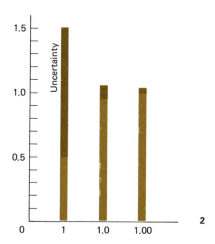

2

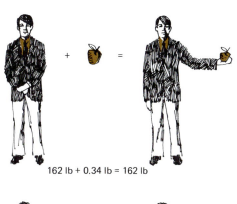

162 lb + 0.34 lb = 162 lb

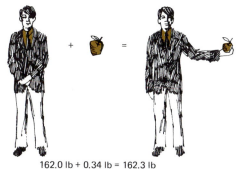

162.0 lb + 0.34 lb = 162.3 lb

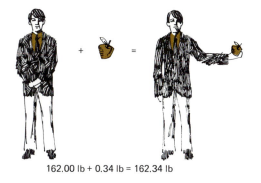

162.00 lb + 0.34 lb = 162.34 lb

3

because all we know of the man's weight is that it is somewhere between 161.5 and 162.5 lb, which means a total uncertainty greater than the apple's weight. If the man's weight is instead 162.0 lb, he and the apple together weigh 162.3 lb; if his weight is 162.00 lb, he and the apple together weigh 162.34 lb. (**3**)

In multiplication and division, the result is entitled to only as many significant figures as there are in the least accurately known quantity. For example, if we divide 1.4×10^5 by 6.70×10^3, we are not justified in writing

$$\frac{1.4 \times 10^5}{6.70 \times 10^3} = 20.89552 \ldots$$

We may properly retain only two significant figures, corresponding to the two significant figures in the numerator, and so the correct answer is just 21.

In a calculation with several steps, however, it is a good idea to keep an extra digit in the intermediate steps and to round off the result only at the end.

$$\frac{5.7 \times 10^4}{3.3 \times 10^{-2}} + \sqrt{1.8 \times 10^{12}}$$
$$= 1.73 \times 10^6 + 1.34 \times 10^6$$
$$= 3.07 \times 10^6$$
$$= 3.1 \times 10^6.$$

The rules for rounding off a number which has more digits than are proper are straightforward. If the digit to be dropped is less than 5, the digit before it stays the same:

$$4.364 \rightarrow 4.36.$$

If the digit to be dropped is 5 or more, the digit before it is increased by 1:

$$2.718 \rightarrow 2.72.$$

If more than one digit is to be dropped, rounding off is done only once. For instance, if only three digits are significant,

$$3.4148 \rightarrow 3.41$$

and not

$$3.4148 \rightarrow 3.415 \rightarrow 3.42.$$

EXERCISES*

1. Express the following numbers in decimal notation:

$2 \times 10^5 =$
$8 \times 10^{-2} =$
$7.819 \times 10^2 =$
$4.51 \times 10^8 =$
$1.003 \times 10^{-6} =$
$10^{-10} =$
$9.56 \times 10^{-5} =$

2. Express the following numbers in powers-of-ten notation:

$70 = \quad\quad \times 10$
$0.14 = \quad\quad \times 10$
$3.81 = \quad\quad \times 10$
$8400 = \quad\quad \times 10$
$1,000,000 = \quad\quad \times 10$
$0.007,890 = \quad\quad \times 10$
$351,600 = \quad\quad \times 10$

3. Evaluate the following:

$(4 \times 10^2)^2 =$
$(2 \times 10^{-6})^3 =$
$(2 \times 10^2)^{-2} =$

4. Evaluate the following:

$\sqrt{4 \times 10^{-8}} =$

$\sqrt{4 \times 10^{12}} =$

$\sqrt{1.6 \times 10^5} =$

5. Evaluate the following:

$$\frac{6 \times 14}{7 \times 10^{-6}} =$$

$$\frac{3 \times 10^4 \times 5 \times 10^{-12}}{10^3} =$$

$$\frac{9 \times 10^{12}}{9 \times 10^{-12}} =$$

$$\frac{8 \times 10^{10} \times 3}{6 \times 10^{-4}} =$$

$$\frac{10^{-3}}{5 \times 10^4 \times 2 \times 10^2} =$$

$$\frac{5 \times 10^5 \times 2 \times 10^{-18}}{4 \times 10^4} =$$

*Answers on p. 399.

APPENDIX C

BRITISH UNITS

In the British system of units, the unit of length is the *foot* (ft), the unit of time is the *second* (s), and the unit of mass is the *slug*. The unit of force is the *pound* (lb). By definition, a body whose mass is 1 slug experiences an acceleration of 1 ft/s when a force of 1 lb acts on it.

The slug is an unfamiliar unit because in everyday life *weights* rather than *masses* are specified in the British system: we go shopping for 10 lb of apples, not $\frac{1}{3}$ slug of apples. In the metric system, on the other hand, masses are normally specified: European grocery scales are calibrated in kilograms, not in newtons.

System of units	Unit of mass	Unit of weight	Acceleration of gravity g	To find mass m given weight w	To find weight w given mass m
Metric	kilogram (kg)	newton (N)	9.8 m/s²	$m\ (kg) = \dfrac{w\ (N)}{9.8\ m/s^2}$	$w\ (N) = m\ (kg) \times 9.8\ m/s^2$
British	slug	pound (lb)	32 ft/s²	$m\ (slugs) = \dfrac{w\ (lb)}{32\ ft/s^2}$	$w\ (lb) = m\ (slugs) \times 32\ ft/s^2$
Conversion of units:		1 slug = 14.6 kg 1 kg = 0.0685 slug		1 N = 0.225 lb 1 lb = 4.45 N	

In order to convert a weight in pounds to a mass in slugs we make use of

$$m\ (slugs) = \frac{w\ (lb)}{g\ (ft/s^2)}.$$

Since g, the acceleration of gravity at the earth's surface, has the value 32 ft/s^2 in British units,

$$m \text{ (slugs)} = \frac{w \text{ (lb)}}{32 \text{ ft/s}^2}.$$

The *weight* of a 1-slug mass is 32 lb, and the *mass* of a 1-lb weight is $\frac{1}{32}$ slug.

Problem. A body that weighs 8 lb is acted on by a net force of 15 lb. What is its acceleration?

Solution. First we find the mass of the body, which is

$$m = \frac{w}{g} = \frac{8 \text{ lb}}{32 \text{ ft/s}^2} = 0.25 \text{ slug}.$$

Now we employ the second law of motion to find that

$$a = \frac{F}{m} = \frac{15 \text{ lb}}{0.25 \text{ slug}} = 60 \text{ ft/s}^2.$$

In the British system of units, the unit of work is the *foot-pound*, abbreviated ft-lb. One ft-lb is equal to the work done by a force of 1 lb acting through a distance of 1 ft. The ft-lb is somewhat larger than the joule:

$$1 \text{ ft-lb} = 1.36 \text{ J},$$

$$1 \text{ J} = 0.738 \text{ ft-lb}.$$

The same formulas of mechanics that hold when metric units are employed hold when British units are employed. Scientists everywhere use metric units exclusively, and most of the world uses the metric system in everyday life and engineering practice as well. A gradual shift to metric units is currently taking place in those English-speaking countries that still hold on to British units, so that in the foreseeable future, conceivably in as little as a generation, the entire world will be using a single system of units. All electrical units are metric already.

A vast number of other systems of units have been used at various times in the past, and some still persist. For instance, there is another British system in which the pound is taken to be the unit of mass and another unit, the *poundal*, is the unit of force. A number of scientists continue to prefer the centimeter and gram to the meter and kilogram; in this system the units of force and energy are the *dyne* and *erg*, respectively. Such scientists often use a complicated set of electrical units in which there are two different units of charge. All these traditional systems are dying out rapidly, however, and the meter-kilogram-second-ampere system, which has received formal international acceptance, will in time be the only survivor.

1. 200,000
   0.08
   781.9
   451,000,000
   0.000,001,003
   0.000,000,000,1
   0.000,0956

2. 7×10^1
   1.4×10^{-1}
   3.81
   8.4×10^3
   1×10^6
   7.890×10^{-3}
   3.516×10^5

3. 1.6×10^5
   8×10^{-18}
   2.5×10^{-5}

4. 2×10^{-4}
   2×10^6
   4×10^2

5. 1.2×10^7
   1.5×10^{-10}
   10^{24}
   4×10^{14}
   10^{-10}
   2.5×10^{-17}

APPENDIX D

THE ELEMENTS

Atomic number	Element	Symbol	Atomic mass*
1	Hydrogen	H	1.008
2	Helium	He	4.003
3	Lithium	Li	6.939
4	Beryllium	Be	9.012
5	Boron	B	10.81
6	Carbon	C	12.01
7	Nitrogen	N	14.01
8	Oxygen	O	16.00
9	Fluorine	F	19.00
10	Neon	Ne	20.18
11	Sodium	Na	22.99
12	Magnesium	Mg	24.31
13	Aluminum	Al	26.98
14	Silicon	Si	28.09
15	Phosphorus	P	30.97
16	Sulfur	S	32.06
17	Chlorine	Cl	35.45
18	Argon	Ar	39.95
19	Potassium	K	39.10
20	Calcium	Ca	40.08
21	Scandium	Sc	44.96
22	Titanium	Ti	47.90
23	Vanadium	V	50.94
24	Chromium	Cr	52.00
25	Manganese	Mn	54.94
26	Iron	Fe	55.85
27	Cobalt	Co	58.93
28	Nickel	Ni	58.71
29	Copper	Cu	63.54
30	Zinc	Zn	65.37
31	Gallium	Ga	69.72
32	Germanium	Ge	72.59
33	Arsenic	As	74.92
34	Selenium	Se	78.96
35	Bromine	Br	79.91

Atomic number	Element	Symbol	Atomic mass*
36	Krypton	Kr	83.80
37	Rubidium	Rb	85.47
38	Strontium	Sr	87.62
39	Yttrium	Y	88.91
40	Zirconium	Zr	91.22
41	Niobium	Nb	92.91
42	Molybdenum	Mo	95.94
43	Technetium	Tc	(99)
44	Ruthenium	Ru	101.0
45	Rhodium	Rh	102.9
46	Palladium	Pd	106.4
47	Silver	Ag	107.9
48	Cadmium	Cd	112.4
49	Indium	In	114.8
50	Tin	Sn	118.7
51	Antimony	Sb	121.8
52	Tellurium	Te	127.6
53	Iodine	I	126.9
54	Xenon	Xe	131.3
55	Cesium	Cs	132.9
56	Barium	Ba	137.4
57	Lanthanum	La	138.9
58	Cerium	Ce	140.1
59	Praseodymium	Pr	140.9
60	Neodymium	Nd	144.2
61	Promethium	Pm	(147)
62	Samarium	Sm	150.4
63	Europium	Eu	152.0
64	Gadolinium	Gd	157.3
65	Terbium	Tb	158.9
66	Dysprosium	Dy	162.5
67	Holmium	Ho	164.9
68	Erbium	Er	167.3
69	Thulium	Tm	168.9
70	Ytterbium	Yb	173.0

Atomic number	Element	Symbol	Atomic mass*
71	Lutetium	Lu	175.0
72	Hafnium	Hf	178.5
73	Tantalum	Ta	181.0
74	Tungsten	W	183.9
75	Rhenium	Re	186.2
76	Osmium	Os	190.2
77	Iridium	Ir	192.2
78	Platinum	Pt	195.1
79	Gold	Au	197.0
80	Mercury	Hg	200.6
81	Thallium	Tl	204.4
82	Lead	Pb	207.2
83	Bismuth	Bi	209.0
84	Polonium	Po	(209)
85	Astatine	At	(210)
86	Radon	Rn	(222)
87	Francium	Fr	(223)
88	Radium	Ra	(226)
89	Actinium	Ac	(227)
90	Thorium	Th	232.0
91	Protactinium	Pa	(231)
92	Uranium	U	238.0
93	Neptunium	Np	(237)
94	Plutonium	Pu	(244)
95	Americium	Am	(243)
96	Curium	Am	(247)
97	Berkelium	Bk	(247)
98	Californium	Cf	(251)
99	Einsteinium	Es	(254)
100	Fermium	Fm	(257)
101	Mendelevium	Md	(256)
102	Nobelium	No	(254)
103	Lawrencium	Lr	(257)
104			
105			

*The unit of mass is the amu. When all the isotopes of an element are radioactive, the mass number of the most stable isotope is given in parentheses.

APPENDIX E

FUNCTIONS OF NUMBERS

n	n^2	n^3	$\sqrt{n}$	$\sqrt[3]{n}$	$1\,000/n$
1	1	1	1.00	1.000	1 000.000
2	4	8	1.414	1.260	500.000
3	9	27	1.732	1.442	333.333
4	16	64	2.000	1.587	250.000
5	25	125	2.236	1.710	200.000
6	36	216	2.449	1.817	166.667
7	49	343	2.646	1.913	142.857
8	64	512	2.828	2.000	125.000
9	81	729	3.000	2.080	111.111
10	100	1 000	3.162	2.154	100.000
11	121	1 331	3.317	2.224	90.909
12	144	1 728	3.464	2.289	83.333
13	169	2 197	3.606	2.351	76.923
14	196	2 744	3.742	2.410	71.429
15	225	3 375	3.873	2.466	66.667
16	256	4 096	4.000	2.520	62.500
17	289	4 913	4.123	2.571	58.824
18	324	5 832	4.243	2.621	55.556
19	361	6 859	4.359	2.668	52.632
20	400	8 000	4.472	2.714	50.000
21	441	9 261	4.583	2.759	47.619
22	484	10 648	4.690	2.802	45.455
23	529	12 167	4.796	2.843	43.478
24	576	13 824	4.899	2.885	41.667
25	625	15 625	5.000	2.924	40.000
26	676	17 576	5.099	2.962	38.462
27	729	19 683	5.196	3.000	37.037
28	784	21 952	5.292	3.037	35.714
29	841	24 389	5.385	3.072	34.483

n	n^2	n^3	$\sqrt{n}$	$\sqrt[3]{n}$	$1\,000/n$
30	900	27 000	5.477	3.107	33.333
31	961	29 791	5.568	3.141	32.258
32	1 024	32 768	5.657	3.175	31.250
33	1 089	35 937	5.745	3.208	30.303
34	1 156	39 304	5.831	3.240	29.412
35	1 225	42 875	5.916	3.271	28.571
36	1 296	46 656	6.000	3.302	27.778
37	1 369	50 653	6.083	3.332	27.027
38	1 444	54 872	6.164	3.362	26.316
39	1 521	59 319	6.245	3.391	25.641
40	1 600	64 000	6.325	3.420	25.000
41	1 681	68 921	6.403	3.448	24.390
42	1 764	74 088	6.481	3.476	23.810
43	1 849	79 507	6.557	3.503	23.256
44	1 936	85 184	6.633	3.530	22.727
45	2 025	91 125	6.708	3.557	22.222
46	2 116	97 336	6.782	3.583	21.739
47	2 209	103 823	6.856	3.609	21.277
48	2 304	110 592	6.928	3.634	20.833
49	2 401	117 649	7.000	3.659	20.408
50	2 500	125 000	7.071	3.684	20.000
51	2 601	132 651	7.141	3.708	19.608
52	2 704	140 608	7.211	3.733	19.231
53	2 809	148 877	7.280	3.756	18.868
54	2 916	157 464	7.348	3.780	18.519
55	3 025	166 375	7.416	3.803	18.182
56	3 136	175 616	7.483	3.826	17.857
57	3 249	185 193	7.550	3.849	17.544
58	3 364	195 112	7.616	3.871	17.241
59	3 481	205 379	7.681	3.893	16.949

n	n^2	n^3	$\sqrt{n}$	$\sqrt[3]{n}$	$1\,000/n$
60	3 600	216 000	7.746	3.915	16.667
61	3 721	226 981	7.810	3.936	16.393
62	3 844	238 328	7.874	3.958	16.129
63	3 969	250 047	7.937	3.979	15.873
64	4 096	262 144	8.000	4.000	15.625
65	4 225	274 625	8.062	4.021	15.385
66	4 356	287 496	8.124	4.041	15.152
67	4 489	300 763	8.185	4.062	14.925
68	4 624	314 432	8.246	4.082	14.706
69	4 761	328 509	8.307	4.102	14.493
70	4 900	343 000	8.367	4.121	14.286
71	5 041	357 911	8.426	4.141	14.085
72	5 184	373 248	8.485	4.160	13.889
73	5 329	389 017	8.544	4.179	13.699
74	5 476	405 224	8.602	4.198	13.514
75	5 625	421 875	8.660	4.217	13.333
76	5 776	438 976	8.718	4.236	13.158
77	5 929	456 533	8.775	4.254	12.987
78	6 084	474 552	8.832	4.273	12.821
79	6 241	493 039	8.888	4.291	12.658
80	6 400	512 000	8.944	4.309	12.500
81	6 561	531 441	9.000	4.327	12.346
82	6 724	551 368	9.055	4.344	12.195
83	6 889	571 787	9.110	4.362	12.048
84	7 056	592 704	9.165	4.380	11.905
85	7 225	614 125	9.220	4.397	11.765
86	7 396	636 056	9.274	4.414	11.628
87	7 569	658 503	9.327	4.431	11.494
88	7 744	681 472	9.381	4.448	11.364
89	7 921	704 969	9.434	4.465	11.236

n	n^2	n^3	$\sqrt{n}$	$\sqrt[3]{n}$	$1\,000/n$
90	8 100	729 000	9.487	4.481	11.111
91	8 281	753 571	9.539	4.498	10.989
92	8 464	778 688	9.592	4.514	10.870
93	8 649	804 357	9.644	4.531	10.753
94	8 836	830 584	9.695	4.547	10.638
95	9 025	857 375	9.747	4.563	10.526
96	9 216	884 736	9.798	4.579	10.417
97	9 409	912 673	9.849	4.595	10.309
98	9 604	941 192	9.899	4.610	10.204
99	9 801	970 299	9.950	4.626	10.101
100	10 000	1 000 000	10.000	4.642	10.000

GLOSSARY

Absolute zero. Absolute zero is that temperature at which random molecular movement in a body of matter would cease. At absolute zero the volume of an ideal gas sample would be zero. Absolute zero corresponds to $-273°$C. The *absolute temperature scale* expresses temperatures in °C above absolute zero; its unit is the *degree Kelvin*, denoted °K. Thus the freezing point of water is 273°K.

Acceleration. The acceleration of a body is the rate at which its velocity changes with time; the change in velocity may be in magnitude or direction or both.

Acceleration of gravity. The acceleration of gravity is the acceleration experienced by a freely falling body near the earth's surface. The symbol of the acceleration of gravity is g, and its value is $9.8 \, \text{m/s}^2$. (In British units, $g = 32 \, \text{ft/s}^2$.)

Alpha particle. An alpha particle is the nucleus of a helium atom. It consists of two neutrons and two protons.

Alternating current. The direction of an alternating current reverses itself periodically.

Ampere. The ampere is the unit of electric current. It is equal to a flow of 1 C/s.

Amplitude. The amplitude of a wave is the maximum value of the wave variable (displacement, pressure, electric field intensity, etc., depending upon the nature of wave) regardless of sign.

Antiparticle. Almost every kind of elementary particle has an antiparticle counterpart whose mass is the same but whose charge and certain other properties are different. When a particle and its antiparticle come together, they both disappear as their mass is turned into energy.

Atom. An atom is the ultimate particle of an element. Every atom consists of a very small positively charged nucleus and a number of electrons at some distance from it. The nucleus contains nearly all the mass of the atom.

Atomic mass unit. Atomic and nuclear masses are expressed in atomic mass units (amu), equal to one-twelfth of the mass of the $^{12}_{6}$C isotope. An amu is equal to $1.66 \times 10^{-27} \, \text{kg}$.

Atomic number. The atomic number of an element is the number of electrons in each of its atoms or, equivalently, the number of protons in each of its atomic nuclei.

Beta particle. A beta particle is an electron (or positron) emitted during the radioactive decay of a nucleus.

Binding energy. The binding energy of a nucleus is the energy equivalent of the difference between its mass and the sum of the masses of its individual constituent nucleons. This amount of energy must be supplied to the nucleus if it is to be completely disintegrated. The binding energy per nucleon is least for very light and very heavy nuclei; hence the *fusion* of very light nuclei to form heavier ones and the *fission* of very heavy nuclei to form lighter ones are both processes that liberate energy.

Bohr theory of the atom. According to the Bohr theory of the atom, an electron can circle an atomic nucleus indefinitely without radiating energy if its orbit is an integral number of electron wavelengths in circumference. The number of wavelengths that fit into a particular permitted orbit is called the *quantum number* of that orbit. The electron energies corresponding to the various quantum numbers constitute the *energy levels* of the atom, of which the lowest is the *ground state* and the rest are *excited states*.

Boyle's law. Boyle's law states that, at constant temperature, the absolute pressure of a sample of a gas is inversely proportional to its volume, so that $pV = $ constant at that temperature regardless of changes in either p or V individually.

Celsius temperature scale. In the celsius (centigrade) temperature scale the freezing point of water is assigned the value $0°C$ and the boiling point of water the value $100°C$.

Centripetal acceleration. The velocity of a body in uniform circular motion continually changes in direction although its magnitude remains constant. The acceleration that causes the body's velocity to change is called centripetal acceleration, and it points toward the center of the body's circular path.

Centripetal force. The inward force that provides a body in uniform circular motion with its centripetal acceleration is called centripetal force.

Charles's law. Charles's law states that, at constant pressure, the volume of a sample of a gas is directly proportional to its absolute temperature, so that $V/T = $ constant at that pressure regardless of changes in either V or T individually.

Coherence. Two sources of waves are coherent if there is a fixed phase relationship between the waves they emit during the time the waves are being observed. Interference can be observed only in waves from coherent sources.

Coulomb. The coulomb is the unit of electric charge.

Coulomb's law. Coulomb's law states that the force one charge exerts upon another is directly proportional to the magnitudes of the charges and inversely proportional to the square of the distance between them. The force between like charges is repulsive, and that between unlike charges is attractive.

Covalent bond. In a covalent bond between adjacent atoms of a molecule or solid, the atoms share one or more electron pairs.

Crystalline solid. Solids whose constituent atoms or molecules are arranged in regular, repeated patterns are called *crystalline*. When only short-range order is present, the solid is *amorphous*.

Density. The density of a substance is its mass per unit volume.

Diffraction. The ability of waves to bend around the edges of obstacles in their paths is called diffraction.

Dispersion. Dispersion refers to the splitting up of a beam of light containing different frequencies by passage through a substance whose index of refraction varies with frequency.

Domain. An assembly of atoms in a ferromagnetic material whose atomic magnetic moments are aligned is called a domain.

Electric charge. Electric charge, like rest mass, is a fundamental property of certain of the elementary particles of which all matter is composed. There are two kinds of electric charge, *positive charge* and *negative charge*; charges of like sign repel, unlike charges attract. The unit of charge is the *coulomb*. All charges, of either sign, occur in multiples of the fundamental *electron charge* of 1.6×10^{-19} C. The principle of *conservation of charge* states that the net electric charge in an isolated system remains constant. Electric charge is relativistically invariant.

Electric current. A flow of electric charge from one place to another is called an electric current. The unit of electric current is the *ampere*, which is equal to a flow of 1 C/s.

Electric field. An electric field exists wherever an electric force acts on a charged particle. The magnitude of an electric field at a point is defined as the force that would act on a charge of $+1$ C placed there. The unit of electric field is the V/m, which is equal to 1 N/C.

Electrolyte. A substance that separates into free ions when dissolved in water is called an electrolyte since the resulting solution is able to conduct electric current.

Electromagnetic induction. Electromagnetic induction refers to the production of an electric field by a changing magnetic field.

Electromagnetic waves. Electromagnetic waves consist of coupled electric and magnetic oscillations. They exist as a consequence of electromagnetic induction, which states that whenever there is a change in a magnetic field, an electric field is produced, and of Maxwell's hypothesis, which states that whenever there is a change in an electric field, a magnetic field is produced. Radio waves, light waves, x-rays, and gamma rays are all electromagnetic waves differing only in their frequency. Electromagnetic waves are produced by accelerated electric charges.

Electron volt. The electron volt is the energy acquired by an electron that has been accelerated by a potential difference of 1 V. It is equal to 1.6×10^{-19} J.

Elements. Elements are the simplest substances encountered in bulk. They cannot be decomposed nor transformed into one another by ordinary chemical or physical means.

Elementary particles. Elementary particles are particles found in nature that do not consist of combinations of other particles. Many elementary particles are known, of which all but the photon, neutrino, antineutrino, positron, electron, proton, and antiproton are unstable outside a nucleus and decay into other particles.

Energy. Energy is that which may be converted into work. When something possesses energy, it is capable of performing work or, in a general sense, of accomplishing a change in some aspect of the physical world. The unit of energy, like that of work, is the *joule*. The three broad categories of energy are *kinetic energy*, which is the energy something possesses by virtue of its motion; *potential energy*, which is the energy something possesses by virtue of its position in a force field; and *rest energy*, which is the energy something possesses by virtue of its mass. The principle of *conservation of energy* states that the total amount of energy in a system isolated from the rest of the universe always remains constant, although energy transformations from one form to another, including rest energy, may occur within the system.

Escape velocity. The minimum velocity needed by an object to permanently escape from the gravitational attraction of an astronomical body is called the escape velocity of the astronomical body.

Exclusion principle. According to the Pauli exclusion principle, no two electrons in an atom can exist in the same quantum state.

Fahrenheit temperature scale. In the fahrenheit temperature scale the freezing point of water is assigned the value 32°F and the boiling point of water the value 212°F.

Force. A force is any influence that can produce a change in the linear momentum of a body.

Force field. A force field is a region of space at every point in which an appropriate test object would experience a force. Thus a *gravitational field* is a region of space in which an object by virtue of its mass is acted on by a force, and an *electric field* is a region of space in which an object by virtue of its electric charge is acted upon by a force.

Frame of reference. A frame of reference is something with respect to which observations are made on something else. When a person standing beside a road sees a car moving along the road, the road is the frame of reference; when a person in the car sees the roadside moving backward past him, the car is the frame of reference. All measurements, including those of time, have meaning only when the frame of reference from which they are made is specified. All frames of reference are equally valid —there is no universal frame of reference. The *special theory of relativity* relates measurements made on an object or phenomenon from frames of reference moving at constant velocity with respect to one another.

Frequency. The frequency of a body undergoing harmonic motion is the number of oscillations it makes per unit time. The frequency of a train of waves is the number of waves that pass a particular point per unit time. The usual unit of frequency is the *hertz* (Hz), which is equal to 1 cycle/s.

Friction. The term friction refers to the resistive forces that arise to oppose the motion of a body past another with which it is in contact.

Fusion, heat of. The heat of fusion of a substance is the amount of heat that must be supplied to change a unit quantity of it at its melting point from the solid to the liquid state; the same amount of heat must be removed from a unit quantity of the substance in the liquid state at its melting point to change it to a solid.

Galvanometer. A galvanometer is a sensitive current-measuring device that is based upon the tendency of a coil carrying a current to rotate in a magnetic field.

Gamma ray. A gamma ray is an energetic photon emitted by a radioactive nucleus.

Generator. A generator is a device that converts mechanical energy into electrical energy.

Gravitation. Newton's law of universal gravitation states that every body in the universe attracts every other body with a force directly proportional to both their masses and inversely proportional to the square of the distance separating them.

Gravitational field. The gravitational field of a body is the alteration in the properties of the region around it caused by its presence. The gravitational field of a body is what interacts with another body brought into its vicinity to produce what we recognize as the gravitational force between them.

Half-life. The time required for half of a given sample of a radioactive substance to decay is called its half-life.

Heat. Heat is a quantity whose addition to a body of matter causes its internal energy to increase and whose removal from a body of matter causes its internal energy to decrease. If the matter does not change state during the process, the change in internal energy results in a corresponding change in temperature. The unit of heat is the *kilocalorie* (kcal), which is that amount of heat required to change the temperature of 1 kg of water by 1 °C.

Heat engine. A heat engine is any device that converts heat into mechanical energy. The laws of thermodynamics govern the behavior of all heat engines.

Hertz. The unit of frequency is the hertz (Hz), which is equal to 1 cycle/s.

Horsepower. The horsepower is a unit of power equal to 746 W.

Ideal gas law. The equation $pV/T =$ constant, a combination of Boyle's and Charles's laws, is called the ideal gas law and is obeyed approximately, though not exactly, by all gases.

Inertia. The inertia of a body refers to the apparent resistance it offers to changes in its state of motion. The first of Newton's laws of motion describes the behavior included in the term inertia.

Interactions, fundamental. There are only four fundamental interactions between elementary particles that are responsible for all the forces in the universe. In order of decreasing strength, these are the *strong nuclear, electromagnetic, weak nuclear,* and *gravitational* interactions. The two nuclear interactions are effective only over short distances.

The electromagnetic and gravitational interactions are unlimited in range, but become weaker with increasing distance.

Interference. The superposition of different waves of the same nature is called interference: *constructive interference* occurs when the resulting composite wave has an amplitude greater than that of either of the original waves, and *destructive interference* occurs when the resulting composite wave has an amplitude less than that of either of the original waves.

Ion. An ion is an atom or group of atoms that carries an electric charge. An atom or group of atoms becomes a negative ion when it picks up one or more electrons in addition to its normal number, and becomes a positive ion when it loses one or more of its usual number.

Ionic bond. Electrons are transferred between the atoms of certain solids so that the resulting crystal consists of positive and negative ions rather than of neutral atoms. Such a solid is said to be held together by ionic bonds. The attractive forces between ions of opposite charge balance the repulsive forces between ions of like charge in an ionic solid.

Isotope. The isotopes of an element have the same atomic number but different mass numbers. Thus the nuclei of the isotopes of an element all contain the same number of protons but have different numbers of neutrons.

Joule. The joule is the unit of work and energy in the metric system. It is equal to 1 kg-m^2/s^2.

Kilocalorie. The kilocalorie (kcal) is the unit of heat in the metric system. It is equal to that amount of heat required to change the temperature of 1 kg of water by 1°C. 1 kcal = 4185 J.

Kilogram. The kilogram (kg) is the unit of mass in the metric system. One kilogram weighs 2.21 lb.

Kinetic energy. Kinetic energy is the energy a moving body possesses by virtue of its motion. If the body has the mass m and the speed v, its kinetic energy is $\frac{1}{2}mv^2$ provided that $m \ll c$.

Kinetic theory. According to the kinetic theory of gases, a gas consists of a great many tiny individual molecules that do not interact with one another except when collisions occur. The molecules are far apart compared with their dimensions and are in constant random motion. The ideal gas law may be derived from the kinetic theory of gases.

Lenz's law. Lenz's law states that the direction of an induced current must be such that its own magnetic field opposes the changes in flux that are inducing it.

Lines of force. Lines of force are means for visualizing a gravitational (or other) field. Their direction at any point is that in which a test body would move if released there, and their concentration in the neighborhood of a point is proportional to the magnitude of the force on a test particle at that point. The direction at any point of a magnetic line of force is that in which a test object (for instance, a compass needle) aligns itself at that point in the field.

Longitudinal waves. Longitudinal waves occur when the individual particles of a medium vibrate back and forth in the direction in which the waves travel. Sound consists of longitudinal waves.

Magnetic field. A magnetic field exists wherever a force acts on a moving charged particle that would not act if the particle were at rest. The magnitude of a magnetic field at a point is defined as the force that would act on a charge of + 1 C moving at a speed of 1 m/s past that point, when the direction of the motion is such as to result in the maximum force. The unit of magnetic field is the *tesla,* equal to 1 N/A-m.

Magnetic force. Charges in motion relative to an observer appear to exert forces upon one another that are different from the electric forces they exert when at rest. These differences are by custom attributed to magnetic forces. In reality, magnetic forces represent relativistic corrections to electric forces due to the motion of the charges involved.

Mass. The property of matter that manifests itself as the inertia of a body at rest is called mass. The unit of mass is the *kilogram*. The *rest mass* of a body is its mass when stationary with respect to an observer.

Mass number. The mass number of a nucleus is the number of nucleons it contains.

Matter waves. A moving body behaves as though it has a wave character. The waves representing such a body are matter waves, also called *de Broglie waves*. The wave variable in a matter wave is its *wave function*, whose square is the *probability density* of the body. The value of the probability density of a particular body at a certain place and time is proportional to the probability of finding the body at that place at that time. Matter waves may thus be regarded as waves of probability.

Maxwell's hypothesis. Maxwell's hypothesis was that a changing electric field produces a magnetic field.

Mechanical equivalent of heat. The mechanical equivalent of heat is the constant ratio between the energy dissipated in some mechanical process and the heat that appears as a result of the process. It is equal to 4185 J/kcal in the metric system.

Metallic bond. The metallic bond which holds metal atoms together in the solid state arises from a "gas" of freely moving electrons that pervades the entire metal.

Meter. The meter is the unit of length in the metric system. One meter is equal to 3.28 ft.

Molecule. A molecule is a group of atoms that stick together strongly enough to act as a single particle. A molecule of a given kind always has a certain definite structure and is complete in itself with little tendency to gain or lose atoms.

Momentum, angular. The angular momentum of a rotating body is a measure of its tendency to continue rotating; it is the rotational counterpart of linear momentum. The principle of *conservation of angular momentum* states that the total angular momentum of a system of particles isolated from the rest of the universe remains constant regardless of what events occur within the system.

Momentum, linear. The linear momentum of a body is the product of its mass and velocity. Linear momentum is a vector quantity whose direction is that of the body's velocity. The principle of *conservation of linear momentum* states that the total linear momentum of a system of particles isolated from the rest of the universe remains constant regardless of what events occur within the system.

Motion, laws of. *Newton's first law of motion* states that a body at rest will remain at rest and a body in motion will continue in motion in a straight line at constant velocity in the absence of any interaction with the rest of the universe. *Newton's second law of motion* states that the net force acting on a body is equal to the rate of change of the body's linear momentum. *Newton's third law of motion* states that, when a body exerts a force on another body, the second body exerts a force on the first body of the same magnitude but in the opposite direction.

Neutrino. The neutrino is a massless, uncharged particle that is emitted together with an electron during beta decay. It possesses energy, momentum, and angular momentum.

Neutron. The neutron is an electrically neutral particle, slightly heavier than the proton, which is present in nuclei together with protons.

Newton. The newton is the unit of force in the metric system. It is equal to 1 kg-m/s^2. One newton is equal to 0.225 lb.

Nuclear fission. In nuclear fission, the absorption of neutrons by certain heavy nuclei causes them to split into smaller *fission fragments*. Because each fission also liberates several neutrons, a rapidly multiplying sequence of fissions called a *chain reaction* can occur if a sufficient amount of the proper material is assembled. A *nuclear reactor* is a device in which a chain reaction can be initiated and controlled.

Nuclear fusion. In nuclear fusion, two light nuclei combine to form a heavier one with the evolution of energy. The sun and stars obtain their energy from fusion reactions.

Nucleus. The nucleus of an atom is located at its center and contains all of the positive charge and nearly all of the mass of the atom.

Nucleon. Neutrons and protons, the constituents of atomic nuclei, are jointly called nucleons.

Ohm. The ohm is the unit of electrical resistance. It is equal to 1 V/A.

Ohm's law. Ohm's law states that the current that flows in a metallic conductor is proportional to the potential difference between its ends.

Periodic law. The periodic law of chemistry states that if the elements are listed in order of atomic number, elements with similar properties recur at regular intervals. The quantum theory of the atom together with the exclusion principle is able to explain the origin of the periodic law.

Period. The period of a wave is the time required for one complete wave to pass a particular point.

Photoelectric effect. The photoelectric effect is the emission of electrons from a metal surface when light is shined on it.

Plasma. A plasma is a gas composed of electrically charged particles, and its behavior depends strongly upon electromagnetic forces. Most of the matter in the universe is in the plasma state.

Polar molecule. A polar molecule is one whose charge distribution is asymmetrical, so that one end is positive and the other negative even though the molecule as a whole is electrically neutral.

Polarization. A polarized beam of transverse waves is one whose vibrations occur in only a single direction perpendicular to the direction in which the beam travels, so that the entire wave motion is confined to a plane called the *plane of polarization*. An *unpolarized* beam of transverse waves is one whose vibrations occur equally often in all directions perpendicular to the direction in which the beam travels.

Pole, magnetic. The ends of a permanent magnet are called its poles. Magnetic lines of force leave the *north pole* of a magnet and enter its *south pole*.

Positron. A positron is a positively charged electron.

Potential difference. The electrical potential difference between two points is the work that must be done to take a charge of 1 C from one point to the other. The unit of potential difference is the *volt*, which is equal to 1 J/C.

Potential energy. Potential energy is the energy a body has by virtue of its position. The gravitational potential energy of a body of mass m at a height h above a particular reference point is mgh. Other examples of potential energy are that of a planet with respect to the sun, that of a piece of iron with respect to a magnet, and that of a body at the end of a stretched spring with respect to its equilibrium position.

Power. The rate at which work is done is called power. The unit of power in the metric system is the *watt*, which is equal to 1 J/s. One *horsepower* is equal to 746 W.

Pressure. The pressure on a surface is the perpendicular force per unit area that acts upon it. *Gauge pressure* is the difference between true pressure and atmospheric pressure.

Quantum theory of the atom. In the quantum theory of the atom only experimentally measurable quantities are considered, and no use is made of mechanical models inconsistent with the uncertainty principle. According to this theory, four quantum numbers are required to describe each electron in an atom. These are the *principal quantum number n*, which governs the electron's energy, the *orbital quantum number l*, which governs the magnitude of its angular momentum, the *magnetic quantum number m_l*, which governs the orientation of its angular momentum, and the *spin magnetic quantum number m_s*, which governs the orientation of its spin.

Quantum theory of light. The quantum theory of light states that light travels in tiny bursts of energy called *quanta* or *photons*. If the frequency of the light is f, each burst has the energy hf, where h is known as *Planck's constant*. The quantum theory of light complements the wave theory of light.

Radioactive nuclei. Radioactive nuclei spontaneously transform themselves into other nuclear species with the emission of charged particle radiation. The radiation may consist of *alpha particles*, which are the nuclei of helium atoms, or *beta particles*, which are positive or negative electrons. The emission of *gamma rays*, which are energetic photons, enables an excited nucleus to lose its excess energy.

Reflection. In diffuse reflection an incident beam of parallel light is spread out in many directions, while in *specular reflection* the angle of reflection is equal to the angle of incidence.

Refraction. The bending of a light beam when passing from one medium to another is called refraction. The quantity that governs the degree to which a light beam will be deflected in entering a medium is its *index of refraction*, defined as the ratio between the speed of light in free space and its value in the medium.

Relativity. The special theory of relativity relates measurements made on an object or phenomenon from frames of reference moving at constant velocity with respect to one another. The *Lorentz contraction* refers to the decrease in the measured length of an object when it is moving relative to an observer. The relativistic *time dilation* refers to the fact that a clock moving with respect to an observer appears to tick less rapidly than it does to an observer traveling with the clock. The *relativity of mass* refers to the increase in the measured mass of an object when it is moving relative to an observer.

Resistance. The resistance of a body of matter is a measure of the extent to which it impedes the passage of electric current. It is defined as the ratio between the potential difference applied across the ends of the body and the current that flows. The unit of resistance is the *ohm*, which is equal to 1 V/A.

Resolution of vectors. A vector can be resolved into two or more other vectors whose sum is equal to the original vector. The new vectors are called the *components* of the original vector, and are normally chosen to be perpendicular to one another.

Scalar quantity. A scalar quantity is one that has magnitude only.

Sound. Sound is a longitudinal wave phenomenon that consists of successive compressions and rarefactions of the medium through which it travels.

Specific heat. The specific heat of a substance is the amount of heat required to change the temperature of a unit quantity of it by $1°C$. The unit of specific heat is the kcal/kg-$°C$.

Spectrum. An *absorption spectrum* results when white light is passed through a cool gas; it is a *dark line spectrum* because it appears as a series of dark lines on a bright background, with the lines representing characteristic wavelengths absorbed by the gas. An *emission spectrum* consists of the various wavelengths of light emitted by an excited substance; it may be a *continuous spectrum*, in which all wavelengths are present, or a *bright line spectrum*, in which only a few wavelengths characteristic of the individual atoms of the substance are present.

Speed. The speed of a moving body is the rate at which distance is being covered by the body. Speed is a scalar quantity, unlike velocity. When the distance is directly proportional to the elapsed time, the body moves at *constant speed.*

Spin. Every electron has a certain intrinsic amount of angular momentum called its spin. The spin of an electron is as fundamental a property as its mass or electric charge. Owing to its spin, every electron acts like a tiny bar magnet. Most other elementary particles also have spin associated with them.

Statistical mechanics. Statistical mechanics attempts to deduce the behavior of assemblies of particles by considering statistically the most probable behavior of their constituent particles.

Sublimation. Sublimation is the direct conversion of a substance from the solid to the vapor state, or vice versa, without it first becoming a liquid.

Superposition, principle of. The principle of superposition states that when two or more waves of the same nature travel past a given point at the same time, the amplitude at the point is the sum of the amplitudes of the individual waves.

Temperature. The temperature of a body of matter is a measure of the average kinetic energy of random motion of its constituent particles. When two bodies are in contact, heat flows from the one at the higher temperature to the one at the lower temperature.

Thermodynamics. The *first law of thermodynamics* states that the work output of any engine is equal to its net energy input plus any decrease in its stored energy. This law is thus a restatement of the principle of conservation of energy. The *second law of thermodynamics* states that no engine can be completely efficient in converting energy to work — some of the input energy must be wasted as heat. This law is a consequence of the tendency of all physical systems to become more and more disordered as time goes on.

Thermometer. A thermometer is a device for measuring temperature. The two temperature scales in common use are the *celsius* (centigrade) scale, in which the freezing point of water is assigned the value $0°C$ and its boiling point the value $100°C$, and the *fahrenheit* scale, in which these points are assigned the values $32°F$ and $212°F$, respectively.

Thermonuclear energy. The energy liberated by nuclear fusion is called thermonuclear energy.

Transformer. An alternating current flowing in the primary coil of a transformer induces another alternating current in the secondary coil. The ratio of the potential differences is proportional to the ratio of turns in the coils.

Transverse waves. Transverse waves occur when the individual particles of a medium vibrate from side to side perpendicular to the direction in which the waves travel. The vibrations of a stretched string are transverse waves.

Uncertainty principle. The uncertainty principle is an expression of the limits set by the wave nature of matter on finding the position and state of motion of a moving body. According to this principle, the product of the uncertainties in simultaneous measurements of the position and momentum of a body cannot be less than $h/2\pi$.

Uniform circular motion. A body traveling in a circle at constant speed is said to be undergoing uniform circular motion.

Van der Waals force. Van der Waals forces originate in the electric attraction between asymmetrical charge distributions in atoms and molecules. Molecular solids and liquids are held together by van der Waals forces.

Vaporization, heat of. The heat of vaporization of a substance is the amount of heat that must be supplied to change a unit quantity of it at its boiling point from the liquid to the gaseous (or vapor) state; the same amount of heat must be removed from a unit quantity of the substance at its boiling point to change it into a liquid.

Vector. A vector is an arrowed line whose length is proportional to the magnitude of some vector quantity and whose direction is that of the quantity. A *vector diagram* is a scale drawing of the various forces, velocities, or other vector quantities involved in the motion of a body. In *vector addition*, the tail of each successive vector is placed at the head of the previous one, with their lengths and original directions kept unchanged. The *resultant* is a vector drawn from the tail of the first vector to the head of the last. A vector can be *resolved* into two or more other vectors called the *components* of the original vector. Usually the components of a vector are chosen to be in mutually perpendicular directions.

Vector quantity. A vector quantity is one that has both magnitude and direction.

Velocity. The velocity of a body is a specification of both its speed and the direction in which it is moving. Velocity is a vector quantity. The *instantaneous velocity* of a body is its velocity at a specific instant of time. The *average velocity* of a body is the total displacement through which it has moved in a time interval divided by the interval.

Volt. The unit of electrical potential difference is the volt. It is equal to 1 J/C.

Watt. The unit of power is the watt, which is equal to 1 J/s.

Wave motion. Wave motion is characterized by the propagation of a change in a medium, rather than by the net motion of the medium itself. The passage of a wave across the surface of a body of water, for instance, involves the motion of a pattern of alternate crests and troughs, with the individual water molecules themselves ideally executing uniform circular motion.

Weight. The weight of a body is the gravitational force exerted on it by the earth. The weight of a body is proportional to its mass.

Work. Work is a measure of the change (in a general sense) a force gives rise to when it acts upon something. The magnitude of the work done by a force on a body is equal to the product of the force and the component of the displacement of the body in the direction of the force. The unit of work is the *joule*.

X-rays. X-rays are high-frequency electromagnetic waves emitted when fast electrons impinge on matter.

ANSWERS TO ODD-NUMBERED EXERCISES

Section 2

1. No. The distinction between vector and scalar quantities is simply that vector quantities have directions associated with them, and both kinds of quantity are found in the physical world.

3. Yes

5. When the sled is pushed by the stick, the vertical component of the force used goes into pressing the sled down, which increases the frictional resistance of its motion. When the sled is pulled by the stick, on the other hand, the vertical component of the force acts to lift the sled, which decreases the frictional resistance. Hence it is easier to pull the sled than to push it.

7. 30 lb; 0

9. c

11. 94 mi

13. 7.6 knots in a direction 22° north of east

15. 64.3 lb vertical, 76.6 lb horizontal

17. 42°

Section 3

1. 16 mi/hr

3. $8\frac{1}{3}$ min

5. 1980 m

7. 0.5 cm/s; 0.167 cm/s

9. 144 mi/hr; 720 mi

Section 4

1. b

3. Yes

5. The squirrel should stay where it is, since if it lets go, it will fall with exactly the same downward acceleration as the bullet and so will be struck.

7. Yes

9. -2 m/s^2

11. The initial speed is 350 mi/hr, which increases to 400 mi/hr and then drops to 300 mi/hr

13. 0.04 m; 4 m

15. 40 m/s

17. 27 m/s

19. 19.8 m/s; 29.6 m/s

21. 0.61 s; 0.67 s; 0.64 s; 0.64 s; it does not reach the bottom while the elevator is in free fall.

Section 5

1. No; yes; yes

3. (a) Each is correct from the point of view of an appropriate observer. (b) The Copernican system is much simpler and easier to analyze. (c) The sun is still the center of the solar system.

5. 236 s (3 min 56s)

7. 6.3×10^4 yr

Section 6

1. The speed of light

3. Zero

5. 2×10^{-11} ft

Section 7

1. If the curtains point directly to the ground, the velocity is constant. If they point forward of plumb, the airplane is decelerating; if aft of plumb, the airplane is accelerating.

3. 2×10^3 m/s   7. $\sqrt{3}\,c/2$

5. 1 m^3; 769 m^3 9. 4.2×10^7 m/s

Section 8

1. Yes, in order that linear momentum be conserved.

3. (a) Yes; (b) in the opposite direction to that in which the man walks; (c) the car also comes to a stop.

5. They have the same recoil momentum—it is the recoil *speeds* that are different.

7. The length of the day will increase, since the water from the icecaps becomes uniformly distributed over the earth, and hence the earth's rotational speed must decrease in order to conserve angular momentum.

9. 40 g; 72 m

11. 8 bullets

13. $1\frac{1}{3}$ m/s

Section 9

1. No. Only a *net* force produces an acceleration; and other forces may come into being, when a force is applied, that cancel it out. Thus pushing down on a book lying on a table does not accelerate the book because the table pushes back with an equal and opposite force.

3. The change in momentum of a person falling onto loose earth is more gradual than if he falls onto concrete, hence the force acting upon him is less.

5. g

7. 0.67 m/s^2

9. 29 m/s

11. 9600 N

13. (a) 250 m/s^2; 25.5 g; 2×10^4 N; 4500 lb
    (b) 50 m/s^2; 5.1 g; 4000 N; 900 lb

15. 8800 N

17. 1080 N; 880 N; 980 N; 980 N; 0

19. 1.96 m/s^2; 1.96 m/s^2

Section 10

1. No. Action and reaction forces act upon different bodies, so a single force can certainly act upon a body.

3. Downward and away from the wall; upward and toward the wall; downward and against the wall; upward and away from the wall.

5. a

Section 11

1. Under no circumstances

3. Yes, because such a laboratory is accelerated and accelerations (though not constant velocities) can be detected by experiments in a closed laboratory.

5. 1.52×10^{-6} m/s^2

7. 384 N

9. 12 N; 52 N

Section 12

1. At the center of the earth the stone has the same mass as at the earth's surface, but its weight is zero.

3. G is a constant of nature with the same value everywhere in the universe.

5. The two are equal, since the centripetal force is provided by the gravitational force.

7. The centripetal force due to the earth's rotation is greatest at the equator, hence the outward reaction forces exerted by particles of the earth are also greatest there. These reaction forces distort the earth's shape. If the earth rotated faster, the distortion would be greater than it is now.

9. The earth rotates from west to east. Hence the satellite sent eastward will have its launching speed increased because of the earth's rotation, and the one sent westward will have its speed decreased. The satellite sent eastward will therefore have the larger orbit.

11. The direction of the force is given by the direction of the lines of force at that point, and the magnitude of the force is given by the density of the lines of force at that point.

13. $g/4$

15. 3.8×10^7 m from the moon

17. $v_{escape}/v_{orbit} = \sqrt{2}$

19. 4.16×10^{23} N; 1.3×10^4 m/s; 26 m/s^2

Section 13

1. No work is done.

3. 3.6×10^6 J/kWh

5. 1040 J; 1040 J

7. 7.52×10^4 J

9. 0.23 hp

11. 6.3×10^{12} W; 1.8×10^3 W; 2.4 hp

13. 2.94×10^6 W; 29,400 light bulbs

Section 14

1. Both are scalar quantities.

3. At the lowest point; at the highest points

5. Under no circumstances; kinetic energy is always a positive quantity.

7. Yes; yes

9. The momentum increases by a factor of $\sqrt{2}$.

11. Much more work is needed to bring its speed from 50 to 60 mi/hr than from 10 to 20 mi/hr.

13. 0.102 m; 1 m

15. 7.07 m/s

17. 24 J; 9.3 J

Section 15

1. Yes; no; no; yes

3. $h/2$

5. 59%

7. 4.85 m/s; 4.85 m/s

9. 1760 J; 1568 J; the "lost" work has been dissipated as heat due to friction in the pulleys.

11. The average force is 21 times greater than the hammer's weight.

13. 6×10^{-11}

15. No

17. $v = (1 + M/m)\sqrt{gh}$

19. 67%; 28%

Section 16

1. The work done goes into the internal energy of the water, as manifested in an increase in its temperature.

3. 626°F; 2138°F

5. -80°C

7. 12,811 m

9. 51.6 kcal

Section 17

1. Aluminum

3. 10 g of ice, because of its heat of fusion

5. The highest temperature that water can have while remaining liquid is its boiling point. Increasing the rate at which heat is supplied to the water thus increases the rate at which steam is produced without changing the temperature of the boiling water.

7. 818 m/s

9. It loses internal energy but its temperature remains constant while the change of state takes place.

11. 620 kcal

13. 0.0616 kg

15. 20.7°C

17. 0.742 kg

19. 60°C

21. 327 m/s

Section 18

1. As the steam inside the can condenses, the internal pressure falls below the external pressure of the atmosphere.

3. 63 atm

5. At constant volume p/T = constant.

7. 4×10^5 N/m^2; 2.67 m^3; 2.67 m^3

9. 1.65 atm

Section 19

1. Because of the volume occupied by its molecules.

3. The internal energy of a solid resides in oscillations of its constituent particles about definite equilibrium positions.

5. Gas molecules undergo frequent collisions with one another, which considerably increases the time needed for a particular molecule to travel from one place to another.

7. 3.33×10^{-9} m; 8.4 diameters

9. Intermolecular attractions decrease gas pressure.

11. 43°K

13. H_2; UF_6

15. 327°C; 927°C

Section 20

1. No, they are quite separate physical principles.

3. More heat

5. 42%

7. 0.5%

Section 21

1. The notion of electricity as a fluid can explain all electric phenomena except those in which the quantization of charge is involved.

3. No; yes

5. The balloon will stick better to the insulating plaster wall because charge would leak from the balloon to the metal wall, which would then repel it.

7. 2 cm

9. 5.7×10^{13} C

11. 1.1×10^{-5} N

13. 0.025 N; 0.001 N; 2.5 N

15. 2.74×10^3 N toward the -1×10^{-4} C charge

17. 5.08 m

Section 22

1. It will not move; it will rotate until aligned with the field.

3. Between the charges, 0.0828 cm from the smaller one

5. 3.6×10^4 N/C

7. 5×10^{-4} N; 5×10^{-4} J

9. 2.4×10^6 J; 1.22×10^4 m; 490 m/s

11. 1.31×10^9 J; 3.12×10^5 kcal

13. 2.84 eV

15. 4.2×10^6 m/s

Section 23

1. The positive charge of an atom is located in its nucleus, and its negative charge is located in its electron cloud.

3. The electric force is stronger than the gravitational force at all separations.

5. No

7. In air, positive and negative ions move in opposite directions when a current flows, whereas in a copper wire only electrons move.

Section 24

1. No, partly because electrons have little inertia and partly because they undergo frequent changes in direction during collisions inside the wire in any case.

3. With a single wire, charge would be permanently transferred from one end to the other, and soon so large a charge separation would occur that the electric field that produced the current would be canceled out. With two wires, charge can be sent on a round trip, so to speak, and a net flow of energy from one place to another can occur without a net flow of charge.

5. $L/2$ and $2A$

7. 6.25×10^{-4} m/s

9. 0.33 A

11. 3.13×10^{18} electrons

13. 1.6 hp

15. 5×10^3 W; 5×10^6 eV; 8×10^{-13} J

Section 25

1. The magnitude of a current can be defined in terms of a more direct and unambiguous experiment than is the case with the coulomb, in part because electrical forces are so strong that electric experiments are difficult to carry out accurately.

3. At first the mutual electrical repulsion of the protons causes the beam diameter to increase, but as they go faster the magnetic attraction increases and the beam diameter decreases.

5. He measures both an electric field and a magnetic field.

7. East; west

9. The magnetic field causes the protons to move in curved paths and hence increases their travel times relative to what they would be if their paths were direct.

11. 8 cm

13. 10^3 N

Section 26

1. $0°$; $90°$

3. Mass and kinetic energy

5. 0.1 N

7. 0.1 N northwest; 0.1 N southeast

9. 3.37 cm

11. 2500 V

13. 2.275 m

Section 27

1. There is never a net force on the loop in such a field. There is no torque on the loop when its plane is perpendicular to the field.

3. Near the equator, where the horizontal component of the earth's magnetic field is strongest. In the polar regions the field has only a very small horizontal component and so exerts relatively little torque on a compass needle.

Section 28

1. Counterclockwise; zero current; clockwise

3. Conservation of energy

5. Diameter, composition

7. When the generator is connected to an outside circuit, work must be done in order to turn its shaft because current then flows in the circuit.

9. Downward

11. 10^3 T/s

13. 4.8×10^{-3} V

15. 10 A

17. 2 turns; 1.1 V

Section 29

1. A magnetic field

3. Electric and magnetic fields

5. Energy

7. The east-west direction

9. Amplitude

11. 455 m

13. 266 vibrations

Section 30

1. For paths or path differences of the order of magnitude of a wavelength or less, the wave aspects of light are conspicuous; for paths much longer than a wavelength, light may be more conveniently considered as a ray phenomenon in many applications.

3. 3 ft

5. Yes. Because the speeds of light in adjacent media of different index of refraction are different, a straight path between a point in one medium and a point in the other is not the fastest. The fastest path will involve a greater distance in the medium of low index of refraction and a smaller distance in the medium of high index of refraction than a straight path would.

7. Light perpendicularly incident is not deflected, so the component colors remain together in the beam.

9. 2×10^8 m/s

11. The frequency is unchanged, but the wavelength is doubled.

Section 31

1. Light from incoherent sources can interfere, but the resulting interference pattern shifts continually because there is no constant phase relationship between the beams from the sources. When the sources are coherent, there is such a relationship, and the pattern is stable and therefore readily discernible.

3. The wavelengths in visible light are very small relative to the size of a building, whereas those in radio waves are more nearly comparable.

5. This is a diffraction effect.

7. Interference, diffraction

Section 32

1. Even a faint light involves many photons per second. In addition, visual responses tend to persist

for a brief time, so successive photons give the impression of a continuous light.

3. The particle aspects of light show up chiefly in its interactions with electrons, which were not discovered until late in the nineteenth century. On the other hand, such wave phenomena as diffraction, interference, and polarization can readily be demonstrated in light with fairly primitive instruments.

5. 3×10^{19}

7. 1.72×10^{31} photons/s

9. 4.63×10^{14} Hz

11. 4.23×10^{21} photons/m^2-s; 1.41×10^{13} photons/ m^3

Section 33

1. The wavelength is inversely proportional to the linear momentum, and is not related to the angular momentum.

3. They are the same.

5. Only if the particle's speed is also known.

7. The dimensions and momenta of macroscopic objects are so large that the uncertainties in their positions and momenta are too small in comparison to be detectable.

9. 2.73×10^{-38} m

11. 2.86×10^{-14} m

13. 6.02×10^{-12} m (relativistic calculation); 6.14×10^{-12} m (nonrelativistic calculation).

15. 6.2%

Section 34

1. The Rutherford model indicates the division of the hydrogen atom into a nucleus and a relatively distant electron, and the Bohr model goes on from there to specify precisely the motion and energy of the electron.

3. The energy difference decreases.

5. A hydrogen sample contains a great many atoms, each of which has a variety of transitions between energy levels which it may undergo.

7. 13.6 eV

9. The average molecular KE at $20°$C is 6.07×10^{-21} J (0.038 eV), whereas the excitation energy here is 1.64×10^{-18} J (10.2 eV).

11. 8.7×10^{-20} J (0.54 eV)

Section 35

1. The uncertainty principle

3. No, because in the Bohr theory the electron must revolve around the nucleus.

5. Such an electron has no vector property associated with it on the average, hence its probability cloud must vary in the same way in all directions.

7. The densest part of a probability cloud, which corresponds to the maximum probability of finding the electron, occurs at a radius proportional to n^2.

Section 36

1. The principal quantum number n

3. All electrons have the same spin magnitude.

5. They are similar.

7. (a) Halogen atoms all lack one electron of having closed outer subshells. (b) Inert gas atoms all have closed outer subshells. (c) Alkali metal atoms all have one electron outside closed inner subshells.

Section 37

1. The existence of such definite orbits is not consistent with the uncertainty principle.

3. 3.5×10^4 °K

5. The gaseous state

7. Positive and negative ions

9. Molecules

11. Solids held together by van der Waals bonds have the lowest melting points since such bonds are weak.

13. The electromagnetic interaction is responsible for all the bonding mechanisms.

Section 38

1. Its atomic number, which is the number of protons it contains.

3.

	$^{6}_{3}$Li	$^{13}_{6}$C	$^{31}_{15}$P	$^{94}_{40}$Zr	$^{137}_{56}$Ba
p	3	6	15	40	56
n	3	7	16	54	81

5. The nuclear density is 8.9×10^{15} times greater than the atomic density in hydrogen.

Section 39

1. The binding energy per nucleon is about a million times greater.

3. 6; 12; carbon

5. $^2_1$H; $^1_0$n; $^1_1$H

7. Because neutrons are uncharged, they are not repelled by atomic nuclei and therefore require less energy to penetrate them.

9. 298 MeV; 8.52 MeV/nucleon

11. 19.8 MeV; 20.6 MeV

13. 6.015 amu

15. 9.6 mg

17. 9×10^{13} J; 21,500 tons

Section 40

1. Helium and radon cannot be combined chemically to form radium, nor can radium be broken down at will into helium and radon.

3. $^{14}_8$O emits a positron, and $^{19}_8$O emits an electron.

5. A deficiency of neutrons relative to the number required for stability

7. $^{206}_{82}$Pb

9. 6400 yr

11. (a) 5.4 MeV. (b) Some of the evolved energy is carried off by the recoil of the $^{228}_{90}$Th nucleus. (c) No, because the products of the decay would total more mass than the original atom. (d) No, for the same reason.

Section 41

1. Both have zero rest mass. The photon is connected with the electromagnetic interaction, whereas the neutrino is connected with the weak nuclear interaction. The neutrino has an antiparticle, whereas the photon is its own antiparticle. The photon can materialize (if it has enough energy) into an electron-positron pair, whereas the neutrino cannot materialize.

3. 1879 MeV; more, because the neutron mass exceeds the proton mass.

Section 42

3. A neutrino

5. 3.78×10^{-25} s; no

7. 142 MeV

9. Such a decay would not conserve baryon number B.

INDEX

INDEX

Physical Constants

Absolute zero	0 K	-273 C
Gravitational constant	G	6.67×10^{-11} N-m^2/kg^2
Acceleration of gravity at earth's surface	g	9.81 m/s^2
Speed of light in free space	c	3.00×10^8 m/s
Boltzmann's constant	k	1.38×10^{-23} J/ K
Planck's constant	h	6.63×10^{-34} J-s
	$h/2\pi$	1.05×10^{-34} J-s
Electrostatic constant	k	8.99×10^9 N-m^2/C^2
Magnetic constant	k'	2.00×10^{-7} N/A^2
Charge of electron	e	1.60×10^{-19} C
Electron rest mass	m_e	9.11×10^{-31} kg